Praise for MILITARY CULTURE SHIFT

"This book is a must read for any military leader serving today regardless of rank. Corie Weathers has done a magnificent job researching in great detail the generational differences that make up our force, and then overlaying on top of them the effects of the twenty-year Global War on Terror, resourcing decisions/cuts over time, and the subsequent impacts on soldiers and families. Written through the eyes of a mental health clinician and Army spouse who has treated hundreds of soldiers and couples who have struggled to keep it all together as they have lived the highest of highs and lowest of lows of Army life. This book exposes the impacts of poorly crafted policy, legislation, and local command policies, and how they have had an oversized effect on an all-volunteer force—Corie speaks truth to power which leaders should heed the warnings. Corie offers recommendations and thoughtful suggestions along the way, including a resounding reminder of what the 26th CSA was fond of telling audiences: 'People are not in the Army, people ARE the Army.' This book should be mandatory reading for anyone going into battalion or brigade command (or other service equivalent), general officers, DoD political appointees, and members of Congress and their professional staffs."
— General Robert B. "Abe" Abrams, US Army (Ret)

"The bedrock of American national security—and the American way of life—are the men and women who serve in our military and their families. No one has greater insight into what's really happening and the challenges they face than does Corie Weathers. Corie takes her own advice: She 'listens to the story' and has done so over fifteen years as a mental health clinician and spouse of an Army chaplain. She is on the front lines of the well-being of those who serve and those who are dearest to them. In this book, Corie tells the story of a proud but weary force and their families. With grace and honesty, she takes a thoughtful look at cultural shifts—generational, technological, and societal—that provide real challenges. Corie gives a clear-eyed diagnosis and a thoughtful, helpful guide for us all, recognizing that culture is in many ways the most important aspect of any group of people working together for a common purpose. If you are currently serving, have served, or care about those who do, this book is worthy reading to help better understand the trials and complexities of our most precious resource in national security."
—Mac Thornberry, former chairman of House Armed Services Committee

"Military culture is an often discussed but rarely understood element of our national defense posture. It is a complex, multi-faceted web of intersecting communities, interests, and subcultures that define a part of our society that stand at the ready to defend our nation in times of crisis. But just as broader cultural shifts impart change on our society, similar seismic shifts tear at the fabric of our armed forces. In *Military Culture Shift*, author and mental health clinician Corie Weathers explores how these shifts are impacting servicemembers and their families and what this means for morale, retention, and leadership. *Military Culture Shift* is a phenomenal study of a vital segment of our society caught in the midst of generational change, reeling from two decades of war, and facing down the greatest threat to our national security in more than three decades."
—Steven Leonard, senior assistant dean of the University of Kansas School of Business; coauthor and editor of *To Boldly Go* and *Power Up*

"Ms. Weathers's book provides invaluable insight into understanding American military culture. The problem is not the warrior, it is society. Her project is a Herculean task, but Ms. Weathers provides us with the essential way forward—which is one our entire society must take together."
—Edward A. Gutiérrez, director of the Center for Military History and Grand Strategy, Hillsdale College

"... professional and personal, often eloquent, and always clear—especially given subject matter such as sequestration. ... [The author's intent is to] show how two decades of a compilation of 'wicked' problems such as war (including the abrupt withdrawal from Afghanistan), a pandemic, climate change, financial quandaries at the Department of Defense, and the stress of relentless service are affecting how and why the US military organization doesn't always work. ... Weathers suggests remedies and tries to 'present a narrative to help you develop empathy for the people involved.' ... The story she chooses to tell is indeed vast—but Weathers uncomplicates things. ... Think of her book as a nuanced, empathetic explanation of how we got to this point, what we can learn from being here, and how we can use the opportunity to improve. The message is dire but the delivery is soothing. ... In addition to the insightful considerations in the narrative, she offers at least fifteen pages of 'Leadership Tips and Questions' that are useful to anyone who deals with other humans."
—J. Ford Huffman, National Defense University Foundation

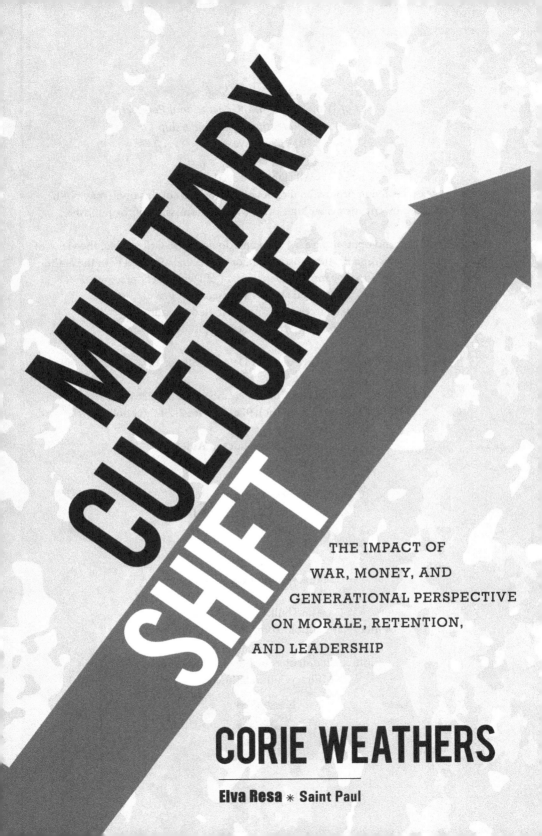
MILITARY CULTURE SHIFT

THE IMPACT OF WAR, MONEY, AND GENERATIONAL PERSPECTIVE ON MORALE, RETENTION, AND LEADERSHIP

CORIE WEATHERS

Elva Resa * Saint Paul

Military Culture Shift:
The Impact of War, Money, and Generational Perspective
on Morale, Retention, and Leadership
©2023 Corie Weathers

All rights reserved.
Except as excerpted in professional reviews, the contents of this book may not be reproduced by any means without prior written permission of the publisher.

Author's research and opinions are her own and do not necessarily reflect the views of the publisher or any organization, institution, or individual affiliated with the author. There is no actual or implied Department of Defense endorsement.
Interviews and quoted articles are used with permission.
Additional sources cited and attributed in endnotes
belong to their respective copyright owners.

Cover design by JuLee Brand of Design Chik for Elva Resa.

Library of Congress Control Number: 2023947269
ISBNs 978-1-934617-78-6 (hc), 978-1-934617-79-3 (ebook)

Printed in the United States of America.
2 3 4 5

Elva Resa Publishing
8362 Tamarack Vlg, Ste 119-106
St Paul, MN 55125

ElvaResa.com
MilitaryFamilyBooks.com

Case and bulk sales available:
MilitaryFamilyBooks.com/Military-Culture-Shift

Additional content available from the author:
MilitaryCultureShift.com

*To the generations of veterans and their families
who endured the longest war in American history.
May hearing your story in a new way
be a balm to unseen wounds,
offer words when you have few,
and instill new confidence to lead the next generation
in an ever-changing landscape.
It has been the honor of my life to serve with you
in the most sacred places of the warrior culture.*

CONTENTS

Introduction 1
    The Military Culture 7
    Shifts Within the Culture 8
    What Can We Do? 10
    How to Engage the Story 11
    Healing Nature of Storytelling 14

GENERATIONAL INFLUENCE ... 17

1 Generational Perspective 19
    Before the Great Culture Shift 19
    After the Great Culture Shift 21
    What Defines a Generation? 23
    Willing to Walk Among the Rubble 29

2 Honor, Duty & Patriotism 33
    The Lost Generation (DoB 1883–1900) 33
    The Greatest Generation (DoB 1901–1927) 37

3 Forgotten & Vilified 43
    The Silent Generation (DoB 1928–1945) 43
    Baby Boomers (DoB 1946–1964) 46

WAR, MONEY & CULTURAL DIVISION ... 55

4 Follow the Money 57
    Congress and Military Funding 59
    Major Factors that Influence Funding 61

5 Law of Diminishing Returns 71
    Recruitment Before 9/11 72
    Generation X (DoB 1965-1982) 74
    9/11 & the Surge 77
    The Power of Imprinting 81

6 The Great Culture Shift 87
    Millennials (DoB 1983–1996) 88
    The Catalyst of the Great Culture Shift 91
    Sequestration 94
    The Apex of Positive Morale 95

MORALE SHIFT...101

7 Broken Promises ...103
    Strength of Our Force Unheard & Dismissed... 105
    The Culture Begins to Fracture ... 107
    "Something Is Making Us Sick" ... 110
    How Did We Get Here? ... 111

8 Constant Change ...115
    Change Is Inevitable... 116
    Time for a Change, Again ... 123
    What Are You Enslaved To? ... 125

SPOUSE CULTURE...129

9 Two for One Model ...131
    Teammate or Second Fiddle? ... 133
    Two for One... 136
    It's Not Personal, It's Business ... 138

10 Trauma Bond ...147
    Looping in Survival Mode... 152
    Trained Passivity... 155
    Is the Military Culture a Welfare State? ... 157

SOCIAL SHIFT...167

11 Social Media & Communication ...169
    Technology Shifts Families Further Inward ... 169
    Shift in Marketing & Communication ... 174
    Online Military Culture a Vulnerability ... 179

12 The Enemy Is No Longer a Country Away ...183
    Generation Z (DoB 1997–2012) ... 183
    Mass-Shooter Generation ... 186
    Rise of the Virtual World... 187

BREAKING POINT...193

13 Cultural Breakdown ...195
    Lockdown for the Military Culture ... 196
    Impact on Employment... 198
    COVID-19 and Gen Z ... 198
    Grief in the Military Culture... 201
    Mental Health Crisis in the Military ... 202
    Measuring a Cultural Breakdown ... 208

14 Shift of Authority & Influence 215
 Education As We Knew It.. 215
 The Way We Learn Has Changed 217
 Reformation of Authority... 221
 Currency of Trust & Authenticity 223
 Win Trust by Redistributing Power 226

SERVICE MEMBER CULTURE ... 231

15 Purpose & Perfection 233
 Piss & Vinegar & Full of Purpose 237
 Dangling Carrot... 244
 Hidden Culture of Fear & Perfection... 247
 People First 252

16 Final Straw or New Beginning?... 259
 Winning Hearts & Minds 260
 Operation Allies Refuge 262
 Aftermath 265
 Quiet Rebellion 271

WHERE DO WE GO FROM HERE? ... 281

17 More Than a Number: Retention & Recruitment 283
 Honoring the Cultural Narrative & Those In It.......................... 283
 Repairing & Restoring Relationships 289
 Leading the Next Generation 294
 More Than Plastic Army Men 300

Leadership Tips and Questions.. 305

Endnotes 319

Acknowledgments 341

INTRODUCTION

"He who wishes to fight must first count the cost."
—Sun Tzu, *The Art of War*

IN AUGUST 2021, STILL REELING from an unresolved global pandemic, the United States military scrambled to leave a country and cause it had invested two decades in. There was a sense of shock at first. When you devote your life and family to a global conflict for that long, war becomes a way of life. Service members and their families stepped into a new void of uncertainty.

American culture, with nearly four hundred TV news channels, half a million podcasts, and more than a dozen social media platforms all with live streaming, distributed opinions to anyone who would listen. As phrases like "This is our Vietnam" and "Was it even worth it?" circulated, I started getting calls from military spouses. Many of them I had worked with throughout the pandemic in 2020 when I had opened up additional counseling appointments and virtual speaking events to serve military families. Others I had experienced life with, worked alongside, or crossed paths with and learned their story.

Up until that point, families had endured a significant amount of stress with the high operations tempo of two wars. The COVID-19 pandemic had been excruciatingly hard for the entire country and world, but I watched the military community go through a psychological breaking point when asked for far more than they thought they could endure.

As I listened, I noticed two things. One, family members were deeply concerned about their service member's mental and emotional state. "It's almost like he is reliving his trauma all over again." Years, decades even, of work they had endured as a couple and family to bring purpose out of trauma seemed to suddenly dissolve. As the news carried on with opinions, interviews, and experts, it seemed like our entire community was paralyzed with the thousand-yard stare. Spouses were scared and didn't know how to support their service members who, once the shock wore off, were angry, devastated, and drinking more, and some were even trying to find their own way back to Afghanistan to help.

The second thing I noticed was a wave of unspoken anger in the military spouse community. It was taboo to talk about, so most pushed it down like every other opinion or need that was secondary to the mission and the pressing needs of their service member. I understood. My first book, *Sacred Spaces*, was about how my husband's deployment to Afghanistan changed our life and marriage. I had experienced the resentment of losing real, tangible time with my spouse only to have him come home changed by the violence and evil of war. Following that first reintegration, it took years for us to work out the fog that was smuggled home in his rucksack and, instead, assign some kind of purpose to it all.

Spouses were now afraid to start all over. They, too, had invested everything in this war and wondered if they were about to lose ten to twenty years of progress fighting for their family. They also harbored their own reaction to the withdrawal from Afghanistan and resented that it was, once again, not their turn to process it. The more I listened, the more I knew this was the sound of grief, exhaustion, and burnout. Like a shaken-up soda bottle, the community was ripe for three possible reactions—implosion, explosion, or both.

The pressure had been building for decades. Despite the best efforts of the Department of Defense (DoD) to address singular issues like mental health, suicide rates, and sexual harassment, families felt largely unheard. Meanwhile, threats of conflict with Russia and China were growing and the new approach to war-fighting involved a more dynamic, quick response force ready to compete with and deter foreign adversaries.[1] It was becoming clear there was no peacetime to look forward to.

For the last fifteen years, I've had the privilege to do life with, work alongside, and study the very tribe I live in. I have listened to and taught families and service members of every rank and season of military experience through various retreat and training environments, and in counseling and coaching. From the youngest and newest military members to our highest-ranking military leadership and their families, I have devoted my career to understanding the complex experience of our warrior culture, especially since 9/11.

As a mental health clinician, I view my role as a first responder to relational chaos. Trained to help people "stop the bleeding," I work with families to look for patterns and problems and then walk them through the healing process. My office over the years has provided a safe place for service members and their spouses to process their thoughts and opinions. Outside the office, military culture has traditionally viewed emotions and convictions as a vulnerability to the mission and a service member's career.

Beginning in 2011, I noticed certain shifts and trends in military culture that were affecting the morale of the force and their families. I also realized that the concerns I was hearing were not just isolated incidents, but common themes shared throughout the culture revealing a quickly growing problem of resentment and discouragement. Individuals and families were fracturing in a community I had come to love. Military spouse unemployment rates were not budging, geo-baching (intentionally living apart from a spouse) was becoming more popular, food insecurity was being exposed, and serious mental health concerns for the entire family (especially military teens) were being published.[2]

An ever-increasing drip of headlines revealed that our culture was not as healthy as it was once portrayed. It started slowly around 2013 with government shutdowns and nervous anxiety around a slimmed-down force. By 2018, headlines like "Military families angry about damage, thefts during moves"[3] and "Military families say base housing is plagued by mold and neglect"[4] started to surface. The murder of Specialist Vanessa Guillen at Fort Hood, Texas, in 2020 made national news, instigating a congressional investigation and a national conversation about leadership and sexual harassment in the military. Headlines like "Fourteen US Army leaders fired or suspended at Fort Hood"[5] further exposed leadership failures to the watching world.

Toxic leadership in locations outside of Fort Hood started to trend on social media as service members reported mistreatment in their ranks. Older generations who valued the institution's chain of command policy were shocked as younger generations publicly exposed internal issues in the force.[6] Suicides, already at concerning levels for both active duty and veterans, were on the rise for officers and military spouses. This headline, "The military has a suicide crisis. Its leaders bear most of the blame,"[7] seemed to connect the dots between service member suicide and something going on with the internal cultural climate.

Still, more headlines surfaced, especially as the withdrawal from Afghanistan came to its tumultuous end:

- Military suicides up 16 percent in 2020, but officials don't blame pandemic. (2021)[8]
- Senate report finds "mistreatment" of military families by housing companies. (2022)[9]
- One year later, troops and veterans involved in Afghanistan exit grapple with mental scars. (2022)[10]
- Navy investigation finds Hawaii water crisis exacerbated by "unacceptable failure of on-scene leadership." (2022)[11]
- Pentagon links leadership failures to violence, harassment at military bases. (2022)[12]

Also in 2022, every branch of the military was struggling to reach its recruitment goals.[13] Headlines like "Lawmakers press Pentagon for answers as military recruiting crisis deepens"[14] highlighted the seriousness of a national defense that was increasingly concerned about the numbers needed to "fight and win the nation's wars."

For those who were part of the military community during this time, it was overwhelming to take in every headline as it was coming out. Doing so would have meant processing what it said about the community we had come to love and depend on. In a world saturated with news, constant crises, opinions, and debate, we learned to tune out the headlines as white noise. Swipe up, scroll to the next story of war, political disagreement, harassment, and celebrity gossip.

When faced with a large and complicated problem, especially one involving an entire group of people, most of us will respond with a variety of reactions including adopting the problem as a new normal, settling for quick wins, or shifting blame. Some may adopt the current circumstances as attributes of the culture, claiming difficulty comes with the lifestyle. The military community, for example, uses phrases like "we chose this lifestyle" or "embrace the suck," even though the circumstances may be unhealthy, undesirable, or even dysfunctional. It is often easier to pacify our gut reaction that something is wrong rather than face the overwhelming task of trying to change the situation.

Some champion smaller, simpler solutions, eating the elephant one bite at a time. Large institutions, businesses, and organizations, for example, may be tempted to apply quick-win solutions to show progress on the larger issue. Many would argue that progress on one area of concern is better than no progress at all. In fact, attempting any solution is a noble effort. However, if the situation involves multiple problems and simultaneous rapidly emerging variables, it is difficult to celebrate progress while the overall situation continues to escalate.

Change, within a system or our part in it, is especially hard when it requires us to work outside our strengths. To keep hope alive and reduce the chance of weaknesses being revealed, individuals, couples, leaders, and even institutions try to solve systemic problems with simple and comfortable solutions. Date nights, for example, can help a couple reconnect, but when betrayal has blown a hole in the relationship's foundation of trust they cannot be the only solution. The couple is going to have to get into the messy, dirty work of repairing the damage or risk losing the relationship. Addressing the underlying root issue that makes a system sick is difficult, overwhelming, and often scary.

Another reaction is to simply point fingers or deflect attention in order to avoid the uncomfortable task of introspection. Some headlines, for example, blame recruitment challenges on Generation Z's inability to meet mental and physical health requirements, or the DoD becoming too progressive, or the nation's citizens being too far removed from the military. When Americans, Congress, interested recruits, and military leaders all contribute to a problem, asking one or two groups to shoulder the blame does not absolve the entire community from self-examination.

The current state of wellness and morale in our military culture is not only a large and complicated problem, it is a "wicked" problem. Design theorists Horst Rittel and Melvin Webber were the first to coin the phrase "wicked problem" in 1973.[15] Wicked problems are problems that are difficult to nail down, have constantly changing or emerging information that is confusing, involve stakeholders who have conflicting values, and are extremely difficult to solve in that they tend to "mischievously" react to solutions with new emerging problems (which is where the negative connotation of "wicked" comes in). In addition, those who do present solutions "have no right to be wrong, in that they are liable for the consequences of the solutions they generate,"[16] especially when those who are touched by the solution are greatly impacted. The social complexity, or variable of people, is what makes these problems difficult to solve.

It is hard to say who started the use of the phrase "wicked problem" in the military culture. It is frequently, and often casually, used to describe problems that are extremely complex, multilayered or dimensional, and presently emerging. General Stanley McChrystal is said to have used it in his 2009 classified assessment of Afghanistan;[17] the popular "spaghetti" PowerPoint slide on the American strategy in Afghanistan is often referenced as a visual example where he was quoted as saying, "When we understand that slide, we'll have won the war."[18] War, the pandemic, and climate change are all examples of wicked problems, where solutions are not only hard to find but simple solutions create even more problems to solve.

The situation the military community is facing after two decades of war, a pandemic, and the withdrawal from Afghanistan is far more complex than the tally of those very big historical moments. It involves additional compounding layers of complexity grounded in cultural tradition, quickly evolving variables of the modern global culture, deep relational wounds, and most importantly, it involves people. Internally, the military culture is fractured by generational differences, exhausted from unrelenting expectations, and challenging the mission-first mentality. The outward manifestation of this wicked problem is a community that is disintegrating while facing the biggest challenge in recruitment since the military moved from the draft to an all-volunteer force.[19]

Since the withdrawal from Afghanistan, the country and DoD have quickly shifted their focus to modernization and new emerging threats to deter, while service members and their families continue to recover from the emotional, physical, and social impacts of the longest war in American history. Without a plan to reconstitute a battle-weary culture, the struggle will continue with the current mass exodus of millennials and Gen X military families and the recruitment of a younger generation weighing the visible cost of enlisting.

We must be willing to intentionally address the systemic root issues that are disintegrating our military culture. A helpful framework for this is Rittel and Webber's design theory, which evaluates and assesses problems before attempting to test solutions. It "revolves around a deep interest to understand the people" for whom we are providing solutions.[20] Design theory invites leaders to regularly empathize with people and actively challenge or examine assumptions and knowledge about a problem, especially when the problem is complex or wicked. This approach reminds us to slow down in order to examine the wicked military culture problem in the context of the people experiencing it.

For the last decade, the military has been exploring empathy in leadership, calling it a soft skill that leaders need to develop. Nothing about war or equipping men and women for war is "soft" and the last thing I believe service members want is to be "soft" or lead in "soft" ways. In fact, Gallup has studied talents and strengths for years and found that empathy is a talent that not everyone can leverage easily. After working with thousands of service members and their spouses on identifying and developing their strengths, I can attest that empathy is not among the most common top five strengths of service members. So while some may struggle to leverage it quickly as a personal leadership strength, no one gets a pass to not try. For now, however, I invite you to leverage curiosity.

Storytelling can be a helpful tool that not only aids our understanding of complex problems, but helps develop empathy as we lean in with curiosity about the people within the story. This book is the story of the people in the military culture, especially after experiencing two decades of war after 9/11. We will pull apart many contributing variables, compounded over time, that resulted in the complex problem we face today and present a narrative to help you develop empathy for the people involved.

If you are or were in the military, it is tempting to believe that you already understand the ins and outs of this tribe. I know I have thought this. When we have lived and breathed the values and traditions of the culture in which we work, it is easy to become overconfident and miss how the culture is evolving right in front of us. Our community has been so overwhelmed over two decades of war, that many military leaders step into their long-awaited leadership role reactive and living in five-meter targets rather than mindful of what has changed.

In order to help you truly understand what brought us to where we are today, let me tell you the story of the people you think you know but could know better. The people you lead want to be seen and are more likely to follow when they experience a leader who shows a deep interest in getting to know them.

I also wrote this book to inspire you to consider the influential part you play as a leader within your own circle of influence. Leaders don't always have to hold office or wear a specific rank, they just have to be willing to be the one to take a step toward change. When that step is in a direction that is beneficial for others around you, people willingly follow. Then, followers step up and want to emulate what you have modeled, turning your brave step into a movement.

THE MILITARY CULTURE

You will find throughout the book, my use of the term *military culture*. For the sake of clarity, I am referring to the collective group of service members and families that have and are serving. There are also subcultures within the military culture, varying in customs, traditions, and experiences, such as special operations, different branches, enlisted and officer, military child and military spouse, and a civilian workforce that includes contractors and policymakers. There is no way to capture the millions of perspectives and unique stories that intimately affect lives and families every day. But I believe there is a larger cultural narrative that can be told in order to help us begin more productive conversations about our individual stories and the ways we can avoid repeating past mistakes.

As part of the process of finding this larger cultural narrative, I studied trends and issues and tested which were isolated incidences and which were culture-wide. I spent countless hours reading research, literature reviews, history books, documentaries, and news articles; talked with military families and leaders; interviewed and consulted experts; and polled the community to capture as accurate a narrative as possible of our military culture.

I gained a first-hand understanding of service members' perspectives by traveling overseas with the US Secretary of Defense to Afghanistan, Iraq, and Turkey, and onboard ships in the Persian Gulf. This is my tribe, my people, and my friends.

Much of what I cover in this book focuses on the majority of the military population: families composed of a service member, spouse, and children. I did my best to mention, where appropriate, the uniqueness of dual-service couples, single soldiers, and couples without children. There are many other specific groups and topics deserving of a separate book to cover them respectfully, including the perspectives of lawmakers, policy experts, same-sex marriages, sexual harassment survivors, female service members, and the health and well-being of our military children. All stories are valuable to the larger cultural narrative.

Ultimately, while there are hundreds of books on leadership and even more on war, military history, and strategy, what I felt was missing was a true understanding of military culture in the context of a deeply layered cultural story. We, as a country and as leaders, cannot make good decisions about the direction and capability of our force without a good culture and climate check. Unfortunately, our culture and climate are in desperate need of attention and repair.

SHIFTS WITHIN THE CULTURE

If you are already a military leader directly serving military members and their families, you've likely noticed a shift in our military culture in addition to the weariness from war and the pandemic. Military leaders say it is harder to "do what has always been done." While the culture indeed has gone through a significant shift from what it once was (hence the title *Military Culture Shift*), this overall feeling that something is different is due to many simultaneous shifts.

Generational Shift

As the world reopened postpandemic, military family events and programs were less successful, leaders were frustrated with changing work dynamics, and approaches that worked before weren't quite working as well. This was partially the military culture mirroring the same ripple effects seen in civilian culture. However, I believe it also marked a moment when there were fewer "old guard" and more "new guard" as a new generation filled a large amount of the force, shifting perspectives.

While we have been fighting multiple global conflicts, millennials have joined, served, and are now filling key positions of leadership. Gen Z is already serving and now building families of their own. The next generation is only a handful of years away from joining as well. Bridging gaps between military generations starts with curiosity and empathy. When we are willing to be curious about another person's story, we are less likely to try to convince them to adopt our perspective formed from our own story, and we instead become willing to look for truth in both. Both contribute to the evolving culture.

The generational shift is just one of many that we will look at throughout the narrative. Seeing the military culture shift as new generations enter the community will help us understand some of the evolving trends that may be very different from what we are comfortable with or how we prefer to lead. We will explore each generation's background, core values, how they shaped the military community, and more importantly how they developed their own unique understanding of what the military culture represents.

Influence of War & Money

While congressional funding may not sound like an interesting topic, the military and all of its programming and resources are funded by Congress, stemming from American tax dollars. This not only ties Americans to the military community in a meaningful way, it also makes the military culture vulnerable based on the availability of funds. Money is a very influential variable in how service members and their families experience even their basic needs, such as housing and food, as well as their experience of deployment and war.

Additionally, World War II revealed that war not only brings the country together, it can be good for the economy as privatized contractors have incredible influence on our capabilities for war and modernization. The defense budget is a means of communication to the nation's allies and adversaries and is a deeply complex topic on its own. Understanding the rise and fall of available funding and how it impacts the military culture, especially during a two-decade global conflict, is eye-opening to the culture's morale and well-being.

Influence of Social Media & Technology

The evolution of technology touches and shapes every generation. Apollo 11's successful mission to the moon, inventions like the computer or graphing calculator, and the rapid expansion of the internet all shifted the way different generations work and learn. While the rise of cell phones and social media has significantly impacted all of society, its unique impact on deeply ingrained social customs and information distribution within the military culture is worth exploring. A community that's often isolated from external family members and friends depends on each other to share in the difficulty and burden of war. Shifts in technology and social media shaped the culture's cohesiveness and communication, and, as a result, shook the very foundation of support the community needs to thrive.

Shift in Authority & Trust

When military leaders describe change in the military culture, often what they are referencing is a shift in the way younger generations view authority. Millennials and Gen Z are demanding significant change from brands, employers, and leaders. Trust in institutions, such as schools, the military, and congress, has been steadily declining for the past forty years. Gallup reports that Americans' confidence was historically low in 2022, then again in 2023.[21] Inspired by the global saturation of information and awareness, Gen Z may be the generational tipping point as it demands transparency over perfection and an authentic relationship based on trust.

In an institution that is strictly structured on a hierarchy of rank and regulated through obedience, authenticity and transparency feel like vulnerability. The military institution must realistically rely on numbers and a mission-first mentality in order to meet its objective to provide national security. Without the human dynamic of leaders inspiring, motivating, and even repairing their relationships with those they lead, there will be no one to fill the ranks.

WHAT CAN WE DO?

There is a palpable tension between mission accomplishment and taking care of people. It's why the 2019 "People First" strategy was received with such a rousing welcome from tired souls who then struggled for years to figure out exactly how to implement it in the face of mission accomplishment. When leaders eventually find themselves pushing paperwork rather than picking up a weapon, it is often easier to simplify their role to its most logical primary purpose: to "deter war and ensure our nation's security." After all, that is what the DoD states on its website and what many envision when they enlist or commission. However, the role of families as the backbone of support and force readiness has also been threaded throughout our history and funded as part of the mission, and families are included in a military leader's responsibilities. In 2014, General Raymond Odierno said, "The strength of our nation is our Army; the strength of our Army is our soldiers; the strength of our soldiers is our families. That is what makes us Army strong." His statement resonated across all branches for a reason.

Most leaders I consult with admit that the day-to-day tasks of working on the mission are not anywhere as difficult as the relationship dynamics of those they lead. Managing a multigenerational team is difficult enough; doing so in the midst of the largest technology shift in five hundred years makes it overwhelming. Most leaders add that the fear of being "canceled" or publicly shamed leaves them paralyzed or wanting to avoid leadership roles altogether.

All of these shifts and how they interact with the culture have been fascinating to me over the years. I love to look for patterns between people and then help inspire new, healthier patterns in order to reduce confusion and conflict. I've applied this same process to teams and organizations and have been observing these larger patterns in the military culture for years. In many ways, stepping back to look for patterns makes the issues seem a little less overwhelming and opens our hearts and minds to the people.

My goal is to share some of these strategies to help you succeed at both the mission and taking care of people. The best we can do is acknowledge this tension and strive to pull the two missions together. The institution can thrive when the people thrive. It is for people that warriors go to war, and for people, they return. Therefore, in order to win the hearts of future warriors, we must be willing to win the people, especially the families these warriors would choose over continuing to serve. We can no longer consider families secondary to the mission, but rather a critical component of mission success. In fact, I believe the essence of the People First strategy is that people ARE the mission.

With this in mind, I implore you to understand the entire culture of warriors and the families they love and support by engaging with their stories.

HOW TO ENGAGE THE STORY

When you finally arrive at a position of leadership, it is exciting to lead people the way you envision. You have likely considered the ways you want to organize yourself and your team, and how you will implement your own leadership style. It can be deeply frustrating to be challenged by a generation that is asking for something different from what you set out to offer or what you experienced from previous respected leaders. Some leaders may dismiss this as immaturity and move forward with what they know, losing trust and loyalty with people in the process.

Listen to the Story

As a clinician, I've learned the value of taking time to hear the story leading up to the presenting problem. You must be willing to hear the story of the people in order to accurately identify and address the problems in front of you. Every great novel builds a relationship between the reader and the characters through a story of significant events that ultimately lead to the climax of the character's plight. At that moment, a reader feels tension and longs for a resolution. If you were to jump into the middle of the story without first understanding a character's motivations and perspective, you would be less invested in the outcome.

Most service members, volunteers, politicians, and civilian contractors rotate in and out of key positions regularly. Leaders have a very small window of time to make an impact, leave a legacy, or amass significant bullet points on an annual evaluation. It's difficult to get continuity from the leader who previously held your position, much less a briefing on the culture you've inherited. So, well-meaning leaders jump into the middle of the story with great enthusiasm and believe they know enough to succeed, only to burn out soon after.

We are at a historical inflection point in our national security and defense. People are leaving the service lifestyle, fewer are joining, and senior leaders who spent two decades at war are flaming out in moral and ethical failures. When you are tasked to lead a culture in this state, it is helpful to know the significant events that built to this moment. Service members and their families have been trying to share their stories for decades. Now, pieces of them are being shared freely over social media, across all platforms. Families are even being called to testify before Congress. By curating as much of the cultural narrative here as possible, I hope to arm you with a better understanding of where you land on the timeline, who is in your ranks, and how to bring order to what feels like cultural chaos.

While it may be tempting to jump to specific chapters on topics of interest, I invite you to experience the narrative. Wicked problems challenge our assertions and invite us to lean in and discover something new within ourselves. Leverage curiosity for the people you work with as we consider solutions.

Engage in Dialogue

The generation of military leaders and spouses before me taught me that I should always highlight the positives rather than expose or complain about the issues. I have gained so much wisdom from them, however, as I raise two Gen Z teenagers, I can't ignore the fact that this generation demands authenticity. They are insisting on change with their feet, their votes, and their voice, both verbally and online. If we want to lead well, we must be willing to focus on the truth of what is in front of us, no matter how complex or disruptive it feels.

It's important to do our best to put aside emotional and experiential bias so we can hear information honestly and clearly. One way to do this is to hear from a variety of service members and families whose experiences are different from ours. Doing so has helped me gain perspective and, ultimately, compassion. Our culture is constantly changing and evolving and so are the topics of importance. Rather than offering solutions that would only lock the relevance of this book into a certain point on history's timeline and rush us into more quick-win solutions, I instead invite you into dialogue.

For example, I'm aware that it is impossible to completely remove my experience as a Gen X officer spouse and mental health clinician as a lens through which I see culture. In some ways, I see it as a strength and a unique perspective that I hope will be an asset to you. Depending on what lens and personal bias you bring, there might be moments you disagree, see another angle, or have information that I did not. Are you an officer? Enlisted? Family member? Policy leader? Lawmaker? Baby boomer? Millennial? With more than two million people in the active-duty culture alone, there is a lot more that could have been added to this project than taken away. Rather than reading from a critical lens, allow yourself to be stretched.

This process opened my eyes to my privileged perspective as an officer's spouse. I've spent years glorifying "the good ol' days" of nice housing and supportive neighbors who helped me survive a very difficult season of deployment and reintegration. Recently, I saw a social media post of an enlisted spouse who had just given birth. She asked if anyone was available to pick up her husband and bring him to the hospital because they didn't have a car. That post was an important reminder to me that we all have different perspectives on convenience, comfort, and community.

Rather than exclude it, I offer glimpses into my bias and also authentically share when I notice it as a humble example of what I am calling each of us to do. I have taken great care to recognize my own bias that could skew my perspective of the bigger story and encourage you to reflect on views that may prevent you from hearing the stories of others and leading well.

Leadership tips and questions for discussion are available at the end of the book to offer additional ways to spark discussion or debate, open dialogue on tough issues, and inform you to make confident decisions. We each have something to offer to the conversation, so I hope this will encourage you and those around you to share your individual stories.

Identify Layers in the Story

As I tested this material over several years in front of service members, leaders, spouses, and organizations, I found that the best way to communicate such a complicated analysis of a culture that spans several generations is to teach it in layers. There are some topics, such as the Vietnam War or US policy, that I could have expanded on more but, since information about those topics can be found in many other resources, I opted instead to focus on presenting issues of our current force and understanding how those topics impact differences in generational leadership styles and family expectations.

In some cases, I have pulled out connected layers to explore in more depth. For example, it is difficult to cover the millennial generation without taking a look at how the introduction of social media changed the way the military community communicates and gets information. The impact of social media on our culture is an additional layer that is worth covering in a separate chapter. While reading this book, and as you apply this information to your leadership practice, recognize that you may discover overlapping and sometimes competing layers of information and cultural influences.

Create Understanding Rather Than Blame

As our presence in Afghanistan came to a difficult end, I reached out to a mentor of mine, retired Colonel Robert "Brad" Brown, and asked if I could interview him. He had commanded our unit through a historic battle that is now in the history books as the "bloodiest and most decorated battle" from America's time in Afghanistan, resulting in two medals of honor, two distinguished service crosses, nine silver stars, and many medals with valor, and the loss of eight of our men. People wanted someone to blame and, even though the finger could be pointed farther up the ranks, we watched as Colonel Brown humbly and with dignity carried the weight of wrongful blame we knew was misplaced.

As Americans and veterans displaced their grief and anger, looking for someone to blame for the way we exited Afghanistan, I asked Colonel Brown how we should handle our desire to blame while processing the difficult truth of our loss.

He shared: "Eventually, more of the facts will be understood of who did what, and why. And the military leadership is in a very difficult position. They have to do their own analysis to ask: What do I own about this? What did I fail to do? What could I have done differently to make this go down in a different way? The same for me as a commander. I was not going to go to *The Washington Post* in person and say, 'Hey, this is someone else's fault.' The chairman of the Joint Chiefs of Staff can't come out and say, 'Hey, we got handed policy constraints that were unworkable to do this in good order and we came up with a plan that was kind of a Hail Mary and we hoped it might work.' This was a constraint. He can't say that. He would never say that. No senior military leader is going to come out and blame civilian officials. Even if the civilian officials were to blame, you're not going to get that. So those that are demanding accountability of the military leadership to come out and throw someone under the bus, that's not how the system works. That's not how the military serves our country or its elected officials. The ballot box holds elected officials accountable and the military leaders are accountable to civilians and their superiors. Over time, the truth will come out and we will see who made decisions where and hopefully learn from it."

With this wisdom, my intention is not to accuse leaders or the DoD for the state of our current force. In fact, it is quite the opposite. Only time and history can reveal the truth of whether there is a place for blame. Colonel Brown's words and example are a reminder that rarely is there one person or entity responsible. I hope this book informs you of the current state of our people, not so we can blame, but so that we can get to work on solutions.

I also hope to encourage you to leverage curiosity, empathy, and a willingness to learn from those in even the highest of leadership roles who also have a story to tell. Together we need to hold two truths: that people are doing the best they can and that the institution must have people in order to function.

HEALING NATURE OF STORYTELLING

The process of finding solutions and healing often requires slowing down and finding the truth, even if it is hard and even when it is complex. Many, many, military families are thriving in the culture. Research has shown, and I have seen it in my own family's life, that the military lifestyle creates grit, character, and skills that families may not have otherwise acquired. The 2021 Active Duty Spouse Survey put out by the DoD reported that 75 percent of those surveyed were food-secure, and 49 percent were satisfied with the military lifestyle. These stories are just as important! However, another way to read the same results is that one out of four families is struggling to put food on the table, and more than half are unsatisfied and willing to leave. We must be willing to hear the whole story, and maybe even lean into the truth we don't want to hear.

Most individuals (and often families, too) join the military because they want to be part of something bigger than themselves. Many join to provide a better life for their family and a future they couldn't otherwise provide on their own. After two decades of high operations tempo, a global pandemic, political and national division, disappointments in the culture, and deteriorating relationships in the home, weary and burned-out leaders are joining the global movement to quietly quit. Some are turning in retirement packets earlier than they intended, while others are choosing to give less in their final years to a career that has taken more than they expected.

Finally, if you are a service member, military spouse, or civilian employee, I hope this book is a balm to the wounds you have accumulated over the years in service to our country. I hope you feel seen and understood, and that you find your voice and your family's voice reflected throughout the pages of this book. I hope you will read and feel less alone. You are not the only one.

I heard K.J. Ramsey, an author and counselor who works with trauma, say in an interview once how important it is for individuals to "see and name rightly what happens to our whole selves in systems that treat people more like products or objects than people."[22]

I like what Ramsey said next and would like to challenge you to adopt it as a way to begin the healing process. She states that the goal is to help individuals "better name the dynamics in their story. Not to blame or point fingers or blow up everything. But if you can name rightly the wound, then you can tend it well."[23]

If that is not you, and you have thrived in every moment of this grand adventure of military life, I hope you will hear the voices of the families reflected throughout and see an opportunity to lead well, whether you are a military leader, spouse, civilian, or policy maker. We cannot continue to look away from the wounding of our nation's service members and families while expecting them to continue to carry the burden of our nation. Families are breaking under the weight of what has been asked of them. Patriotism is at an all-time low and, even if it was right for national security, modernization has been prioritized over people. I hope you will be able to bring your unique strengths and talents to the table and cast vision for the people around you.

Now more than ever, loyalty is won or lost in how we care for our most valuable asset: people. It is possible to positively shift military culture while serving the mission, and we have to do both.

GENERATIONAL INFLUENCE

"I have had enough experience in all my years, and have read enough of the past, to know that advice to grandchildren is usually wasted. If the second and third generations could profit by the experience of the first generation, we would not be having some of the troubles we have today."

—Harry S. Truman

Chapter 1

GENERATIONAL PERSPECTIVE

FOR MANY YEARS, THE CULTURE of military family life was driven by deep bonds formed over a love of country and the other military families you experienced life with. Those relationships were further forged in the fire of shared unspeakable combat trauma and a tribal mentality that held our community together and connected us to the veterans of the past. The American culture reflected some of what we were experiencing as studios put out movies like *The Hurt Locker*, *Black Hawk Down*, *We Were Soldiers*, and the popular *Band of Brothers* series, just to name a few. After 9/11, the entire nation seemed ready to support the mission of defending our country, whatever that meant. As late as May 2011, I remember being glued to the television as President Barack Obama announced the death of Osama bin Laden, and our entire neighborhood came out of their homes to celebrate in the streets.

But then something changed. I call this the "Great Culture Shift."

BEFORE THE GREAT CULTURE SHIFT

Like many military families, joining the military for us was not only about serving our country; it provided financial security for our family. We were delayed in following hundreds of thousands of service members after 9/11 due to graduate school, but by the time we joined in 2007, families had already gone through two or three deployments. For many new families, sign-on bonuses at the time made the decision to join easier and financially promising. It was not uncommon to hear of $25,000 bonuses, educational degrees paid off, and new soldiers driving brand-new trucks and sports cars to celebrate.

When we arrived at our first duty station at Fort Carson, Colorado, we were offered a newly constructed home in a new neighborhood on the installation. The force was growing significantly, and housing demands were more than housing contractors could keep up with. After barely escaping the housing crash, we felt great about our decision. The military lifestyle was already proving better than anything we could have afforded in the civilian world.

When we joined active duty, we were told immediately about deployment cycles already in motion. Accepting this assignment meant my husband would deploy to Afghanistan a short nine months from our move-in. For conventional forces, deployment cycles included one year deployed, then one year home. While we didn't see it at the time, that level of predictability gave us time to wrap our minds around what to expect for the next two to three years of this lifestyle.

During my husband's deployment to Afghanistan in 2009-2010, almost all the service members on our street deployed as well, leaving military spouses outnumbered by their children twelve and under. We had one neighborhood dad who graciously offered to hang Christmas lights or help wrangle kids after work. I relied heavily on my neighbors during that time for my sanity and survival. I did my best to be reliable for them as well. We took turns running to the grocery store, watching each other's kids, hosting birthday parties, and celebrating holidays together. On days we were especially tired or needed a break from chicken nuggets or cooking an entire dinner, we shared a potluck dinner.

Our mother hen of the neighborhood was Mama Pam, a loud Aggie whose hugs were bigger than her Texas pride. Her kids were the oldest on the street and we learned much of our parenting and navigation of military life from her. She taught us how to potty train, how to teach respect and selfless service to our children, and how to get involved with our assigned units. She would remind us, "Don't complain if you aren't willing to do something about it." When tragedy hit a little too close to home, she and her kids led the way as we huddled together and served each other.

During the deployment, our squadron had regular family readiness group (FRG) meetings where we were briefed on any and all new information. A paid position called a family readiness support assistant (FRSA) was assigned to our unit and was often filled by a military spouse. This person would distribute information through phone calls and emails, coordinate and organize meetings, arrange childcare, and even organize family events. Unless you wanted to be alone, you always had a place to celebrate a holiday dinner, and even then, most wouldn't let you be alone for long. With an entire brigade or division deployed at once, everyone was "in the suck" together. So together was how we did life.

I remember thinking that first year that it felt like I was going back in time when I drove through the gate from Colorado Springs. Like a portal to the 1950s, most military spouses were female and the most popular careers were selling Pampered Chef, Mary Kay, and Tupperware. As archaic as it sounds now, these were the most portable careers at the time and could be accomplished while children napped and went to school. The spouses who worked outside the home were in professional and highly portable fields, such as nursing and education. However, the majority of spouses opted not to work and stayed home with their kids because of the high degree of flexibility required by the unpredictability and work hours of their service members.

The ability to communicate with our spouses overseas was better than anyone before us had ever seen. Skype wasn't perfect, but it was better than the tools of the generation prior: limited to writing letters or going months with no word from their service member. Email was mostly reliable, and depending on where your spouse was deployed, you could almost guarantee an email every day. If your videos were short and low enough in quality, you could email a video message back and forth. If you were lucky enough to have a BlackBerry, you were learning how to message each other in real time, even from the playground.

At the end of that first deployment, everyone's husband on the block came back almost at once. This is where I learned a new cultural lesson about military life: families seem to disappear when the service member comes home. Family time was precious. You had a year or less to reintegrate, recover, and rebuild before they were off again for another scheduled deployment. If you weren't prepared for it, it was quite a shock to have the village of military spouses you leaned on suddenly devoting all their energy to their marriages. Some families relocated not too long after, but if you were able to stick around long enough, the service members also got to know each other and the neighborhood would come back to life. Clearly, there was something transformative about a community having a shared cause, values, and a common enemy. But another component was a shared need that we could fill for each other.

Chances are, if you came into the military community before 2011, your experience of the culture was similar. Service members from that time describe deep bonds formed and strengthened amid shared deployment experiences.

AFTER THE GREAT CULTURE SHIFT

I first noticed the Great Culture Shift around 2011, after moving and settling into another new house on post at Fort Stewart, Georgia. People weren't coming out of their homes. Playgrounds weren't busy. At first, I figured there was something special about my experience at Fort Carson, or perhaps I had looked at those first years of military life through rose-colored glasses. But as I spoke with others who had relocated across the globe, they described the same thing. Positions no longer had funding, and people were laid off. Anxiety started to rise as rumors spread of service members getting handed "pink slips" (notices of dismissal) while still on deployment. I noticed other changes, too. As we began a second deployment, our neighbors did not. They were still closed inside with their families intact or, in some cases, never introduced themselves at all.

Working part-time as a mental health counselor off post, my waitlist of military families grew. Soon, I heard the same concerns behind confidential doors. Something very big was happening and my gut told me to pay attention. It was then that I started to lean in to even the smallest shifts happening in and out of the military culture. Sometimes, it would start with a topic brought up by a military spouse in a counseling session only to pop up soon after in other sessions, at group events, and even playing out in my own military family.

By 2015, the military's family support system seemed to be falling apart across the globe. People weren't showing up for in-person events, even if it was to distribute critical information or enrich their marriages. Facebook pages and virtual town halls became a way to get information and find resources after moving. Military spouses were quickly gaining a reputation as negative and dramatic. And soon, cyberbullying increased as the derogatory nickname "dependa" was used to bully spouses who seemed entitled or negative. This kept some families off social media altogether.

Around 2016, generational differences emerged as baby boomers took the most senior leadership positions in the branches. Gen X executed large-scale military movements, while millennials stepped into boots-on-the-ground command roles. Communication and leadership definitions were evolving while older generations continued to lead as they had been led. Families asked for more support while an oversaturation of programs and resources continued to be underutilized. As the world rapidly changed with information at everyone's fingertips, politics splintered the country, and the line between foreign and domestic enemies blurred. Cancel culture was only a tweet away.

As my career evolved to include consulting with large-scale organizations and niche groups like special operations forces, I tested some of the trends I was witnessing to see if they resonated in other groups and branches. Across the board, whether it was military spouses, couples, service members, or leaders, the top concerns were resentment, identity and purpose, and burnout. We were losing the will of the people directly tasked to execute the mission.

In the years since, I've seen a steady decline in what once made this lifestyle endearing and enticing for the next generation. In decades past, around 80 percent of those who served directly knew a family member who had served before them.[24] Today, many veteran and active duty parents discourage their kids from joining. Senior military leaders, command teams, and organizations share that:

▷ Generation Z military families are less eager to self-identify as military families.

- Those that do are less likely to show up for events that support and/or develop well-being and resiliency.
- Traditional methods of distributing information are no longer efficient or effective.
- In-person social events are suffering or stagnant.
- In 2022, each military branch struggled to reach recruitment numbers[25] with the Army ending the fiscal year close to twenty-one thousand under.[26]
- Fewer families are living on installations than ever before.
- Traditional autocratic and hierarchical leadership that depends on trust and respect is being challenged by younger generations.
- Experienced and seasoned service members are leaving as they reach the twenty-year retirement mark rather than accepting promotions.

What happened? How can the DoD sustain a willing and ready force that is dissolving at this rate? How concerned should we be that the family culture and traditions as we knew them are disintegrating?

The most common explanation is that generational differences must be getting in the way.

WHAT DEFINES A GENERATION?

Generational labels are referenced in media more than ever as the DoD seeks to relate to its newest generation of recruits, many of whom have different values and expectations from leaders of other generations. Social media has provided a forum where generations openly jab at each other in jest, while differences contribute to very real barriers in communication, confusion, and a widening gap between those who have experienced wartime and those who have not. Though it is just one of many variables that affect current circumstances, learning what defines each generation is a great place to begin.

Generational differences are embedded in our everyday relationships, including family, work, and social encounters like waiting in line at the grocery store. As a new generation enters young adulthood, its members offer innovation and a new perspective that changes human culture. Businesses and institutions, inspired to remain competitive, face the difficult task of adapting their branding and internal culture to recruit this new generation while still encouraging a productive multigenerational environment.

Conflict and misunderstanding can exist between any generations. New generations tend to swing in the opposite direction from their upbringing. The dilemma, for any institution or business, is establishing a culture that is flexible while honoring the most important components that define each generation.

Communicating through these differences is imperative if we expect a healthy work or social environment. Some people may cling tightly to their generational category while others resist a generalized label that threatens their uniqueness. Similar to personality tests, generational labels are not meant to confine, but rather to provide insight into cultural changes, increase understanding, and act as a springboard for further dialogue about individual perspectives.

Part of what we are seeing, and will continue to see, in the military culture is the demand for flexibility in a culture that has previously succeeded with, and been defined by, inflexible structures and systems. To cater to that request entirely would be to change the DNA of the organization, which is one of the many reasons why the DoD is struggling with retention and recruitment. Leveraging more curiosity around generational differences and expectations can help discern which areas are more adaptable and which must remain inflexible.

Regardless of which generational category you are in, you likely feel the tension of either wanting to bring about change or to prevent it. Young generations, inspired to make their mark on the world, bring new strategies and ideas, delivered with a sense of urgency and experimentation. Meanwhile, older generations tend to be more opposed to large organizational shifts, defending their years of experience with hard-earned knowledge, position, and working style.

Recently, I was brought in to consult a multigenerational team of active duty military and civilians (mostly retired military) on communication and teamwork. The boomers and millennials were clashing over different styles of communication and work ethic, and they were hoping to reduce conflict and work more efficiently. To break the ice and introduce a level of play, I gave them all Play-Doh and asked them to make something that represents who they are. One older gentleman who was a former special forces operator made a ball and called it a rock. He announced, "This is me. I'm not moving. People will come and go, they may even stub their toe on this rock, but I've worked too hard, and I know what I believe in. There's not much more I'm willing to change."

He was direct but endearing, molded by years of purposeful military service, constant change, and striving through promotion lists. He was also motivated by the strong work ethic and values boomers are known for. He spoke of the years of structure that shaped his career and positively etched his soul. All of us will find ourselves at one point or another wanting to be valued and respected for the years we've put in. After all, isn't that how leadership is earned?

This resistance is normal. In *The Fourth Turning: An American Prophecy*, William Strauss and Neil Howe examine generations throughout history to find predictable patterns of cultural change. Strauss and Howe share that each generation exerts its own version of dominance over the two that are younger as they enter adulthood. No consecutive generations are alike. "Your generation isn't like the generation that shaped you, but it has much in common with the generation that shaped the generation that shaped you."[27] I see this in family systems work as well. Values that drive one generation are challenged in the next. Then, the pendulum shifts again as the next generation challenges the one before.

To learn how to truly work together, we must be willing to value innovation, healthy debate, and creativity as necessary tools for growth, success, and relevance. Leaning into how each generation brings these tools to the table is crucial to being a successful leader and working on a team. Yet, much of the generational discussion, especially around the need for empathy in the workplace, doesn't easily fit into the military institution, where hierarchy, rank structure, and obedience to orders are critical and culturally ingrained. Slowing down to appreciate, much less listen to, generational differences seems counterproductive.

While each individual and family has a story, each generational cohort has a common story, and the military culture as a whole has a story. By learning each generation's cultural narrative, we gain insight into motivations and values. We also begin to identify our own story, helping us distinguish between reactions clouded by our own generational perspective versus gained wisdom. Several factors influence how generations are defined and shaped over time.

Geographic Location of Upbringing

Typically, generational categories are organized by birth year, however, it is quite common for someone to fall into a category that does not quite describe their worldview. Jason Dorsey, president of the Center for Generational Kinetics, researches generational characteristics in the workplace. In order to help distinguish between the generations, Dorsey describes several life markers and experiences that should be factored, in addition to date of birth.

For example, the geographic location you were born in as well as your culture shapes your worldview. Were you born in an urban or rural environment? The conservative Bible Belt or the Pacific Northwest? What country? Did that country have access to technology and steady information distribution of world events? Urban locations often surge with the first wave of new technology, while rural areas may not have the same advantage until years later.

Season of Life

In 2023, the Pew Research Center stated that it planned to end its use of generational categories when conducting and reporting on generational differences. They stated, "We'll only do generational analysis when we have historical data that allows us to compare generations at similar stages of life."[28] What Pew and many other social scientists conclude is that there are traits that have more to do with the season of life, such as young adults gravitating toward more liberal views. With as much polarized division the country has endured, blurring hard lines between generations helps "avoid reinforcing harmful stereotypes or oversimplifying people's complex lived experiences."[29] It is an important distinction.

With the military continuing to reach the youngest generation, generational labels are still used. It's important to consider the experiences of each person as well as what behaviors and beliefs are influenced by their present life stage that might change as they enter the next. In this book, we'll consider traits and beliefs that tend to stick with the cohort over time to determine their contribution to and our response to their retention.

Parenting Style

Dorsey, similar to Strauss and Howe, contends that parenting plays an important role. For example, the Generation X nickname "forgotten generation" stems from baby boomer parents who pursued careers and financial security throughout the 1980s and 1990s, leaving their latchkey kids to play outside till dark and cook their own TV dinners in the microwave.

My birth year is right on the Gen X/millennial line. My cousin, just three years younger, was raised much differently by his parents, who were on the tail end of the boomers. They parented intentionally, following their sons from Indiana to Georgia when they graduated college. For them, parenting was about being attentive and heavily involved. My cousin is a millennial through and through, while I identify much more strongly as a member of Gen X.

Technology

Technology has an incredible way of shaping culture. Every generation has been influenced by some level of advancement in technology and has experienced its impact on education, communication, and connection. Social media, for example, drastically changed the way we view human connection, as well as marketing and information distribution.

Recently, I was visiting a college with my son where a baby boomer was briefing an audience of parents. Wanting to make a point about Gen Z needing help to manage screen time, he pulled out his cell phone and announced, "Can you believe we now have a computer in our pocket?" Parents in the audience were mostly Gen X and only somewhat related to his enthusiasm. While baby boomers had witnessed a time of slide rules and mechanical calculators before electronic versions existed, these Gen X parents knew about portable phones by their twenties. Millennials experienced an era of the expansion of social media and the iPhone, while Gen Z will have grown up with virtual reality headsets and the beginnings of artificial intelligence. Technology is often a marker of generational differences and, depending on your season of life, may drastically impact your perspective and approach to life choices, military or otherwise.

Historical Markers

In addition to where and how you grew up, major life events have a way of marking key moments in your timeline and shaping your values. The assassination of President John F. Kennedy is a memory that almost all baby boomers share. The same is true for Gen Xers remembering the catastrophic mission of NASA's *Challenger*. The events of September 11, 2001, brought the entire country together with a new sense of patriotism yet ignited different responses from boomers, Gen X, and millennials. The COVID-19 pandemic is another historical marker for all generations. It dramatically altered Gen Z's most formative years and opened up new opportunities for innovation.

Military Subculture

While overall generational descriptions may be congruent between civilians and military families, the military as a subculture is an additional lens through which we should see generations. There are common values passed down or embraced by almost every military family generation. The willingness of service members and their families to accept significant sacrifice—family time, personal autonomy, one's life—is a shared burden and expectation for all.

There are also considerably more life stressors, such as frequent relocations, deployments, isolation from extended family members, and potential combat trauma. Families are expected to absorb the complexity of these stressors and many do so out of their own sense of service and sacrifice.

The camaraderie, however, is most often referenced as the benefit that sets the military community apart from the civilian community. Traditions, ceremonies, specific social norms, and a deep connection to history solidify a strong sense of "oneness" that is difficult to explain but easy to spot in other military families. Each branch maintains its own set of values and mottos, but they all share a sense of duty, honor, respect, selfless service, discipline, and responsibility. Family members who may never see combat but offer support from the homefront adopt the same values and teach them to their children. Despite the stressors, or perhaps because of them, a deep bond and mutual sense of identity are shared within the culture.

The connection between veterans and current service families is especially endearing. I have always known that when I travel, the USO or any installation is a safe place to be "among the tribe." The same is true when I meet a fellow military spouse. I recently met an eighty-three-year-old military spouse at the Frankfurt airport in Germany. As soon as we realized what we had in common, the generation gap melted away and we talked for hours. It doesn't matter what branch of service or current status, there is an unspoken bond we share and the relationship generally starts from a place of trust. While in the civilian world, a person averages more than 140 hours of investment before being considered "a good friend,"[30] the very real experience among military spouses is that most will say they can identify a new emergency contact for their kids in less than sixty minutes, especially if it is another military family.

It is easy to assume that if we share the same values, even a similar lifestyle within the military, then everyone's experience must also be the same. This assumption has significantly contributed to the generational confusion. It's important to understand that each generation experiences military culture differently. Depending on when you enter the community and the generational perspective you bring, you are likely to experience historical events, social norms, and even the concept of community differently from others around you.

Historians have tried to capture the spirit of soldiers and war heroes and nail down what gives an individual the courage to fight some of the most horrific battles. Many agree that you must understand who they were coming in to understand who they became. According to World War I historian Edward Gutiérrez, "The pre-war life experience and personality of a soldier dictate how that soldier will react to battle. Individual predispositions shape a soldier's experience. When you gather that en masse, these individual voices become a collective narrative of warrior motivation and reaction to war."[31]

In order to examine why a service member would be willing to go into battle, or even join the military and bring their family into this culture, we must examine who (collectively) is joining. Similarly, an individual or family's pre-military experience and generational personality dictate how that family will assimilate into the culture. This matters because many leaders are frustrated with why the younger generation is not as interested in some of the traditions and social customs many before them have come to love. We cannot assume that what worked for military families in World War II, the 1980s, or even shortly after 9/11 will work for this generation.

Applying Gutiérrez's words even further can help us understand why each generation has reacted differently to post-9/11 combat. The more we are able to listen to the voices of each generation as distinctive from other voices within the same tribe, the more we will find what motivates them toward productivity, community, and wellness.

While whole books have been written about a single generation, we will be looking instead at key markers that shaped each generation and how that impacted their experience of the military culture. Likewise, we will look at how the military subculture had a hand in shaping each generation. Because the current makeup of the force includes baby boomers and younger, we will dive much deeper into their experiences over the last twenty years. However, as Strauss and Howe emphasized, we must go back a bit further to understand the cohort's experience and the military they walked into.

Finally, untangling the problems faced today means understanding the larger military story. The story involves a deeper exploration into how each generation has shaped the culture we experience today. As each new generation enters, it inherits the progress and the burdens of previous generations.

WILLING TO WALK AMONG THE RUBBLE

Wicked problems within a system can be overwhelming. Today's military leaders on every level are faced with the near-impossible task of recruiting the next generation while healing the current one. We need leaders who are willing to walk through the rubble and get to know the people they serve alongside. It may require putting aside the comfortable script, including the way you see life, work, and this culture. It will also require your willingness to look for and be more aware of cognitive biases that inform how you see a problem. What may seem like a simple problem, may, in fact, be far more complex when you see it through the eyes of another generation.

On September 14, 2001, President George W. Bush stood on the rubble of what was the New York City Twin Towers, dressed in a button-down shirt and khakis. By then, we knew the enemy had come to us and that something would need to be done about it. The US people, especially those at Ground Zero, were in a state of shock, exhaustion, grief, and paralysis. We were not ready to go to war.[32]

Bush's visit to the collapsed towers started with walking around the rubble. The stench of death, fire, and ash filled the air. The first responders around him had been there for days, exhausted from their efforts to find survivors. Spontaneously, Bush was asked to give remarks and someone handed him a bullhorn. He recalled, "My first instinct was to reprise parts of the speech I had given at the National Cathedral earlier that day."[33] And he did until someone in the crowd yelled, "We can't hear you!"

Bush reflected, "It became clear that my messaging at that point in time was not hitting the mark." The people in the middle of the war zone needed a different message than the people in the church who were there to grieve. The president quickly shifted away from the script and simply said, "I can hear you! The rest of the world hears you. And the people who knocked these buildings down will hear all of us soon." The weary first responders erupted in cheers, chanting out their anger and frustration with a united "U-S-A! U-S-A!" The president's words, spoken into a bullhorn, won the will of the people.

In the days that followed, America continued to unite as one. Memories of

9/11 hold both tragedy and an incredible sense of patriotism. We had something valuable to protect. In the aftermath, commissioning and enlistment for Gen X into our all-volunteer force skyrocketed. Millennials had just become eligible and were already joining to serve their country for the very first time.[34]

While there were many moments of inspiration surrounding 9/11, I've thought a lot about why Bush's speech, now known as the "bullhorn address," was so powerful. Our nation was in a vulnerable state of complex grief, a mixture of anger, sadness, confusion, and fear. Perhaps we would have received anything he was able to offer. However, he had already delivered two other speeches, including one I call the "American Dad" address: the official well-crafted speech delivered from the Oval Office. In that speech, it was a message to everyone, including allies and terrorists, of just how seriously he was going to take this attack. He spoke with a somber, firm tone that conveyed someone was in trouble, he was just deciding who would get punished first. We needed a protector and provider and Bush delivered that for us in a suit and tie behind the Resolute Desk. But the American Dad address was not what stood out to most of us who sat glued to our televisions all week.

Bush had also delivered the National Cathedral remarks earlier that day where he spoke to those who had lost loved ones. He took the role of a comforter, offering sympathy and validating the grief-stricken with the wisdom that time and history would bring purpose from their pain. He also named the attack, calling it pure evil. Those present that day may remember parts of that speech, but once again, it was not what stood out most that day. It was the words yelled through a bullhorn just hours later that ignited patriotism and renewed the spirit of an entire country.

Looking back, and knowing what I now know about communication, all three speeches were necessary. Each message was congruent with what the American people needed to hear. Speeches of comfort and protection revealed his deep loyalty and commitment, expected of a leader, yet the unscripted bullhorn address captured the moment Bush led from the trenches. President Bush didn't realize he was impulsively writing a historic speech, he was simply answering the man from the crowd by saying, "I can hear you!" However, his willingness to meet the people where they were established trust in the relationship for his words to answer something deeper that the crowd didn't even know they needed. They were tired, angry, devastated, and traumatized. Their ash-covered faces and eyes spoke words they didn't have yet. The first message they heard was that he genuinely heard (and saw) them.

Metaphorically, our military culture is in a similar state. The people are looking for leaders who will, of course, fulfill their commitment as a comforter and protector and who will also look into the eyes of those who carry the cost of service. We can learn a lot from this moment, when a leader won the will of the people with just a few words. At a moment when morale was low and people were angry, Bush's leadership strengthened their resolve to give more when they were ready to give in, and in doing so, he won their loyalty. Moments like these make or break leaders, businesses, and institutions because they expose the connection, or the disconnect, they have with the people they serve. Looking deeper at this moment also shows how we can be more intentional as we read through the cultural story and wicked problem that impacts millions of people.

As the crowd erupted in a cheer signaling to the president that his words hit something deeper, the president yelled, "The world hears you" and a second message was delivered. This one is interesting. On one hand, I believe the crowd heard that they were not alone. The rest of the country (and people around the world) desperately wanted to be there to help carry the burden they shouldered. We also shared their grief and anger. They were not alone or wrong for feeling what they were feeling because others did too. The president closed the gap between the people at Ground Zero and those outside the city, further validating their experience but also igniting advocacy in others who were listening.

But the deeper, and more important, message was that he acknowledged that they had an understandable desire for justice. Instead of quelling it or minimizing it, it was as if he was giving their voice the bullhorn and validating their right, as humans, to have thoughts, feelings, and opinions on the matter. We are held accountable for how we choose to communicate them, but thoughts and feelings don't have to be threatening or intimidating. It is what makes us human, even when we disagree. By validating that the world could hear their anger, the president acknowledged that he heard it, saw it as part of being human, and allowed it. The crowd responded even more loudly, their resolve building.

When the president ended with, "the people who knocked down these buildings will hear all of us soon," he delivered a final powerful message of commitment to action. He knew the situation called for a response, and the people wanted to know what that response would be. Although his words didn't share the details of the mission ahead, he answered the unspoken question of whether the people had a leader who valued them enough to defend them. In those few words, this leader won the people, strengthened their resolve to give more when they were ready to give in, and won their loyalty.

No one could have known just how overwhelming the decades ahead would be, nor how long we would commit ourselves to it. What was powerful about Bush's bullhorn speech was not just that he communicated the mission to defend the country, but that he acknowledged the experience of the American people first. He recognized the importance of winning the will of the people, and when it came time to communicate, he publicly acknowledged that he understood their needs and would be their advocate. Only then, amid cheers and chants that communicated he had won their trust, did he announce the mission.

Especially in the military, we are trained to avoid when possible commitment, engagement, or the mission without a full analysis and courses of action assessed for vulnerabilities. Yet, as you walk into the rubble who can you commit to being as a leader within your circle of influence? Consider one action you can start today, even if that is just being more intentionally present with the next person in front of you.

Within the military culture, there is no mission without trust and loyalty.

AN EXAMPLE WORTH REMEMBERING
Everyone can change and grow.

The boomer who saw himself as the immovable rock later brought me back to lead his directorate through the same training. Empowered with a new perspective of generational strengths, he started the training with vision casting how important it was to productivity and morale to not only understand your own generational story, but to truly listen to others' stories. While he stood by his direct, focused, get-the-job-done leadership style, he explained that he understood it was his style and not necessarily theirs. At the end of the training, he asked the team not only where leadership could improve, but for their accountability as they made adjustments. Likewise, he asked them to provide feedback on areas they could improve on and how he could best hold them accountable to change. By leveraging the compassion and curiosity he naturally had for the team and work he valued, he was able to translate his generational strength from an immovable rock to a leader who provides a secure foundation of honesty and support they can trust and rely on.

Chapter 2
HONOR, DUTY & PATRIOTISM

ONE OF THE BIGGEST QUESTIONS the Department of Defense faces is how to recruit the next generation of service members. For decades, honor, duty, and patriotism were strong motivators to serve the country. The meaning of these terms, however, has shifted as younger generations form their own opinions on government, service, and global conflicts. These words have also been tied to political division and even violence, requiring us to reconsider the assumption that everyone joins because of a sense of honor, duty, and patriotism.

It is important to understand where our values originate and how they shape our perspectives of leadership and service. Many values are passed down or influenced by family. Exploring past generations also gives insight into trends that are likely to circle back and return in future generations.

Historians and storytellers have long captured the timeline of wars, the history of governments, and the stories of war heroes, but seldom has our military cultural story been given the same level of insight. Exploring the motivations and values of past generations paints a picture of how our military culture came to be and what decisions contributed to its evolution. We can learn how leaders motivated people during crises, throughout significant political division, and how they foreshadowed some of the issues we see today.

THE LOST GENERATION (DOB 1883–1900)
Honorable & Dutiful

The lost generation was born after the Industrial Revolution and fought in World War I. The youngest of this generation served in World War II as well. Nicknamed by writer Gertrude Stein, the generation's name described the "disoriented, wandering" spirit of veterans who survived World War I only to return to a completely different American culture. It was a time of booming literacy in a divisive, politically chaotic, and media-saturated time.

Before the First World War, Congress wanted to avoid the likelihood of families becoming financially dependent on the military and passed legislation in 1847 stating that married men were ineligible to enlist. Some of the wives of those already serving or those who married after enlisting followed their service members, setting up camp and living outside of installations. Known as "camp followers," they helped with critical tasks like sewing, cooking, providing for horses, and more. In some cases they earned rations for their labor. Service members did not earn enough to support the entire family. These rations, although minimal, eased the financial burden on families and allowed spouses to stay close to their service members.

Leading up to the Great War, President Woodrow Wilson made every effort to avoid the conflict. However, on April 2, 1917, he asked Congress to declare war on Germany.[35] Allies needed fresh troops to replace those exhausted on the battlefield. Six weeks after the US joined, Wilson signed the Selective Service Act into law, requiring all single men in the US between the ages of twenty-one and thirty to register for military service. By the end of World War I, twenty-four million men had registered and more than half of the 4.8 million Americans who served in the war had been drafted.

Throughout World War I, the military continued to discourage married military members from re-enlisting. Camp followers were not allowed to follow their service members overseas and draft registrants were eligible for a deferment if they were married. Enlistees who decided to marry needed permission from their superiors, leading to the old adage, "If the military wanted you to have a spouse, they would have issued you one." The phrase has widely circulated since.

In addition to needing troops, the US would need a national military. According to Brian Neumann, a historian at the Army's Center of Military History, "The United States was in it, but they had to define what 'in it' meant."[36] Building a national military would take building an entire infrastructure to support it in comparison to the large military forces of other countries. This would include barracks, mess halls, hospitals, roads, and training facilities. Congress approved $3 billion to build a "million-man army" and the infrastructure needed to support it. Many Americans, who were also new immigrants, were not too keen on the idea of a national army, much less being drafted in to support it.

Historians describe those who enlisted, and many who were drafted, as doing so out of a sense of duty and honor. These values were key motivators for much of American culture. Serving others and your country came from a sense of duty and was one of the highest callings you could follow. Literary works of that time praised the heroism of men who fought in the Civil War, encouraging future generations to live up to this sense of greatness and honor. In President

Wilson's 1914 Fourth of July address, he used some of this language to bring support for the war, recruit more to serve, and unite the country: "The most patriotic man, ladies and gentlemen, is sometimes the man who goes in the direction that he thinks right even when he sees half the world against him. It is the dictate of patriotism to sacrifice yourself if you think that that is the path of honor and of duty. Do not blame others if they do not agree with you. Do not die with bitterness in your heart because you did not convince the rest of the world, but die happy because you believe that you tried to serve your country by not selling your soul."[37]

Wilson's successful use of messaging is worth noting. Although his communications were much slower and limited compared to today's media and press, his consistent campaign messaging of values that were already enforced and used in everyday American life reached into homes, churches, and public squares and ultimately shaped the nation.

As women fought for full citizenship in the women's suffrage movement, men marching off to war meant that women could enter manufacturing, shore-duty positions, and other ways of supporting the war. The Red Cross, Salvation Army, and many other organizations provided a place for women to volunteer. While this contributed to a perspective shift in women's value in the workplace and foreshadowed the future of Rosie Riveters in World War II, men returned and reclaimed jobs, forcing women back into traditional roles in the home.

Before World War I, benefits and pay included pensions and bounty land applications for military service and were available to widows of those killed in the line of duty. Early pensions were only available to those who were injured. In 1916, the military instituted an "up and out" policy, forcing out servicemen who had not been promoted by a certain amount of time and paying retirement compensation by calculating base pay multiplied by years of service, a policy very similar to what is used today.

On November 11, 1918, a cease-fire and armistice were declared, and just a few short months later, the Spanish Flu was discovered in the States. Because America had been at war and troops were coming home from training or overseas, the flu spread faster. Worldwide, the Spanish Flu killed more people than the Great War, between twenty and forty million; 675,000 of those were Americans[38] (in comparison, COVID-19 killed 676,000 Americans by 2021).[39] Many women were widowed from the war, followed by the Spanish Flu, and this made it more difficult for single women to marry. Many of them gravitated toward employment opportunities created by the incredible loss.

The country was understandably in a turbulent time. Censored media in different countries shared limited or sometimes false information. The source of the pandemic is still unknown, but due to Spain's ability to report on it freely, it was named the Spanish Flu.

The lost generation veterans experienced considerable mental health struggles and physical disabilities after the war. Considering many had gone to war with the thought that entering battle was a path to honor and greatness, many felt a sense of disillusionment and dread when they returned, contributing to their generational nickname. The American civilian population had evolved without them while they "had themselves changed ... most had seen more of the world than had their elders and had been exposed to beliefs and customs strange to them and to those they left behind."[40]

With the ratification of the 19th Amendment in 1920 giving women the right to vote, America entered the Roaring Twenties. Women deviated from the socially acceptable standards of dress and emerged as "flappers" with bobbed hair and short skirts. Poets and authors like Gertrude Stein, Ernest Hemingway, and F. Scott Fitzgerald shaped the country with stories that spread the tone of materialism, expression, and rebellion. They also spread the stories of disillusionment and wrote characters with a "lost" spirit in works like *The Great Gatsby* and *The Sun Also Rises*.

In the arts, Dadaism emerged. The movement was born out of a response to the modern age and often in protest to the horrors of war. Usually sarcastic, humorous, and labeled "anti-art," some of the pieces from that time mirror some of what we see today in Gen Z memes.

In addition to a world war and pandemic, the lost generation experienced significant job and income loss as the economy crashed during the Great Depression. Just before that in 1924, Congress had approved Adjustment Compensation Certificates, or bonuses, for veterans of World War I. They would be paid $1.25 for each day they served overseas and $1.00 for each day they served in the US. However, the payout would not be until 1945. Since World War I ended in 1918, this was a significant amount of time to wait.

As the Great Depression continued into 1932, somewhere between ten thousand and twenty-five thousand World War I veterans marched into Washington, DC, demanding an early lump sum payment. Called the Bonus Expeditionary Force, they set up tents and encampments and waited for Congress to decide. When the House passed but the Senate did not, some went home discouraged. Thousands more, however, chose to remain and protest. In a horrific display of policing their own, troops led by Brigadier General Perry Miles and Army Chief of Staff General Douglas MacArthur used tanks, tear gas, and the burning of tents to dissuade protesters. Congress eventually appropriated $100,000 to send protesters home.[41]

In an effort to adjust the unemployment rate, President Franklin D. Roosevelt later created the Civilian Conservation Corps, a relief project that created jobs for environmental projects. He then issued an executive order allowing twenty-five thousand veterans to enroll, exempting them from the previous requirements of being single and under age twenty-five to enlist. In 1936, Congress finally passed a bill to disburse around $2 billion in veterans' benefits. The Bonus Expeditionary Force laid the foundation for future bonuses for military service and the GI Bill of Rights that support veterans and their families today.

THE GREATEST GENERATION (DOB 1901–1927)
Committed & Patriotic

When an entire generation is motivated by honor and duty and then forged through war, a global pandemic, and financial disaster, their children emerge with a sense of steadfast determination. Those children are the greatest generation, a term coined by well-known NBC journalist Tom Brokaw. His book, *The Silent Generation*, details the generation's stories of commitment and sacrifice. Inspired by their parents' service and motivated by patriotism, many from the greatest generation served through World War II, the most costly war to date.[42] Having a distinct memory of the Great Depression, they were also known for personal responsibility, frugality, and a work ethic that kept many loyal to one career for a lifetime.

Commitment and selfless service were threaded throughout every area of life, much like honor and duty were for their parents. Many Americans at the time held to traditional values in religion, marriage, and roles in the home, making the progress women had gained during World War I challenging to sustain. Divorce rates were extremely low. Integrity and honesty were character traits deeply valued in family, work, and entertainment.

Since 1918, America had been in a time of peace. Before World War I, the US War Department served as a modest version of the Department of Defense we have today. As the stock market crashed in 1929 and the Great Depression began, military spending was less than 1 percent of the US gross domestic product (GDP). Compare that to 11.3 percent during the Korean War, 8.6 percent during the Vietnam War, and roughly 3 percent in modern peacetime years.[43] While there were efforts to modernize the military throughout the 1930s, much of that was difficult under low budget constraints and economic hardship. Roosevelt focused primarily on the New Deal and work programs to help the economy and bring down the 17 percent unemployment rate.

On September 8, 1939, just days after the Nazis invaded Poland, Roosevelt proclaimed a limited national emergency in the safeguarding and strengthening of the national defense. In 1940, Congress increased the military budget and the president signed the Selective Training and Service Act, the first peacetime draft requiring all males ages twenty-one to thirty-five to register and serve at least one year. At the time, married men were still given a deferment unless they had enlisted before marrying.

On December 8, 1941, the US declared war against Japan in response to the attack on Pearl Harbor. On December 11, Nazi Germany and Fascist Italy declared war against the US. As many young men heard of the attack on Pearl Harbor, they quickly enlisted to do their part out of patriotism and duty. In the predicament of needing a much larger presence to fight a global war, the force could no longer reach numbers while giving deferments to married men.

Family stability was now a factor: "The effectiveness of war operations depends in large part upon civilian and military morale. A vital factor in upholding this morale is some reasonable maintenance of families of men engaged in military service."[44] A family allotment was prorated to service members based on the size of their family. Whereas the previous draft created a loophole for men to be excused from the war by getting married, there was now concern that couples would marry for government stipend financial security.

With the new Selective Training and Service Act, registrants grew from sixteen million to an impressive forty-two million by 1942. The sheer size of the US force during World War II is significant compared to 2021's 2.1 million, which is only 1 percent of the population and an all-volunteer force. In 1942:

- 38.8 percent (6,332,000) of US servicemen and all servicewomen were volunteers
- 61.2 percent (11,535,000) were draftees
- Average duration of service: thirty-three months
- Total US casualties: 407,316; injured: 671,278[45]

To help with the financial needs of military families, Henry Stimson, the Secretary of War, instituted Army Emergency Relief that would support families in financial distress or need. Congress later followed up that year with the Servicemen's Dependents Allowance Act, which would adjust wages based on the number of dependents in the household. These benefits now offered financial security to families who considered military service. To help with the new family allotment, the War Department created the Office of Dependency Benefits (ODB). Over $13 billion would be paid to four million military dependents between 1942 and 1945. Three million of those were army wives.[46]

The war, in many ways, was great for the economy as Americans mobilized back into the workforce to help modernize and expand the force's capabilities.

The economy soon began to recover, and unemployment was reduced to 1.2 percent by 1944.⁴⁷

Some of the most inspirational stories of World War II military culture describe the American people coming together to help win the war. If you were not already in a war-related occupation, there were opportunities to buy war bonds, stamps, or work for a business that invested in another way. Much of American life was changed by war production efforts. Women and minoritized groups again filled labor shortages created by the men who left for war.

Messaging and propaganda effectively saturated media, print, and the radio, the cheapest entertainment form, reaching a massive audience. Some of the most famous campaigns continue to inspire marketing efforts today. Ads highlighted masculinity, women's opportunities, faith, and patriotism to create buy-in for America's families and civilians. Posters like Rosie the Riveter and Uncle Sam's "I Want You," "Loose Lips Sink Ships," and others not only surged recruiting efforts but gave citizens specific ways to get involved.

The idea of military family was not only reserved for those who had husbands, sons, or other family members in the service. Everyone knew someone serving or had a family member in war-related work. Fresh off of economic hardship, Americans were willing to live with rations, shortages, and price controls. Deployments were not six months or even a year. Instead, soldiers were gone for years at a time with limited (if any) contact with family back home.

It is important to note the incredible investment, modernization, and advancements made during World War II. The US spent 4.1 trillion dollars between 1941 and 1945, and 37.5 percent of GDP at the peak of the war. It was the most the country has ever spent on defense. War bonds covered 63 percent of that and were sold to citizens to redeem after a maturity period of thirty or forty years. Even children bought war stamps. In that time, in addition to producing for allies, the US produced:⁴⁸

- 303,000 aircraft
- 100,000 tanks and armored vehicles
- 82,000 landing crafts
- 41,000 guns and howitzers
- 907 cruisers/destroyers/escorts
- 211 submarines
- 27 aircraft carriers
- 10 battleships

As businesses came together to help war production, they discovered the economic benefit of working with the government. By providing and innovating the latest and greatest technology, it also gave our military a competitive edge against adversaries. Seeing how this could evolve into something much bigger, President Dwight D. Eisenhower warned Americans in his 1961 farewell address: "A vital element in keeping the peace is our military establishment. Our arms must be mighty, ready for instant action, so that no potential aggressor may be tempted to risk his own destruction ... American makers of plowshares could, with time and as required, make swords as well. But now we can no longer risk emergency improvisation of national defense; we have been compelled to create a permanent armaments industry of vast proportions ... This conjunction of an immense military establishment and a large arms industry is new in the American experience ... Yet we must not fail to comprehend its grave implications ... In the councils of government, we must guard against the acquisition of unwarranted influence, whether sought or unsought, by the military-industrial complex. The potential for the disastrous rise of misplaced power exists and will persist."[49]

As the war came to a close, the military faced the incredible task of deciding what to do with the enormous number of service members deployed with no enemy to fight. Not wanting to repeat the mistakes of post-World War I demobilization failures, the Army devised the Adjusted Service Rating Score. This new rating system awarded soldiers points for how long they had been overseas and in the service, the number of dependent children (up to three), and the decorations they received. Once they accumulated eighty-five points, they could return home.

The point system created misinterpretation and was perceived as unfair. Those who had more children or served in specialized units earned points faster and were able to come home earlier, causing other service members to take out their frustration on leaders. Eventually, the Army ended its point system after soldiers completed two years of service but it succeeded in bringing home more than seven million service members in two years at a manageable pace. The passage of the GI Bill also kept troops from flooding the job market and sent many to school, something World War I veterans resented being left out of.

African American veterans returned to racial violence and were also denied many of the benefits guaranteed under the GI Bill, including mortgage assistance and vocational training. Although they weren't specifically written out, the bill was structured in a way that limited access and allowed white veterans the opportunity to prosper. Historian Ira Katznelson writes that there was "no greater instrument for widening an already huge racial gap in postwar America than the GI Bill."[50] Full integration would not occur until after the Korean War.

As the greatest generation war heroes returned, they brought back an even stronger set of values and morals and found new ways to serve by reintegrating into key leadership roles in the community and government. While some call this generation more patriotic than any other, others like Tom Brokaw point out the areas in which they were slow to make progress. Brokaw notes the generation's failure to acknowledge the importance of women's rights, even after women had proven critical to war efforts, twice. They were also slow to value the importance of racial equality, especially considering that 13 percent of all military members in the war were minorities serving with a similar sense of patriotism. They would similarly struggle with a country that was beginning a shift with a new generation motivated by something other than patriotism.

As the draft expired in 1947, Congress passed a peacetime draft. The National Security Act of 1947 reorganized foreign policy and military establishments of the US government, including the creation of the Central Intelligence Agency (CIA). A single Department of Defense under a Secretary of Defense would oversee the War Department, the Navy Department, and the new Department of the Air Force. It also paved the way for future "think tanks," filled with civilian experts on war strategy, that would later become war colleges and universities. During World War II, Congress had given women full status in the Women's Army Corps (WAC) due to recruitment concerns. In 1948, President Harry Truman signed the Women's Armed Services Integration Act into law, allowing women to compose up to 2 percent of personnel in each branch.[51]

The military family culture was beginning to form as the Army allowed families to move with their service members to new military bases across Europe and Asia. Wives were now seen as "assets" who could "improve the readiness of men" and "positively sway" them to reenlist or make the military their career.[52]

The greatest generation's experience of technological advances like the radio and telephone changed communication drastically in their early years. Much of the efforts that went into war production became commercial products in homes after the war. The cavity magnetron, for example, was a small palm-sized device that improved radar technology with the use of shorter micro wavelengths. Advancements in radar improved meteorology and the ability to study weather patterns. By the 1970s, this technology would make its way into homes in the form of microwaves.

The most well-known scientific advancement was the atomic bomb, dropped on Hiroshima and Nagasaki, marking the end of the war in the Pacific. This discovery, heavily debated and evaluated today, propelled the US as a dominant figure in the world and began the competitive arms race.

By 1948, tensions were already growing between the US and the Soviet Union, marking the beginning of the Cold War and related conversations of anti-communism. Just five years after the end of World War II, the US rushed to ready and train exhausted veterans to fight a completely new enemy with an entirely different terrain in the Korean War.

The Korean War, lasting three years (1950-1953), was much different from World War II. More than 6.8 million men and women served, many of whom were sons and daughters of the greatest generation. This force was entirely different from those who had trained and were activated for World War II. Also different was the lack of media attention that the Korean War received. Training was often rushed while service members and their families were ready for peacetime. Since World War II, Congress has not offered another declaration of war, even during the War on Terror.

AN EXAMPLE WORTH REMEMBERING
Sometimes, going against cultural norms changes history.

During World War II, Oleta Crain was one of three African American women to enter officer training and the only one to be retained. Despite her own experience of segregation and challenges, Crain exhibited optimism and used her influence as a commanding officer to ensure that all of her troops were treated fairly. She persuaded the commandant to end segregation in the pool and trained other female drill instructors on how to build confidence in their voice to earn respect and loyalty. When Crain retired in 1963, she spent the rest of her career working for the US Department of Labor where she eventually became the regional administrator of the Women's Bureau for Colorado, Montana, North Dakota, South Dakota, Utah, and Wyoming. There, she focused on "balancing work and family responsibilities, reforming welfare, and improving women's employment opportunities."[53] An employee who worked for Crain at the Department of Labor remembered her passion for women's rights. Crain suggested that men "ease up on the macho, breadwinner role," saying, "We'll all live longer."[54]

Chapter 3

FORGOTTEN & VILIFIED

KNOWING AND UNDERSTANDING THE STORIES of the generation before us can inspire compassion, explain how our own circumstances evolved, and reveal trends that may pass down to future generations. Generations, like periods of history, develop themes that repeat or circle back around, providing new insight into those around us.

THE SILENT GENERATION (DOB 1928–1945)
Quiet & Reserved

The oldest members of the silent generation entered the Second World War but their enlistment peaked throughout the Korean and Vietnam Wars. *Time* magazine was the first to name this group in its 1951 article "The Younger Generation," attempting to capture the focus of this new generation: " … waiting for the hand of fate to fall on its shoulders, meanwhile working fairly hard and saying almost nothing. The most startling fact about the younger generation is its silence. With some rare exceptions, youth is nowhere near the rostrum. By comparison with the Flaming Youth of their fathers & mothers, today's younger generation is a still, small flame. It does not issue manifestos, make speeches, or carry posters. It has been called the 'silent generation.' But what does the silence mean? … Hardly anyone wants to go into the Army; there is little enthusiasm for the military life, no enthusiasm for war. Youngsters do not talk like heroes; they admit freely that they will try to stay out of the draft as long as they can."[55]

As members of the silent generation came of age, they experienced the horrors of World War II through young eyes, many of them losing fathers or siblings to the war. They saw the horrific truth of Nazism and the destructive capability of the atomic bomb. They would grow up with communism as a new enemy and the possibility of it crossing into their own country. Strauss and Howe refer to this generation as "adaptive" in that they grew up overprotected by their parents, leading to their tendency to become mediators of conflict rather than signing up to fight. They were known for taking fewer risks and keeping their heads down.

Although labeled silent and risk-averse, they, like every generation, evolved as they entered adulthood and decided to challenge the decisions of generations before them. They did not experience a global conflict uniting the country or an economic crash, in fact, they were known to devote their talents to rebuilding and reshaping almost every other area of culture. While the greatest generation before them would contribute to seven presidents and prioritize service in their communities, the silent generation would explore the moon, introduce rock and roll with Elvis Presley, and bring laughter and entertainment to the television.

Many critics of the name silent generation remind us that most of the rebellion of the 1950s through the early 1960s came from this not-so-silent generation. Martin Luther King Jr. was a part of this generation, as well as many other activists who advocated for changes in education, civil rights, and equality for women. They also changed the stigma of divorce as they reformed marriage laws.

For families during the "forgotten" Korean War, much was unspoken and left undone by the government. American forces "buried their dead in temporary cemeteries, assuming they could go back and claim the bodies once the war was won, as they had in World War II."[56] But since there was no peace treaty or official end to the Korean War, access was denied to those sites.[57]

According to a RAND study, up to a third of prisoners of war (POW) Americans died while in captivity.[58] Silent generation Senator John McCain, having survived as a POW, would later join with other families to form The National League of Families of American Prisoners and Missing in Southeast Asia to help release prisoners, account for those lost, and bring answers to waiting families.

Military family programming continued to evolve as well. By the 1960s, family members outnumbered service members in the DoD. In 1969, a group of military spouses advocated for the financial security of their friends who were widowed after their service member's deaths. Out of this effort, the Survivor Benefit Plan became law, allowing for continued monthly payments over the life of the beneficiary with some limitations that are being challenged even today.

Throughout the Korean War and after, registration of men ages eighteen to twenty-six continued, this time allowing for deferment if a student was enrolled in school. Recruiting focused on health professionals. Lobbyists continued to build a list of other deferments, such as scientists and farmers. Even though the government advocated for volunteer enlistment, there was much debate on whether an all-volunteer force was even possible.

In an effort to keep the force from dwindling, a new strategy was implemented, allowing service members to possibly avoid combat placement by selecting their service and specialty. This volunteer strategy was successful for recruiting throughout Vietnam as the Army reached 95 percent of its quota this way and 50 percent of overall military recruits in 1970.[59] This also highlighted the military's value of education and specialty in the forces with recruiting aimed at those with degrees. Military pay also increased and the Army began recruiting through television ads.

While more than ten million baby boomers would serve between 1960 and 1970, it was the silent generation that held a heavy presence in the ranks throughout the Vietnam War. As Americans protested America's involvement in Vietnam, President Richard Nixon saw the opportunity to end the draft as a way to discourage the protesting. With opposition from Congress and the DoD on the effectiveness of an all-volunteer force, he delayed the effort until 1973, when Congress officially ended the draft shortly before Vietnam ended. It would be the last time until 1980 that draftees would be called.

The importance of the switch to an all-volunteer force cannot be stressed enough. It is a major historical marker for the silent generation and boomers, and the ramifications of this decision remain significant. The decision to move to an all-volunteer force model was not made lightly. The military would now need to attract and recruit, compete with the civilian sector, and revisit the conundrum of trying to limit the dependency of families on the military.

Not much has been written about the silent generation, other than the important part it played in the Korean and Vietnam Wars. Many call them the luckiest generation because their families were more intact, they gained more education than generations before them, and they were able to capitalize on veteran benefits while experiencing fewer casualties and more peacetime. They are also credited for turning World War II enemies (Germany and Japan) into allies.

They were known to inherit their parents' frugal thinking and traditional values of marriage and family, with many marrying and starting a family quite young. Raised in a "children should be seen and not heard" environment, they continued that narrative when raising their kids, the baby boomers. Baby boomers, however, would challenge this approach as young adults in their thoughts on family and the Vietnam War.

Although President Joe Biden is often mislabeled as a boomer, he is actually a member of the silent generation. Often frustrated that their accomplishments have been credited to boomers, members of the silent generation are known for "their ability to make things work by bringing people together in common cause" and "manage conflict without being disrespectful, to compromise in the name of making progress, and to avoid confrontation for the sake of confrontation—all while making those organizations and institutions more successful than they had been before, or since."⁶⁰ Many silents feel this is what Biden did following the national and political division of Donald Trump's presidency.

In 1970, a silent generation journalist then in his thirties revisited the first *Time* article's attempt to describe his generation. He wrote: "Our silence on campus had its price, no doubt, but it also had its rewards, not the least of which was the chance to grow at our own pace and to pursue, with no guilt whatsoever, the totally irrelevant. ... Above all, we were the last generation to accept without question—or to pretend to accept—the traditional American values of work, order and patriotism. ... Our views on religion are scarcely less confused. Is God dead? Don't ask us. For the majority of us, religion is merely a word, sometimes honored, sometimes not. ... Though we bear no scars from the Depression and 'the war'—their twin traumas—both are the vivid memories of childhood to us, rather than cold, historical incidents in a textbook. We can understand, as the young cannot, why the older generation is afraid, and more sadly, why it is resentful of those who seem to have everything but gratitude. To both young and old, we are almost invisible. The young often see us as the cop-outs—as the shorthaired, button-down junior exec or the suburban housewife in a station wagon—and many of us are. Our parents and older brothers and sisters often see us as the fellow travelers of the youthful enemy, which many of us are too. ... Like any generation, we contain contradictions and exceptions, including those, particularly among the blacks, who want to burn and bury the system. But the revolutionaries among us, political or cultural, are a minority; reform, not revolution, is our aim. As a generation, we are distinguished by our lack of anger. Circled by fury, we are the unfurious; surrounded by passion, we are the dispassionate. Most of us by this time have made a commitment to the kind of country we want to live in, and often that commitment is pursued with all the energy and talent we possess."⁶¹

Many have compared Gen Z values and behaviors to those of the silents. Perhaps this is because the climate in which both generations were born and raised has notable similarities. As the Global War on Terror slowly ended and older generations are still recovering from their part in it, Gen Z has been labeled as "keeping their heads down," delaying their thoughts and beliefs on joining the military, and questioning what they believe about US involvement in global conflicts. In many ways, they are much quieter than the millennials before them.

BABY BOOMERS (DOB 1946–1964)
Hard Working & Keeping Up With the Joneses

Baby boomers today make up a small portion of the military force, but those still in the force today hold senior leadership positions. General Mark Milley, chairman of the Joint Chiefs from 2019 to 2023, as well as many other branch chiefs and much of the DoD leadership, are baby boomers. Understanding the impact boomers had on the culture as well as how the military culture impacted them is vitally important to understanding the current leadership climate and direction. Every generation has an opportunity to shape the trajectory and vision of the institution, and baby boomers have a story of their own to tell.

Baby boomers were given their nickname from the "boom" of births to the younger greatest generation and silent generation between 1946 and 1964. The 1946 publication of the book *The Common Sense Book of Baby and Child Care* by Benjamin Spock challenged the "children should be seen but not heard" concept, drawing more attention to the emotional needs of children and steering parents away from strict schedules and verbal tones. Spock's book sold half a million copies within the first six months and influenced the childhood of boomers, but even more how boomers would later parent.[62]

Many silents argued that this parenting style permitted too much individualization, beginning a generational gap between them and their children. Boomers embraced other values from their parents, such as a strong sense of patriotism and work ethic. Their parents, after all, were the heroes of World War II.

The Vietnam War lasted from 1955 to 1975, with the silent generation making up the initial force as the greatest generation began their retirement. By the end of Vietnam, boomers were a large portion of young draftees (more than 10 percent of boomer men served[63]) leaving a significant number of them deeply affected by their involvement. The average age for those who served in Vietnam was nineteen years. The average age for World War II was twenty-six years.

Vietnam was the first war or conflict to be televised, igniting protests and political division throughout the nation. The question "Why are we in Vietnam?" sparked much debate from Americans, and events like the 1968 My Lai massacre portrayed America as the perpetrator rather than the victim or peacekeeper.

Walter Cronkite, an American journalist and member of the greatest generation, was a correspondent during World War II, flying in bombing raids over Germany and covering the Allied landing on the beaches of Normandy. After a successful move to CBS's *The Morning Show*, he became the longtime anchor of *CBS Evening News with Walter Cronkite* from 1962 to 1981. He covered the assassination of John F. Kennedy and the Apollo 11 moon landing. In 1968 he went on a two-week fact-finding trip to Vietnam, calling it the "Report from Vietnam." Back in New York, Cronkite swayed from his approved script to offer his personal commentary on what he had seen:

"We have been too often disappointed by the optimism of the American leaders, both in Vietnam and Washington, to have faith any longer in the silver linings they find in the darkest clouds. ... For it seems now more certain than ever, that the bloody experience of Vietnam is to end in a stalemate. ... To say that we are closer to victory today is to believe, in the face of the evidence, the optimists who have been wrong in the past. ...

"To say that we are mired in stalemate seems the only realistic, yet unsatisfactory conclusion. On the off chance that military and political analysts are right, in the next few months we must test the enemy's intentions, in case this is indeed his last big gasp before negotiations.

"But it is increasingly clear to this reporter that the only rational way out then will be to negotiate, not as victors, but as an honorable people who lived up to their pledge to defend democracy, and did the best they could."[64]

Cronkite's actions and words were powerful. Although Americans had already started to turn against the war, this moment is remembered as one that validated much of the country's feelings about the war and the government. It has been reported that President Lyndon Johnson, who was considering running for a second term, said after the broadcast, "If I've lost Cronkite, I've lost Middle America." Cronkite and other journalists' more opinionated coverage continued through Nixon's presidency and the Watergate scandal, shaping key historical events.

For those who served in Vietnam, it was more about surviving the war and a viscerally opinionated and unwelcoming country.[65] Vietnam continues to be a painful lesson of a failed global conflict and was very much part of the conversation as America faced future conflicts. Countless books, movies, documentaries, and other content have covered the complexity of this conflict and its impact on generations. Rather than focusing here on the details of the war itself, consider how the war impacted the military culture and leaders within. How would it shape that generation to lead the next?

Although Vietnam veterans experienced a sort of break or peacetime after Vietnam, many continue to walk around with the emotional and physical wounds of a war that included chemical weapons (like Agent Orange) that are still being evaluated and investigated today. The moral and spiritual injury of a war that was so negatively backed by their country continues to be an open wound that may never fully heal.

One veteran who shares the "too little too late" sentiment said, "they still do not want to hear our stories. More recently I am hearing 'Thank you for your service.' ... When I hear those words, they really seem hollow. I wish they would just not say anything."[66]

Perhaps there was some mentoring from those who served in Korea, but Vietnam vets had little help transitioning back into society. Reintegrating from

World War II was certainly different, and previous generations did not openly talk about their experience of war and returning home. Many veterans reported that even their own parents who had served in World War II avoided asking them about their time in Vietnam, some for as long as forty years.

This America did not acknowledge Vietnam veterans or make room for their injuries. Instead, Americans wanted to move on and forget, seeing these veterans as drug addicts rather than heroes. My father, a pilot who served in Vietnam and specifically with Operation Babylift out of Saigon, said that while he was in school, they were ordered to not wear uniforms because service members were being beaten by protesters.

According to James Wright in *Enduring Vietnam: An American Generation and Its War*, "One reporter wrote of the returning [Vietnam] veteran, 'Silently he is slipping thru the back door of the nation which sent him to war.' There were no parades, 'no frenzied homecoming celebrations.' Instead, the veteran has been 'vilified, condemned, ostracized. He has been branded a murderer, a junkie, an undisciplined disgrace.' Perhaps most cutting, 'for the first time in American military history, he has been labeled a loser.' The stories of 'heroism and dedication' had 'been lost under a sea of public disgust.'"[67]

Today, it is hard to write or think about Vietnam without considering the war in Afghanistan as well. The Vietnam conflict lasted twenty years, as did the Afghanistan conflict, both having eerily similar endings. Perhaps that is why in 2021, as America withdrew from Afghanistan with the chaos of evacuation flights, statements like "This is our Vietnam" and "Was Afghanistan a waste?" triggered a large reaction, especially from boomer veterans who refused to repeat the experiences of Vietnam.

Much of the comparison of Afghanistan to Vietnam came about as early as 2001 from concerned leaders. Colin Powell, Secretary of Defense Donald Rumsfeld, President George W. Bush, and National Security Advisor Condoleezza Rice expressed the comparison. President Barack Obama would later bring it up again, influencing President Biden's direct decision on the Afghanistan withdrawal. "I wasn't going to ask [US troops] to continue to risk their lives in a military action that should have ended long ago," Biden said. "Our leaders did that in Vietnam when I got here as a young man. I will not do it in Afghanistan."[68]

Although the Vietnam and Afghanistan conflicts were drastically different, fought in very different landscapes, for different reasons, and on different scales, the aftershocks of the exits were similar enough that they created notable responses from the community. For many Vietnam veterans, the path to healing has always been to offer the support they did not receive to those who have continued to serve in global conflicts since Vietnam.

In this regard, boomers were brought up in their own version of the country at war, externally and internally as the nation wrestled through opinions on war and civil rights. Older boomers were influenced by the assassinations of Martin Luther King Jr. and President Kennedy, Watergate, the March on Washington, Rosa Parks's activism, and the Civil Rights Act of 1964, as well as the rise (and later fall) of the Berlin Wall.

They were also the young adults of the '60s and '70s leading the sexual revolution (e.g., Woodstock) and making significant contributions to music (e.g., The Beatles), feminism, and women in the workplace and education. The family, while still leaning traditional, began to break down barriers of diversity and individual freedoms with the legalization of interracial marriage and abortion. These dynamics, according to sociologists, shifted marriage and home life, as individuals who had higher education were attracted to those who shared those achievements, making it more acceptable for both spouses to work outside the home.

In the military, the '70s brought expanded benefits to the dependents of female service members to match the benefits of male service members. Families, not just wives, were now seen as central to the success of an all-volunteer force. Benefits and provisions were seen as a way to create buy-in for families considering the culture and profession. Bonuses and money for education continued as recruitment strategies. However, now that family was front and center, housing, childcare, health benefits, advocacy programs, the commissary and other family-friendly provisions made the military career more enticing to service members with families.

Jonathan Pontell coined the term "generation Jones" for the younger boomer generation who were largely influenced by the idea of "keeping up with the Joneses" based on a comic strip of that name. They largely felt left out of opportunities like Woodstock and the advantages older boomers received, and they struggled with wanting the same successes. They would go on to parent the younger Gen X and older millennial generations and are known for pessimism and a more cynical approach to politics. Boomers remain, however, the most influential and involved generation in politics, the economy, and the workplace.[69]

During the Cold War, the alliance between China, Japan, and the West to counteract the threat of the Soviet Union was a significant shift in the '80s. President Ronald Reagan leaned toward modernization of the force as well as researching and developing nuclear arms. The pressure to make the all-volunteer force successful turned recruiters toward poverty-stricken or low-income populations that would benefit from education benefits and career advancement opportunities. By 1982, the proportion of recruits with high school degrees reached a record number of 80 percent or higher in all branches.[70]

The 1980s also escalated the message that women were welcome in service and that the institution could provide all the support a family needs. Families entered the military not only for a career, but for a lifestyle that gave military wives, especially, an opportunity to stay home with the help of childcare, housing, and medical benefits reducing the financial burden on the family. Wives who struggled to find employment or did not want to work could leverage their desire to make an impact by volunteering while also influencing their spouse's chances of promotion.

Much of the criticism toward boomers comes from their commitment to work ethic and highly competitive motivation to "make a difference" in their surroundings. When it comes to communication and work style, boomers still very much appreciate face-to-face conversations rather than texts or emails.

The online conflict between millennials and boomers is real and fierce with younger generations making fun of boomers' slow adaptation to social media and technology. However, boomers adapted well to the internet for information, news, and content. As social media came on the scene, boomers took over the early platforms like Facebook, which was more conversational than later platforms like Instagram and Snapchat.

Boomers contributed heavily to technology with space exploration and inventions like the computer. Most of their adult life was a world without the internet, so the main source of content and news was either in print through magazines and newspapers (still delivered to their door) or the evening news on their television. They remember television changing to color and have seen content expand from four major networks to satellite and now streaming.

Even after Walter Cronkite's pivotal moment in broadcasting, much of the news and media before the internet was considered "vetted." With NBC's *Nightly News with Tom Brokaw* or CBS's *60 Minutes*, very few Americans questioned the news they were receiving. In 2022, five hundred hours of new content were uploaded onto YouTube and 380 new websites were uploaded every sixty seconds.[71] Speaking as the Gen X mediator, perhaps boomers deserve a little empathy for the transition they have had to experience from "all news is vetted news" to the current concerns regarding "fake news."

While the boomer views on government and the institution are understandably cynical and anti-establishment, their view of authority comes down to respect, competition, and hierarchy. Throughout the '80s and '90s, with both men and women able to thrive in the workplace, they strived for achievement and career advancement, leading to other labels like "privileged and entitled" and "absent" due to their hands-off approach to parenting.

Similar to the generations before them, most baby boomers valued staying with one career over their lifetime if possible. For that reason, promotion was gained by keeping your head down and doing the job. As our "rock" boomer describes, "the bullseye for success is to make the boss happy." When it comes to feedback, no news means good news. The only feedback expected from superiors is critical feedback. In other words, boomers do not expect accolades or positive feedback, although it is appreciated if they receive it.

Boomers expect to work until the job is finished, even if that means after-hours. Loyalty and commitment to the job translate to promotions and a future in that career. This is likely why boomers have been described as less supportive of the flexible work policies advocated for by millennials. They did not have the opportunity for remote work available to them throughout the majority of their careers and made significant sacrifices to climb the ladder.

In a recent workshop, I arranged participants into small groups based on generations. Boomers, when tasked with an assignment, were quick to finish, aiming for a consensus that they had "done enough." Interestingly, they had a more logical, reserved approach and quickly finished the task before other generations, who were still collaborating, talking through opinions, and even deliberating on who would speak for the group. Boomers described their value system as one defined by right versus wrong and loyalty to the job. After listening to other generations who expressed their strong values around family, the boomers (mostly men) admitted they struggled with work/family balance but also felt they were raised during a time more accepting of that.

In the military culture, this '80s work style, combined with the fact that the military culture lagged behind progressive civilian culture, translated to men working long hours outside the home while wives remained at home. Throughout the Cold War and even well into the first decade after 9/11, it was not uncommon for service members to be expected at the office after-hours, willing to do whatever was necessary to get the job done.

Generation Jones (younger baby boomers) and Generation X were met with a new era of marketing for the military that highlighted adventure and the ideological world of military service. Movies like *Top Gun*, released in 1983, spiked enrollment for the Navy and opened the door for a new military-entertainment complex between Hollywood filmmakers and the Pentagon.

Even after the hurt of Vietnam, patriotism is still a value and motivation for boomers: "91 percent of Vietnam veterans of actual combat and 90 percent of those who saw heavy combat" were proud to have served; "66 percent of Vietnam veterans said they would serve again if called upon."[72] Today, a majority of the public is appreciative of Vietnam veterans, likely due to boomers.

In a recent poll of what values boomers, Gen X, millennials, and Gen Z respect the most in a leader, a few keywords stood out across all generations. All age groups most valued integrity, compassion, humility, and empathy. Although these are not surprising values to expect from a leader, the fact that every generation references the same words is important. The real question is how each generation defines each of those traits. Empathy, for example, is a word that has created a generation gap between millennials and boomers. Millennials have asked for empathy in the areas of diversity and policy change. Boomers, however, now entering later years of retirement and sandwiched between aging parents and adult children, are asking for empathy themselves.

Boomers, along with the generations before them, paved the way for the current military infrastructure, rich traditions, sense of community, and benefits. Raised on the heroic and often painful war stories of generations before them, boomers are patriotic and stern about what the military stands for. They expect gratitude, service, hard work, and unquestionable loyalty.

While praised for their contributions to music and social reform, boomers are highly criticized in the civilian world for their impact on family, economics, and commercialism. Boomers are also often compared to their parents, the greatest generation, for how few served in Vietnam. Boomers would say it is an unfair comparison, considering Vietnam did not need as many in the draft as World War II did.

In the military culture, boomers cultivated the experience of community and advocated for programming that meets so many of the needs of military families today. They mentored the generation following them. Although they likely worked Gen X into the ground with long work hours and doctrines of respect, loyalty, and responsibility, boomers were also known for setting the same standard for themselves.

AN EXAMPLE WORTH REMEMBERING
Kindness differentiates a leader from a bully.

Army Colonel Ian Palmer, chief of staff of III Armored Corps, Fort Cavazos, and a member of Generation X, shares:

"When I arrived at my first assignment, as is true with most second lieutenants, I didn't know much about the Army other than ROTC and my time at what was then called the Officer Basic Course. I observed a lack of values and a lack of personnel respect that forced me to question whether I really had a future in the Army. If this is what the Army is about, I thought, then I need to rethink my future plans. I loved being a platoon leader and being around soldiers, but it appeared that leadership meant something different than what I thought.

"Fortunately, after about a year, I was placed in a new position with an entirely different chain of command. My new boss, Captain Thom Sutton, changed the trajectory of my career by demonstrating that one could be a successful and demanding leader in the Army while also promoting the positive and healthy treatment of people. He held us all to very high standards, but was also a teacher, mentor, and someone I could look up to and model my leadership style after without compromising my values. God places people in your life when they are needed the most and Thom Sutton rekindled my passion for the Army, for leadership, and taught me so much about how to treat people.

"The most important thing I derived was that we all choose what kind of leader we want to be. We can treat people with respect or we can lazily rely on old stereotypes that tell us that we need to break people down in order to lead them. I learned that leadership is service; it's teaching, coaching, and modeling. Leading well means creating a legacy of people who take the positive lessons you teach them and perpetuate them. I am grateful that I learned very early in my career that I wanted to be like Thom Sutton."

WAR, MONEY & CULTURAL DIVISION

"We will bankrupt ourselves in the vain search for absolute security."

—Dwight D. Eisenhower

Chapter 4

FOLLOW THE MONEY

THE PROCESS OF GOVERNMENT FUNDING is a complicated topic for even the most knowledgeable analysts and policymakers. However, understanding how it touches every aspect of military life will shape a leader's perspective on many challenges in the force today. Leaders must be willing to hear the needs of the people in front of them and know how and when to connect their presenting concerns to larger, more systemic issues, something not possible without following the money. Some concerns exist today as direct consequences of financial decisions made decades ago. Maintaining awareness of the process and systems associated with the defense budget helps leaders perform better today and prepare for better decisions tomorrow. It may even prevent them from repeating some of history's prior mistakes.

After a speaking event in the fall of 2015, I was invited to lunch with a group of seasoned Army and Air Force spouses. I remember looking around the table, noting that all of them had a good twenty years of life experience on me, which meant they were likely baby boomers. I listened as they reminisced about the good ol' days when the military community came together, especially during deployments. It sounded similar to my experience in Colorado a few years prior. They connected with each other over a time when communication with their service member was through letters in the mail and when spouse groups and social events were a thriving part of the culture.

As our conversation unfolded, I also noted an underlying resentment. They told stories of the effort and energy it took to build consistent support within family programs and said my generation seemed to take it for granted. For a moment, I felt a bit embarrassed that my inner posture was to sit and learn from this generation when they clearly were expressing frustration toward those of us who had not "taken the baton and run with it." I admitted openly that I had, indeed, benefited from mentors like them, and I had not hosted, volunteered, taught, or supported others in the way they had modeled. Internally, I wondered why that was true.

My taking ownership of my generation's lack of action must have made the general's wife to my left more at ease. She had hardly spoken but then leaned in and said, "I'm so tired. I have aging parents and adult children to now take care of. I need to let go but it's hard to watch what we've built fall apart." The others sitting around the table all quietly nodded in agreement. Then, she turned to me and said something that sparked a new curiosity in me. She said, "If you really want to understand what's happening to our community, follow the money."

Only a small portion of the defense budget is appropriated to military leaders, making their pressing responsibilities more deserving of their time and energy. Therefore, it's tempting to dismiss understanding the larger budget's systemic reach. In addition, managing appropriated (and nonappropriated) funding is difficult, connected to laws with real consequences if used inappropriately, and consumes whole teams of leaders, civilian employees, and volunteers.

On top of how funding is allocated, the complexity of how to spend money is discouraging when red tape debilitates or even sabotages attempts to maintain force readiness or improve morale. Specific buckets of funding are allocated for a specific use and sometimes must be spent within a certain fiscal period. This creates a fear-based environment of "use it or lose it." Leftover funds at the end of the fiscal period are therefore spent quickly or sometimes frivolously. Otherwise, the budget risks being reduced or cut completely.

Families are even more limited in their understanding of government funding, even though their very livelihood depends on it. Assimilating into the military culture does not include training or an explanation of how the institution functions. When there are playgrounds in every neighborhood, grocery stores, maintenance for housing, and a variety of programs from childcare to discounted amusement park tickets, it is easy to make assumptions that there is an endless supply of resources available.

The intent of this chapter is not to dive into the details of economics, politics, or debates around military spending. Entire libraries have been filled with all the rabbit holes one could go down on this topic. This will also not be a line-by-line evaluation of the defense budget. The Congressional Appropriations Committee for 2023 proposed $1.7 trillion in a bill 4,155 pages long,[73] covering far more than the defense budget, of course. It would be overwhelming at best.

Instead, our focus will be a high-level understanding of how the military receives funding, some of the important topics that impact military readiness and families, and the major historical financial markers that have shaped today's military culture over the last twenty years. The goal is to give you a foundational understanding of how money gets to the military and how specific decisions have impacted families in the military community today.

CONGRESS AND MILITARY FUNDING

Long before US independence from Great Britain, English kings had the power to initiate war and to raise and maintain a military presence, much to the detriment of the people and without accountability. The English Declaration of Rights of 1688 provided that the king could not maintain standing armies without first gaining consent from Parliament, thus securing a system of accountability. As the king and Parliament continued to disagree on the use of force and who had the influence to wield it, colonists watched as the king's armies (often corrupt) infringed upon the liberties of innocent lives both in Britain and in the new colonies.

This deep concern over the power of government authorities to raise a standing force became the focus of much debate in the creation of the US Declaration of Independence and Constitution. Ultimately, it was agreed that there was value in the national security of a standing army, but only with checks and balances between Congress and the President of the United States.

Article I, Section 8, Clauses 11-14 of the United States Constitution:
> The Congress shall have power …
>
> To declare War, grant Letters of Marque and Reprisal, and make Rules concerning Captures on Land and Water.
>
> To raise and support Armies, but no Appropriation of Money to that Use shall be for a longer Term than two Years.
>
> To provide and maintain a Navy.
>
> To make Rules for the Government and Regulation of the land and naval Forces.

The "no appropriation of money to that use shall be for a longer term than two years" was added to further limit the government or president from investing in a standing army indefinitely without accountability. Today, special committees of Congress in both the House and Senate oversee the Department of Defense, the Armed Forces, and portions of the Department of Energy. One time during their administration, the president sends their National Security Strategy (NSS) to Congress to communicate the executive branch's vision of national security. In response, the Department of Defense issues its National Defense Strategy (NDS), which outlines how it will execute the objectives outlined by the administration's NSS. Each year, budget requests to implement the strategy are sent to Congress.

Committees from both the House and Senate analyze and then attempt to agree on the defense budget and then appropriate funding through two bills. The first bill, the National Defense Authorization Act (NDAA), was first passed in 1961 as a way to outline the annual DoD budget, nuclear weapons programs, and other defense-related programs. The bill must be passed by the House and Senate before making it to the president for signature into law.

The passing of the NDAA does not actually appropriate the funds suggested in the budget; it only authorizes the programs, projects, and policies that should be used for congressionally-appropriated funds. A second bill, an appropriations bill, goes through a similar process of passing the House and Senate to also be approved into law in order to direct the actual funds. The House Armed Services Committee and the Senate Committee on Armed Services try to work with the House and Senate appropriations committees, which write and have jurisdiction over the appropriations bill, to have as much in common between the two bills as possible before it goes to the president for signing.

Boomers and Gen X likely remember the *Schoolhouse Rock* "I'm Just a Bill" cartoon from the '70s. In it, "Bill" describes to a young boy the long process of a bill becoming a law only to get to the president's desk where it can be vetoed and returned to Congress to start over. A long and often political process, the amount of time it takes to authorize the NDAA can vary, but generally it adheres to a consistent schedule in order to keep agencies running. The relationship between lawmakers, Congress, and the president can impact not only what is approved, but how long it takes to approve it.

For example, the 2021 NDAA, named the William M. (Mac) Thornberry National Defense Authorization Act for Fiscal Year 2021 in honor of Representative Mac Thornberry, chair of the House Armed Services Committee, totaled $740 billion. The bill budgeted for increased modernization, military pay raises, military base realignment, and much more. President Trump threatened to veto the bill as it included a provision to limit the president's use of emergency declarations to divert DoD Homeland Security funding to finance the expansion of the Mexico–United States barrier.

There has always been disagreement over domestic versus defense spending and debate around what amendments should be made to the annual NDAA The defense budget is one of the largest portions of the overall federal budget, hovering around 15 percent during the four years before the COVID-19 pandemic. Considering a good portion of federal spending is on auto-spend (Social Security being the largest portion), defense spending is a considerable amount of spending that must be approved each year. With federal funds needed for other national initiatives like education and healthcare, there is great emphasis and debate on spending money wisely. Some analysts would say that in order to get through the bureaucratic process, ultimately the NDAA needs to become the lowest common denominator between military and civilian leadership to pass into law.

The NDAA covers an incredible amount of information; the most recent 2023 NDAA fills 670 pages.[74] Some of the biggest categories included in the NDAA are funding for all branches of the force; salaries for service members and civilian employees; military training; maintenance and modernization of technology and arms, equipment, and facilities; healthcare for civilian employees, service members and beneficiaries; and DoD operations. However, topics embedded within the NDAA, such as the US-Mexico border wall and 2023's rescinding of the COVID-19 vaccine mandate for service members, can slow down the process to the point of a potential government shutdown. The 2023 NDAA also included military aid to Taiwan and additional funding to Ukraine[75] and addressed continued concern over support of the Navy's number of maintained and ready ships.

As the House and Senate work to find agreement with the DoD every year, one question the parties approach differently is, "How much money does a standing and ready force actually need to be able to defend the nation while also remaining competitive against its adversaries?" This one question invites additional questions:

- How large should the nation's force be?
- What adjustments should be made, if any, during peacetime?
- How do we as a country define peacetime?
- How much should be devoted to modernization and advancement?
- Where does the Department of Defense's responsibility for the care of its service members and families begin and end?

MAJOR FACTORS THAT INFLUENCE FUNDING

Lawmakers, analysts, and policy experts can spend a lifetime studying all the variables that impact federal spending and the defense budget. While it is impossible to address all of them here, some variables that affect force readiness and the daily lives of families are worth a high-level overview.

Money During War, Peace & the Gray Zone

Considering the history of World War II and Vietnam, it may seem logical to assume that defense funding will go up during wartime and down during peacetime. In some ways, this was the initial pattern, with extremely limited funding during peacetime after World War I versus the beginning of World War II, when nearly 40 percent of the GDP was devoted to national defense. Yet, in order to think constructively about the allocation of money, we must consider several important factors.

First, the delineation between wartime and peacetime is not always clear. During the two world wars, Congress made a clear declaration of war. However, that has not happened since. While Congress has the sole authority to declare war, the president has the authority to leverage military force when needed and Congress is designed to authorize such uses of force. A modern example is President George W. Bush's force against Iraq after 9/11. The president is also able to use military force in more informal ways. For example, as commander in chief, the president can deploy US troops in situations that do not specifically amount to war: the support of allies, peacekeeping attempts (as President Bill Clinton did in Bosnia), or low-level hostility involvements such as the contested bombing campaign in Libya in 2011 (President Obama). The United Nations also provides authority for the use of force, as President Truman "argued that his use of force in Korea was a 'police action' to enforce the UN Charter, not a war."[76]

Second, money is spent differently during wartime than during peacetime. During the Vietnam War, military spending increased but the allocation of that spending was on consumable war items, like ammunition, compared to budgets focused on modernization during peacetime.[77] So as we discuss numbers, especially how the defense budget number has changed over the years, it is important to differentiate that the $230 billion budgeted during Vietnam or the Afghanistan conflict, for example, would be spent very differently from the same number during a time of fewer operational conflicts.

Third, national defense has become increasingly more complex. As the DoD's scope and reach have expanded due to changes in warfare, care for military families, and evolving recruitment strategies, so has spending.

This all makes the topic of how the government should budget and allocate money much more complicated than building up for wartime or scaling down for peacetime. The word *war* has been especially emphasized during the two decades following 9/11 to describe the consistent wartime operations tempo felt within the military community. If we were to get into semantics, however, the War on Terror (officially the Global War on Terrorism or GWOT) was an international counterterrorism military campaign started by the US that included what have been called the Afghanistan and Iraq wars. These global conflicts led to Operation Enduring Freedom (OEF) and Operation Inherent Resolve (OIR) but, just as with Korea, Vietnam, and Desert Shield/Desert Storm, Afghanistan and Iraq were considered global conflicts with congressional involvement, but no official declaration of war.

This was also the beginning of Congress providing funding designated for emergency requirements and later for Overseas Contingency Operations (OCO). A separate OCO fund was originally designed to finance operations in Afghanistan and Iraq and has less oversight than other funding areas. Some refer to this "war fund" as a "slush fund" for the Pentagon. It has become a topic of much debate and will be important as we look into the impact of budget cuts beginning in 2011.

As the US withdrew from Afghanistan in 2021, it marked an interesting moment for military members who had joined since 9/11. To most, it felt the "war" was over. Perhaps US citizens felt the same and many assumed the US would go into some sort of peacetime such as after the Cold War and before 9/11. Military families expected a slower pace, fewer deployments, and a ramping down of training. However, instead, conversations surrounding Ukraine, Russia, and China escalated and the operations tempo did as well. Today, the Center for Preventive Action is tracking more than thirty ongoing global conflicts that are of concern to the US.[78]

The point here is that the US declaring war is a significant statement, to its own people, allies, and adversaries. Defending the nation as outlined by the president's NSS, the defense budget, and the NDAA includes a complex strategic combination of supporting current operations and deterring ongoing future threats. Advancing the modernization of the force for worldwide competitive advantage not only prepares for future conflicts but also sends a message to allies and adversaries. The NDAA, then, is also a tool of communication in as much as it is a financial budget.

During the Cold War, pentagon leaders relied on a "two-war standard" in determining the size and strength of the force. The premise was that at any given time, the force needed to be prepared to fight two wars simultaneously. However, Afghanistan and Iraq demonstrated that it's not an easy or inexpensive task when there are additional ongoing operations that promote global stability such as strategic posturing, training other militaries, and more.[79]

The military's role in providing global stability has also been one of debate. In 2018, it was estimated that "over 6,000 US troops [were] currently stationed throughout the [African] continent in approximately forty-six sites, which include forward operating bases, cooperative security locations (such as drone installations), and contingency locations."[80] Even though military leaders maintain they were not engaged in a war with Africa, managing ongoing threats in other countries often means operating in a "gray zone" between war and peace.

This, again, brings up difficult questions we must consider when looking at defense spending, questions that policy leaders and scholars also grapple with. Steven Feldstein, in an article for Carnegie Endowment for International Peace, wrote, "As the roles and boundaries of what the military ought to be responsible for have become increasingly murky, the US military has become a 'Super Walmart' that offers vast resources and economies of scale to address any situation. Former US ambassador to Nigeria John Campbell describes this phenomenon as the 'securitization' of US foreign policy, arguing that since the September 11, 2001, terrorist attacks, huge increases in Pentagon resources have led successive US administrations to entrust the military with solving a growing array of nonmilitary problems."[81]

These questions are important to ask but also reveal the complexity behind the national defense strategy and the DoD's approach to meeting those goals. Does the defense budget reflect the amount needed to protect and defend against actual threats against the US? What should our role be in global stability, especially in a world of finite resources? Does operating in gray zone environments reduce threats or create more conflict? Do our activities and presence justify our need for sustained and increased funding when there would naturally be a conversation around reduced spending in the absence of war?

Answers to such questions impact recruitment and readiness. If there is no end to global conflict and no peacetime for a country and its defense, then there is no respite for those whose jobs and livelihoods exist to defend the nation. What role, then, does respite and recuperation play in maintaining a ready force?

Today's oversaturation of real-time information, global conflicts, and concern about US involvement create a war-weary state. The warning of King Solomon's proverb, "Where there is no vision, the people will perish"[82] emphasizes that, without understanding or clarity, people will move into a state of chaos. What we have seen since the Afghanistan withdrawal is the emotional and organizational chaos of a generation that has not known a time of peace in their lifetime. Communicating clarity and vision as a leader is paramount. Although respite or recuperation may not be specifically written into the defense strategy and budget, leaders on every level have the responsibility, and in many cases funding, to make sure it creatively happens.

Military-Industrial Complex

As Vietnam ended and a time of relative peace began, the defense budget continued to steadily rise until 2011. Rather than face budget cuts, this was an opportunity for modernizing programs and acquiring major weapons systems. For industry sectors that supported the military, it provided significant opportunities for government funding, not limited to wartime. It also opened funding for innovation and research and provided more jobs. The Defense Department today would cease to operate without the production and work of government contractors. Whether it is designing and producing the latest advancements in ammunition and weaponry or running the payroll for personnel, this has become the way the military runs the business of defending the nation.

Today, many believe that what President Eisenhower prophesied about the military-industrial complex (MIC) has come true. Large-scale industries like Raytheon Technologies, Lockheed Martin, Boeing, Northrop Grumman, and General Dynamics (nicknamed the Big 5) are the biggest and wealthiest of the more than four million government contractors making up the MIC. The term MIC was not used until the second half of the 20th century, when concerns grew over the level of influence these larger industries might have in shaping national and global warfare, considering the amount of the budget that went specifically toward them. Lockheed Martin has been the largest of the Big 5, reported in 2020 to have $48.3 billion in obligations to the government.[83]

Another concern about the MIC is that many of the leading industries maintain a level of secrecy around their innovation and research in order to keep a competitive advantage in securing future contracts with the government. It also raises concerns that the competition around their innovation encourages the escalation of capabilities of warfare among countries.

While having a war to fight (or being ready for one) benefits these industries, there is also concern that the lines have been blurred between the military and its contractors. Some 80 percent of three and four-star officers are often provided second careers within these industries. Networking in and out of the Pentagon is not encouraged but happens.[84]

The fiscal year 2023 NDAA (like previous years) included growing attention and concern about cybersecurity. This was particularly true within the US Army, as the operating model of warfare shifted from antiterrorism and counterinsurgency operations to "large-scale combat [and] multidomain operations as part of a joint and multinational force. Funding around research and development also shifted to include space and cyber-space technology."[85] Research and development (R&D) has long been a heavy part of the NDAA since the Reagan administration, however, in the last few years, cybersecurity has received more R&D funding, creating new opportunities for cybersecurity companies.[86] The need for this technology has driven many private companies to innovate and compete for R&D government contracts. Many are now calling it the cyber-industrial complex.[87]

As data breaches, digital weaponry, and the influence of cyber-attacks continue to threaten military personnel and American citizens, the government, lawmakers, and Congress must rely on industry leaders for information, capability, and guidance. According to former House Armed Services Committee

Chairman Mac Thornberry, "Adversaries continue to try to disrupt our social cohesion, our elections, and support for a strong military. Others are against terrorist groups that continue to plot against us. All of that makes the line between war and peace fuzzier than ever, which has all sorts of implications for the military and for the civilian-military relationship."

The question surrounding this topic is the classic "chicken or egg" debate. Does the defense budget for modernization reflect the needs identified by the government or does it stem from the influence of industry leaders who create needs only they can fill?

Allies & Adversaries

Happenings in other parts of the world can directly impact the US defense budget and its priorities. Altogether, US allies spend less than half of what the US spends on defense and therefore depend on the US for "essential capabilities, regarding for instance, intelligence, surveillance, and reconnaissance; air-to-air refueling; ballistic missile defense; and airborne electronic warfare."[88]

In 2006, the North Atlantic Treaty Organization (NATO) defense ministers agreed to spend 2 percent of their countries' GDP toward military defense. While the 2 percent expectation has encouraged allies to increase their own spending on defense, many have done so slowly. Russia's invasion of Ukraine in 2014 and escalation in 2022 ignited many countries to quickly catch up on their commitment and increase their spending, especially on new major equipment.

Some experts predicted this situation in what they call the security vacuum of Europe. If allies are unwilling or unable to invest in their own security and defense, they are more likely to remain dependent on the US. As conflicts arise, who then will protect Europe? While the US could stay out of global conflicts between other countries, NATO agreements as well as relationships with allies are just as important for US security. Theoretically, providing support to allies is a preventive means to deter future conflicts or another world war. But it's an expensive and complicated position. When providing intelligence and military capability support to Ukraine, for example, the US had to carefully avoid its own conflict with Russia. Relationships like these create complex decisions around US defense funding, not only for the support of allies (such as the 2023 NDAA support for Ukraine and Taiwan) but in modernization to deter other conflicts.

In addition to cybersecurity changing the landscape, nuclear missile programs have continued to be a priority from the Cold War to the 2023 NDAA. As Russia and China continued to advance their own nuclear capabilities, the US defense budget directly reflected its defense strategy. For example, in the FY23 budget, the Sea-Launched Cruise Missile-Nuclear program (SLCM-N) was zeroed out after President Biden wanted to "reduce the role of nuclear weapons in the defense strategy."⁸⁹ However, when asked by Congress his opinion on the administration's view, Chairman of the Joint Chiefs General Milley disagreed saying, "My general view is that this president or any president deserves to have multiple options to deal with national security situations."⁹⁰

So while one would think that defense spending would go down as war or global conflict ends, peacetime (or reduction in global conflicts) offers just as much of an opportunity for the DoD to "pursue a qualitative military advantage over potential adversaries through a robust and well-directed modernization program."⁹¹ To global allies and adversaries, the investment in modernization and quality of life is an indicator of national resolve to remain dominant in warfare. This means that decisions on how defense money is allocated and spent are just as much a means to deter foreign adversaries from inciting conflict as it is a way to prepare to decisively win in potential future wars.

Use It or Lose It

As stated earlier, the fear of losing money if it is not spent in a timely manner is a palpable concern, and the flow, or lack thereof, trickles down to impact service members and their families throughout the year. Budgets for various accounts can last for one fiscal year or as many as five or even eight fiscal years into the future. This means the DoD must budget not only for future threats of global conflict but also consider economic inflation and other variables. By the time funds are delivered, the needs of the defense may have changed.

Specifically for the category of operations and maintenance, Congress tasks the defense to provide a quarterly report showing that as close as possible to 25 percent was spent in that quarter. In addition to the quarterly expectations, there is a legislative provision that states that no more than 20 percent can be spent in the last two months of the fiscal year. Considering some funds expire at the end of the year, pressures and anxiety around how money should be spent create the "use it or lose it" fear described earlier.

The means by which this is accomplished at the eleventh hour is through prioritized "unfunded requests" (UFRs: pronounced yoo-fers).⁹² These requests are mission essential or mission enhancing items, for example, from a combatant command that did not make it through the budget process to get to Capitol Hill. Or they may have necessities that have arisen during the current fiscal year which, as such, were not budgeted and therefore not funded. The higher headquarters prioritizes ("racks-and-stacks") these UFRs and requests that any unspent or unobligated funds be redirected to fulfill this requirement.

Although the use-it-or-lose-it fear is not entirely based on actual legislation, the concern is that if money is not used or returned to Congress it will be assumed that too much money was appropriated and therefore cut for the next fiscal year. It is not entirely an irrational fear. "Between fiscal year (FY) 2013 and FY 2018, the Department of Defense had over $81 billion canceled, most of it—$49 billion—from operations and maintenance (O&M) accounts."[93] Regardless of whether the funding in question is for family programming, research and development, or weaponry, this mentality and fear of losing funding can become invasive throughout the organization. When looking at the defense budget, the question analysts ask is "Does the defense budget reflect what is actually needed, or what the defense has had and is afraid to lose?"

In some cases, red tape and paperwork for approving the spending of appropriated money delays the actual spending of those funds further into the year causing even more stress. The timing of money spent and the justification of money spent on government contracts that are difficult to explain (such as the Air Force's famous $10,000 toilet seat)[94] has created much debate by analysts and legislative leaders on whether there is wasteful spending in the defense.

Commitments to Personnel and Families

While many other factors are worthy of exploration, the DoD's commitment to service members and families is one that is not only the most expensive category of the defense budget but also one of the most highly debated. The defense has been known to provide "gold standard" benefits to service members and their families in return for their commitment and sacrifice. But it has also been largely criticized for doing so. While operational costs related to civilian and service member personnel include salaries and job security, there is also the promise of healthcare, retirement, other benefits (such as housing and installation management), and programming that supports every stage of the career. This area is considered "cradle-to-grave" benefits that make up the most expensive part of this category.

In 2022, about 7 percent of the federal budget went to "benefits to veterans and former career employees of the federal government, both civilian and military, totaling $420 billion. About nine-tenths of the benefits available to all veterans are either disability payments or medical care, which is often specialized to deal with the unusual conditions that military service may impose."[95] Inside the defense budget, close to 12 percent was allocated to healthcare, and $9.2 billion was budgeted for family programming. In 2023, a 4.8 percent pay increase for service members and an increase in Basic Allowance for Housing (BAH) was also included.

To maintain a ready force, the government has an established relationship with personnel as well as their families. This commitment requires a significant portion of the defense budget to provide benefits and provisions as the needs of families change and recruitment incentives evolve.

Though members of both the House and Senate express concerns about spending, the last thing representatives want is to be one who isn't willing to support veterans, service members, or their families. So, historically, Congress has provided more than what is requested by the president. In December 2022, the NDAA for 2023 was passed with $45 billion more than what was requested by President Biden. Some of this, of course, included support for Ukraine and accounted for inflation, however, it should be noted that even when there is a need to cut government and defense spending, doing so is hard.

More than ever before, access to legislative and military leaders is a mere social media post away. As service members and their families feel the weight of the operations tempo, they are taking their needs directly to those who are making financial decisions. Some of their needs originate from the consequences of prior budget cuts. A broader awareness of those consequences is being shared by families who have better digital access to their congressional representatives and are unafraid to share it with the world digitally. Meanwhile, Gen Z is watching as military families openly express just how far defense spending goes to support the families who are already serving.

AN EXAMPLE WORTH REMEMBERING
Congress was built for advocacy.

Between 1991 and 2016, a DoD "accounting problem" caused the IRS to "incorrectly withhold taxes on disability severance payments to combat-injured veterans" rather than make nontaxable payments. The number of veterans impacted was initially unknown, with some within the government minimizing the extent of the problem. In March 2016, Senator John Boozman (AR) and Senator Mark Warner (VA) introduced a bipartisan bill that was unanimously passed that same year called the Combat-Injured Veterans Tax Fairness Act of 2016 (114-292). Thanks to the passing of this law, the DoD was directed to identify veterans and work with the IRS to help inform those affected. By 2018, the number reached 130,000 with payouts amounting to as little as $1,750 or as much as $10,000.

Military spouse and millennial Heba Abdelaal was a staffer in Senator Boozman's Washington, DC, office at the time. She shares, "I'll never forget the immense amount of national interest this legislation received back in 2016. It was one of the last pieces of legislation President Obama signed into law. It was passed unanimously. If it hadn't been for the advocacy of a legal organization and impacted veterans, Congress would have never known this had happened. It was one of those moments where the system worked to right a wrong, just like it's supposed to. And it was one of the most inspiring things I've ever worked on as a staffer. It was personally a lesson for me that the institution of Congress can right wrongs in society. As a result, I now feel that helping our military community elevate their advocacy at the national level and in front of Congress is what I can continue to give back to our military family."

Chapter 5

LAW OF DIMINISHING RETURNS

WHEN THE FOUNDING FATHERS wrote the Constitution, they did so with the power of a king in mind. It was important that one person not hold the power over providing for people's needs, rather that "the people ought to hold the purse strings." Congress, as representatives of the people, manages the purse strings on our behalf. For the military community, this means that our most basic needs depend on the ebb and flow of defense dollars and the focus of each year's NDAA. Key decisions indeed hold power to the community's needs and also its morale.

Just as each generation remembers how their personal story shifted with the introduction of new technology, each generation experiences its own version of the military culture as military life shifts over time. While the presence of global conflicts definitely shapes a generation's experience in the military lifestyle, so does the presence or absence of funding. As funding rose and fell throughout the two decades of war, many mistakenly believed that their neighbors of another generation had the same experience.

Understanding how money shaped how someone experienced the culture is not as easy as it sounds. Service members and their families may not connect their current positive or negative experience in the military lifestyle with changes in funding. Additionally, most naturally assume their generation and values reflect the majority.

As an example, according to the 2021 DoD demographics of active-duty personnel, 12.1 percent[96] of the force was forty years and older (Gen X and boomer). This number rose from 7 percent in 2020 due to the COVID-19 pandemic's impact on rising inflation, fragile housing markets, and changes to the career search process in a volatile economy facing a possible recession, suggesting that many from these generations at military retirement age chose to remain rather than retire, even as morale was low.[97]

Each generation holds some level of influence, whether it be through wisdom or the powerful voice of the majority. Gen X and boomer leaders tend to overestimate their generation's population in the force. While they have incredible influence as leaders, they are outnumbered.

The military offers a strong hierarchical and authoritarian framework for leadership, yet cultural leadership involves the art of leading people just as much as the science. Culturally, technology and social media have flipped the totem pole upside down, giving younger generations access to the top as well as the voice to influence major decisions for brands and institutions.

As we explore how money shaped the culture and how each generation experienced that culture, it will be important to remember this seemingly obvious but crucial point. There are entire chapters of the military narrative that current generations have never known and yet we are attempting to lead them as if they have. Understanding these generationally different military narratives is key to truly hearing what people are trying to say.

RECRUITMENT BEFORE 9/11

When we left off with boomers it was the '80s and '90s, the peak of their service. Understanding the boomer narrative is crucial in that they were leading a large portion of the DoD and military branches by the time of the Afghanistan withdrawal in 2021. Their leadership style was shaped by their experiences of the Gulf War, budget cuts, and of course, 9/11.

The older cohort of boomers may remember a majority of the Cold War, an aggressive and costly arms race between the US and the Soviet Union. However, a majority of boomers would serve during the Gulf War (1990-91). The silent generation was in middle-to-senior positions of influence at the time when President George H. W. Bush, part of the greatest generation, leveraged military force into Kuwait. The Gulf War would be televised to the American people on a new level compared to the Vietnam War.

Thanks to targeted recruiting and advertising leading into the '90s, women made up 11 percent of the military. By 1991, combat aviation opened up to women as well and two years after that, combat ships also opened up to women.[98] While the '80s brought movies like *Full Metal Jacket*, *Red Dawn*, and *Top Gun*, the '90s brought *Crimson Tide*, *A Few Good Men*, *The Hunt for Red October*, *Saving Private Ryan*, and *Forest Gump*, among many others. The military-entertainment complex continued to successfully aid in recruitment for all branches. According to Marine veteran Alex Hollings, "*Saving Private Ryan* was the first movie I ever saw that made me consider the sacrifices inherent to service in a context other than death. These men had left their lives behind to go fight, and in that moment, watching from my couch, I admired them all that much more for it."[99] Still, only 3.5 percent of the US population was in active duty, National Guard, or Reserves (today, 1 percent).

After the Cold War and Gulf War ended, the military leaned out with significant budget cuts that reduced spending to "less than one-third of what it was during the peak of the Vietnam War."[100] The military also faced personnel cutbacks for the Department of Veterans Affairs (VA), and defense "roles were reduced to peacekeeping missions and disaster relief."[101] Many remember friends getting letters that read, 'As of July 1, your services are no longer needed.'

Deployments were less frequent and success was measured by a "lack of violence and no casualties." With no major war to fight, the military "engaged in nation-building activities to help restore order and improve life in war-torn areas."[102] A spouse whose husband served in Desert Storm described what many have echoed about the leadership climate then, "There was so much trust and respect [for leaders] at that time. It was a great military career for us. I will say, we would have never left Afghanistan like we did."[103]

According to others who served in the Army in the '90s, "doctrine, discipline, respect for rank, army values, rules, and regulations were not just known, but heavily enforced."[104] Knowing what we know of the relationship between the silent generation and boomers, this work climate makes sense. Both heavily value work ethic and respect within a hierarchical system, although each from a different motivation. What it was that made this season of leadership so endearing is difficult to define. It may have been the silent generation's leadership style or perhaps the lack of technology that created a more in-person work environment. Perhaps it is the amnesia of time that helps people look back on only the best moments of their experience. Yet, many look back and remark on those years as positive leadership years.

The culture at the time was heavy in military tradition, social events, and support, but those who served during that time reflect back saying there was "no funding for family and marriage [morale programs] or support for families separated by solo rotations to places like Korea." Housing and maintenance were governed by units, and families were expected to keep their lawn "fifty feet in any direction from home mowed to shorter than four inches" and units would inspect with a "three strikes you're out"[105] for non-compliance. Installation security was relaxed and minimal compared to what it would be after 9/11.

Spouses continued to invest in the community with social events and programming that fulfilled a sense of duty and purpose. Spouses married to service members in command roles became "key-callers," who regularly called spouses on the roster as a way to check on their well-being and distribute information about the unit, important dates, or activities. Although there were male military spouses at the time, they were the minority and less acknowledged. Spouse events remained themed toward women's interests and were mostly scheduled during the day. Those who were in same-sex relationships were also in the minority and existed under President Clinton's Don't Ask, Don't Tell (DADT) policy. Most same-sex couples, therefore, hid their relationships or isolated themselves from community involvement.

As the military increased its demands on the family and the spouse throughout the '80s, specifically to help run programs and offer social support in the culture, spouses became increasingly intolerant and dissatisfied. The result was families demanding additional resources from the DoD to support the military's expectations, even as the Defense Department faced budget cuts. Mady Wechsler Segal, PhD, a sociology professor at the University of Maryland, provided a good description of the tug-of-war between the military and families, saying both are "greedy" institutions that make "great demands on individuals in terms of commitment, loyalty, time, and energy."[106]

The events of 9/11 rocked the nation and the military culture and resulted in a surge in funding, recruitment, patriotism, and reenlistment. One spouse shared, "As a Navy wife, life didn't really change much. Sure, it was a little scarier, but the tempo of training and deployments didn't change much at all. The Navy has always been an actively deploying force with relatively few losses. My Army and Air Force friends' lives [however] were absolutely upended. They went from fewer deployments to being deployed all the time. Their deployments were generally quite dangerous. And Army and Air Force families sustained catastrophic losses that impacted the whole community."[107]

GENERATION X (DOB 1965–1982)
Forgotten & Independent

Although the first of the Gen X cohort turned eighteen in the early '80s, the youngest entered adulthood around 9/11. This means that while older Gen X members may have served in the Gulf War, the rest of the cohort was the perfect age to respond and enlist as the US Armed Forces invaded Afghanistan in October of 2001 against al-Qaeda terrorists and their Taliban supporters.

Referring to themselves as "latchkey kids," Gen X spent much of their childhoods with autonomy and freedom. With divorce numbers at their highest and women entering the workplace more than ever, it was not uncommon for both parents to work and Gen X to be trusted at home alone or with siblings. Largely independent, they were introduced to the convenience of microwave TV dinners and were independently responsible for homework and house chores. This early sense of responsibility resulted in Gen X becoming an entrepreneurial generation, preferring to work independently or run their own businesses. A recent study showed 47.2 percent of US small business owners are Gen X.[108]

Playing outside until the street lights came on was a common memory Gen X shared given that neighbors and neighborhoods were generally considered safe. Gen X spent a considerable amount of time playing outside, riding bikes. Many remember Hasbro's GI Joe inviting children into the world of special forces action figures and their enemies who were trying to take over the world.

Created in the 1970s by Army Vietnam veteran Larry Hama, GI Joe and a team of other veterans (including some from the British Armed Forces and a few who served in World War II) introduced patriotism, military terms, military gear, and martial arts into American homes.[109] In 1985, the first two animated seasons of *GI Joe: A Real American Hero* launched just in time for Gen X's Saturday morning television cartoon programming.

Television, movie rental stores, movie theaters, video games, and the computer became the technology of Gen X's childhood and adolescence, email and the internet not coming into their world until young adulthood. Gen X would be the last generation to remember adolescence or adulthood without the internet. Meanwhile, entertainment made the shift from outside the home to inside, with most of it on screens. Atari could turn a television into a personalized arcade. VHS tapes rented from the local video store allowed home movies on-demand. Concerns from older generations circulated about how much screen time Gen X was absorbing, and advertisers escalated their ads and commercials toward latchkey kids at home.

This brings up another generational dynamic worth exploring. Ask most Gen Xers to finish these sentences and you'll get quick answers with a smile:

"My bologna has a first name ..."

"After these messages ..."

While generations before them had the same news channels and media, Gen X was marketed to specifically during certain times of the weekday and on Saturdays. Advertisers knew that many families were sitting down in front of their televisions at the same time watching the same shows. An entire generation was imprinted with the same information during their most formative years. Entertainment and content have expanded exponentially since then, creating so many platforms that people now get different angles of the news, content curated based on their interests, and likely have different experiences than their peers. This might have ended the era of marketing and messaging to large audiences through limited pathways.

Another huge shift toward the American culture moving inward was the murder of Adam Walsh in 1981. Boomers' hands-off parenting style began to shift when Adam was abducted from a Sears department store at the Hollywood Mall in Florida. As Adam's parents grappled with his disappearance and death, his father, John Walsh, became an advocate for other missing children. His 1983 TV film *Adam* was watched initially by more than thirty-eight million people. Walsh later became the host of another TV show, *America's Most Wanted*. Throughout the '80s and '90s, the faces of missing children would be aired on television and printed on the back of milk cartons, introducing fear and insecurity to Gen X youth and boomer parents. Before this, malls had been the safe hangout of choice for most adolescents and families.

Over the course of Gen X's young adulthood, neighborhoods, shopping centers, and neighbors felt less safe. "Stranger danger" was taught at home, in school, and during TV afternoon specials, telling kids to watch out for white vans with tinted windows and strangers luring children with candy. At school, the Cold War triggered nuclear fall-out drills, raising anxiety for children. According to Greg Lukianoff and Jonathan Haidt, authors of *The Coddling of the American Mind*, this rise of paranoia and fear in the '80s and '90s led to the beginning of overprotective parenting, which in turn impacted the generation that followed, the millennials.

Sandwiched quietly between the boomer and millennial generations (who are in constant conflict) plus the new curiosity surrounding Gen Z, Gen X is quick to point out that they are the "forgotten generation." However, they would also become known for rebellion through punk rock; political satire in shows like *Saturday Night Live*, *The Daily Show,* and *The Colbert Report*; and the launch of music-video television with MTV.

Gen X also had its own pandemic to navigate. The AIDS epidemic and the initial mystery that surrounded it filled their adolescent and young adult years with anxiety, specifically around sexuality, sexual expression, and social interaction. Again, mass media advertisements targeted young people's dating choices and anti-drug campaigns.

Patriotism, likely thanks to the boomer generation, still took center stage in films during Gen X's formative years with movies that reflected Cold War fears and wars of the past. American flags covered almost everything, including Rocky's shorts in *Rocky*, *Rocky II*, and *Rocky III*, as well as movies like *Independence Day*, *GI Jane*, *Glory*, and many more.

In addition to an independent work style, members of Gen X are similar to their parents in that they believe a strong work ethic leads to promotion or advancement. Although they similarly expect little feedback unless it is constructive criticism, it comes from a different motivation of autonomy and independence. Where they diverged from boomers is their value of a consistent family-work balance. After watching their parents sacrifice family time for their jobs or careers, Gen X wanted work to end at a reasonable hour so they could be more invested in their home lives. The strengths and differences between these two generations could not be better put to the test than after 9/11.

9/11 & THE SURGE

September 11, 2001, was the first time our country had been attacked on American soil since Pearl Harbor, igniting a similar patriotic response from the country. It was also our country's first-ever live-streamed surprise experience of war; however, most didn't realize we were watching war unfold until it was too late and too riveting to turn off our televisions.

The patriotic response from the entire country, let alone those within the military, was palpable. We watched a united Congress sing "God bless America" together on the Capitol steps. Civilians eagerly offered help in various ways throughout the aftermath, while many enlisted and commissioned into the force, including older millennials. One Harvard study found that shortly after the attack, "79 percent of American college students [millennials] supported air strikes in Afghanistan and 68 percent would support the use of ground troops. Three in four said they trusted the military to do the right thing all or most of the time. Ninety-two percent considered themselves to be patriotic."[110]

By October 7, 2001, President Bush announced a US military response in Afghanistan, beginning with al-Qaeda terrorist training camps and Taliban military installations. By March 20, 2003, the US led a coalition that included the United Kingdom, Australia, and Poland to invade Iraq, calling it Operation Iraqi Freedom. Starting with airstrikes on Sadam Hussein's palace and other military targets, eighty-two thousand service members were deployed in response to intelligence reporting the stockpiling of "weapons of mass destruction," a topic still debated today. After just five weeks, Iraqi forces were overwhelmed but an insurgency of al-Qaeda fighters poured into the country "sparking guerrilla warfare tactics against US troops and civil war between the Sunni and Shia tribes."[111]

With two major conflicts in two different countries, every generation within the force (and their families), was willing to sacrifice until the job was finished. Even if it meant longer work hours and extended deployments, military families felt that "holding down the homefront" for their deployed service member was their service to the country. Boomers, including some who had experienced the Gulf War, now had the opportunity to lead the next generation through an important global conflict backed by the country and its allies. The trust and value in the hierarchical structure by all gave a sense of order during intense trainings and deployments to multiple countries.

Within the first decade after 9/11, military spending increased 50 percent, adjusted for inflation,[112] doubling its spending and obligations to companies and contractors. While much of this was building the infrastructure for homeland security, focus was also placed on rebuilding the force for a response to Afghanistan and Iraq, with additional incentives and new programming to support the demands the military was placing on families.

With so much of the culture evolving between 2001 and 2010, Gen X was being introduced to a military life that looked far different from what boomers had experienced. While the military had leaned out and built back up before, this climate was different. There was a sense of patriotism in the country and an awareness of what was being protected. There was also conversation about what military families were sacrificing to make that possible. Those enlisting for the first time could receive bonuses up to $20,000 and those reenlisting could receive bonuses in various amounts as well.[113]

Addressing installation housing issues had already been part of the conversation since the '80s. By 1995, funding and attention went to maintaining and revitalizing single-soldier barracks. However, despite the reduction in force (RIF) during the '90s, the need for more and better housing for families only grew. In some locations, housing dated back to World War II and was no longer suitable for service members and families. Yet with budget cuts in the late '90s, addressing the housing crisis was put on the back burner and families had the option of living off the installation in the civilian community.

One answer to this dilemma was the DoD's plan in 2006 to privatize 87 percent of military-owned housing by 2010. By doing so, the DoD could lean on the expertise of civilian construction and maintenance, and build or renovate homes faster. By 2005, more than 112,000 family housing units had been privatized and there were plans for 76,000 more units in four years. The oversight of these programs by each of the branches, however, was still under review by Congress in 2005, including the concern for how families would communicate feedback to the DoD on how the new privatized experience was going.[114]

For those (like our family) who were assimilating to the military culture during those years, the military was a provider of new or newly remodeled homes. If you were coming in after the housing crash of 2007, gaining new housing was a potential relief from the economy. Why invest in a home that could become a financial risk or burden when the military could give you a brand new home and respond to any maintenance request you had?

Shortly after we arrived at our first duty station, as we walked through our new house, a housing office employee told me that even if I needed help changing a lightbulb during the deployment, I could call maintenance. My first impressions of military housing were not exactly shared by boomers. While they were grateful to have the promise of better quality housing, their experience was that the military wasn't quick to respond to their housing and childcare needs. It would take consistent funding and support over several years to improve the quality for families and to rescript the preexisting negative narrative.

It is difficult to capture the sheer amount of support that developed in those beginning years after 9/11. An online search brings up overwhelming amounts of programs, benefits, and nonprofits that developed options and care for service members, veterans, and their families after 9/11. The most utilized may be the Post-9/11 GI Bill, which provides education benefits to service members or their dependents if they choose to pass it down. There was an immense amount of monetary support that flooded in from civilians to nonprofits after 9/11, including the USO, National Military Family Association, and other nonprofits serving military or veteran families.

New family support programming through the DoD made a memorable impact, especially on the active-duty culture. Though not an exhaustive list, the following additions significantly enhanced the quality of life for families.

Family Readiness Programs

Each branch has a different name for its unit family readiness programs (e.g., Army family readiness group, Marine/Navy/Coast Guard ombudsman groups, Air Force key spouses). These groups originated with military spouses offering coffees and social events to distribute important information. The military acknowledged these groups in the '80s as a valuable resource for morale and connection to families. After 9/11, spouses of leaders and other volunteers had a difficult time managing the stress of intense operations tempo as well as efficiently reaching out to other family members. In response, the military branches adopted the spouse-run group as a command-sponsored program (and thereby funded it) as an official tool for communication. With the help of paid staff positions, there was a liaison between the command and families.

These family readiness support administrator (FRSA) or key spouse manager positions were often filled by a military spouse who took the administrative burden off the volunteers. In 2007, FRSAs became available Army-wide.

General George Casey Jr., then Army chief of staff, wrote in *Army Echoes*: "This is not a quick fix; the Army is planning for the long-term. ... These battalion-level family readiness support assistants are as important after a deployment as they are prior to a deployment. ... We will continue to look for more ways to help. We owe this to our families, and these steps are just a 'down payment.'"

There is simply no longer any question that in an all-volunteer force, family readiness equates to readiness of the force itself."[115]

Strong & Ready Teams (Formerly Strong Bonds)

In 1999, US Army chaplain Jason Duckworth started a program to invest in the morale and readiness of military couples and families. It quickly spread throughout active and reserve components of the Army. Events supported stronger relationships, communication, and marriage enrichment. In 2004, the law was amended to allow command funding for chaplain-led programs to assist in building and maintaining a strong family structure. This meant there was similar programming available and funded across every branch. During the build-up, Strong Bonds[116] served thousands of service members and families through weekend retreats with food, lodging, childcare, and training.

Medical Benefits

Before 2001, "retirees lost all medical benefits once they became eligible for Medicare due to age or disability."[117] With the development of TRICARE for Life, veterans are provided full medical and pharmacy coverage. The impact this had on retention was significant as this coverage was not available to those who separated from active duty without retiring. The VA has benefits for veterans as well, however, coverage is dependent on disability rating. With this addition to the budget, it became the "fastest growing portion of the defense budget more than doubling in real (inflation-adjusted) terms since 2001."[118] The argument today is that 20 percent or more of the military's health care spending is for working-age retirees, meaning they do not qualify for Medicare yet. The thought is that these retirees could, in theory, benefit from a second career and qualify for benefits through that employer, which would reduce federal spending.

Childcare Services

In order to support the demands of the military deploying with more frequency, more childcare support was needed for families at home. The DoD had been offering what many had called the gold standard of childcare since the start of the all-volunteer force in 1973. With some major revamping of oversight and safety in the '80s, it was (similar to housing) privatized to civilian contractors in 2000. By 2005, there continued to be waitlists and limited hours, so recommendations were made to invest even more funding into childcare. By 2011, President Obama launched an initiative that included ensuring excellence in childcare and increasing childcare availability. By 2020, childcare costs had risen to up to $1 billion a year.

Military OneSource & Behavioral Health

In 2004, the DoD started the Military and Family Life Counseling Program, which provides licensed counselors (MFLCs) contracted through privatized companies to service members and their families. These clinicians provide confidential, nonmedical counseling as an alternative to families using TRICARE benefits for medical-based behavioral health. Another DoD program, Military OneSource, became available in 2004 to active-duty, National Guard, and Reserve members and their families and now includes veterans as well. Privatized through outside companies, it provides extensive support 24/7 through nonmedical coaching, tax help, spouse employment services, and much more.

In 2009, the Master Resilience Training Course was created in conjunction with the Walter Reed Army Institute of Research and the United States Military Academy at West Point. Based on research from the University of Pennsylvania on positive psychology's impact on depression and post-traumatic stress, the program was developed to teach resilience and coping skills to service members and eventually family members. As of 2022, fifty-five thousand Army soldiers and spouses have become trainers in the curriculum and gone on to teach resilience skills to other soldiers and families.

THE POWER OF IMPRINTING

Research suggests that key political events, especially those experienced between the ages of fourteen and twenty-four, "have the most powerful influence in shaping one's lifetime political attitudes. Events occurring at age eighteen have three times as much of an impact on one's worldview as an event that occurs at age forty."[119] Perhaps this explains the strong reaction of older millennials and Gen X who experienced 9/11 during their formative years. Considering most older Gen X and boomers experienced a sort of peacetime during their own formative years, the trajectory that each would take in their military and political views would turn out to be very different.

This idea of early imprinting can be applied to each generation's early experiences of the military culture as well. In my book *Sacred Spaces: My Journey to the Heart of Military Marriage*, I cover how the brain forms memories (both positive and negative) in the hippocampus. The five senses play an important role in anchoring those memories, but when paired with our human need for connection, safety, and belonging, strong imprintings can occur.

There are some experiences, especially moments that have a way of changing the trajectory of our lives, that are so strong they become what I call sacred spaces. Sacred in this case does not necessarily imply spiritual, rather it means "set apart" or different from the everyday moments of life. These kinds of sacred moments, such as the loss of a loved one, a battle that only a few experienced, or a sunset where you felt the presence of God, are to be treated with great respect.

You need others to "tread lightly" when you speak of these moments because no one else can understand what that moment was like in your narrative.

There are plenty of moments throughout the military lifestyle that are sacred spaces for both service members and families, many that are not even shared between a military couple. I share in *Sacred Spaces* my husband's experience of his first deployment. His support of his buddies during a historic battle in Afghanistan imprinted on him his expectations of military camaraderie and what he believed the military culture would be like from that point forward.

Our first assignment had imprinted so strongly on both my husband and me that I was curious how other families would describe what defines their expectations of the culture. Almost everyone I asked responded that their strongest imprintings came from moments that included two variables. The first was that they felt vulnerable in some way, such as experiencing a first assignment, being stationed overseas where they had limited support and were far from family, or during a deployment or other vulnerable life stressor. The second variable was that the culture strongly (positively) met or (negatively) did not meet basic needs critical to their well-being. Two spouses shared their imprints:

"I lost my opportunity to work in my career with this overseas move. ... It was also my first experience interacting closely with senior leadership and senior spouses ... who really showed me what it meant to mentor and lead in a military community. My biggest impression for me was that it was the first time I felt someone acknowledge that military members could not do what they do without a supportive spouse and family who sacrifices so much too."

"At the time GWOT dollars were pouring into the installation, FRGs were thriving and on-post housing had just been updated for my husband's ranked zone. We were babies, learning how to exist in the world and the structure from that unit helped us."

This was true for me as well when our supportive neighborhood and unit overwhelmed me with their generosity and kindness. Also during that first assignment and deployment, I distinctly remember the housing office giving away items to neighborhood families—backpacks stuffed with school supplies in August, pumpkins in October, Christmas trees in December, and flats of petunias in the spring. Each time, I was surprised, grateful, and overwhelmed by the generosity that made deployment and raising two toddlers on my own a little easier.

As a family assimilates into the military culture, they enter in a vulnerable state, with great need and often no family or friends nearby to lean on. There are new ways to do even the smallest tasks (most requiring a government ID card), and enrolling for medical care or childcare is dependent on your official relationship with your service member (spouse, child). Even single soldiers arrive at basic training with the vulnerability of little to no competence of what it means to be a service member. Basic training introduces new ways of understanding community, discipline, order, and authority. The early days (or years) of assimilation are most impressionable for the entire family because it is when they are at their most vulnerable. During this time they are likely to draw significant conclusions about the culture and form expectations of others for the future.

The same imprinting can happen for large cohorts of people at once. Anxiety around planes or security since 9/11, or the sound of a cough while out in public since the pandemic, are two examples. Both of these are imprinted memories of emotions, connections to loved ones, and thoughts held physiologically in the bodies of those who went through those experiences.

Boomers had experienced winning the Cold War and for most, the Gulf War as well. They experienced peace during the following years, so the imprinting they brought into the post-9/11 build-up was a sense of patriotism, positivity, and mixed gratitude. Although they may have endured budget cuts, funding was then flowing and new programming answered some of the biggest needs they had been advocating for.

For a majority of Gen X who entered service after 9/11, there was no connection or personal memory of budget cuts, pink slips, or having their lawns measured. Their assimilation into the culture looked a lot like mine. Even though my experience was from a privileged officer family's perspective, most families of all ranks and branches assimilating after 9/11 remember bonuses, new housing, new playgrounds, home maintenance, and marriage retreats. If your spouse was deployed, there were even more benefits available. For example, with the new privatized housing support, a family was only responsible for lawn care inside the fenced area of the backyard. During a deployment, you could request lawn service, get deployment bonuses and receive your pay tax-free, as well as take advantage of additional childcare and free sports and recreation classes for your children. In some ways, families were better off financially to volunteer for deployment, and many did.

The imprint of the military culture and what it meant to be part of it was that the military was able and ready to provide for your every need. From sign-on bonuses to petunias, the culture could even provide places for your spouse and child to find purpose. You were not alone. It was okay to leave your family of origin behind; you were gaining a new family within the military culture. The experienced generation before you was ready, willing, and able to host information sessions, coffees, and social events to mentor you during a difficult time.

What also made Gen X's imprinting different from boomers was that they entered a new culture while operations tempo steadily escalated, making them vulnerable in another way. Boomers experienced deployments (even consistent deployments), but deployments after 9/11 came with heavy intensity, visibility, and in multiple countries. Many families, like ours, entered the service with only enough time to stabilize their families before quickly deploying. Some deployments, scheduled for nine months, were extended to a year and some one-year deployments were extended to fifteen months. This created a vacuum of need for new families entering the lifestyle.

An Army spouse shared: "[The operations] tempo was so high and … because deployment was so hard, our unit was very close and very good friends. I wanted to think that was all units. And I wanted to think that [Child & Youth Services] would always be as useful."

By 2010, Afghanistan and other global conflicts were still going and there was no end in sight. The tempo did not stop over those first ten years. If anything, it escalated. Most Gen X families were young with small children. Their upbringings instilled their value of being present at home. While women wanted to work, many had always had the option to stay at home. With so much of the military providing basic needs, families were willing to live on one income. This dependence on the military served young families well at first. Although Gen X got involved, volunteered, and even created new programming to support the needs of families throughout GWOT, their level of need did not change and they took on fewer leadership opportunities as the years went on. As a Gen X military family member, I saw and can acknowledge that on the whole we consumed as much, if not more, than we gave.

At this point, the entire military community, including boomers, had been in a constant state of need for a decade. For many who served in combat tours, there was now a decade's worth of trauma and loss below the surface. While the institution continued to meet basic needs, the community was beginning to show cracks in its resiliency. It was simply too much to take care of themselves while also trying to serve others in great need around them.

The law of diminishing returns was originally an economic principle but has since been applied to a variety of environments, including the workplace. The *I Love Lucy* episode in the chocolate factory is a humorous example. As the conveyor belt speeds up, Lucy and Ethel try to work faster, only to become so overwhelmed that it turns into a hilarious disaster. The law states that if a variable is increased with the hope of increasing the return, after a certain point, if all other variables remain the same, the return rate will start to decline.

The DoD's provision for military families, although significant, was nowhere near enough to help families get back to a baseline of wellness. The continued demand for resiliency, energy, and availability from families instead resulted in exhaustion, new bio-psycho-social issues, and burnout. Without changing the demand, the situation was destined to decline.

AN EXAMPLE WORTH REMEMBERING
Empathy is a call away.

My husband, Matthew Weathers, a chaplain and member of Gen X, shares: "From 2020 to 2022, I served as the deputy division chaplain for the US Army's 1st Cavalry Division, the largest division in the Army, with more than twenty thousand soldiers, not including their families. During that time, while dealing with the COVID-19 pandemic, our headquarters was incredibly busy, implementing the findings of the Fort Hood (now Cavazos) Independent Review Committee, maintaining a large division-forward element for Operation Atlantic Resolve, deploying two brigades, and preparing for another to respond to crises in Eastern Europe. Accomplishing the mission required very long hours and a deep abiding trust in one another to 'box above their weight.'

In 2022, the division headquarters was shaken by the devastating news that the fourteen-year-old son of one of their beloved and central officers had completed suicide. Because the staff is an older population, many other leaders within the headquarters had teenagers of their own, triggering shock, sadness, and concern. One of those was the division chief of staff at the time, Colonel Kevin S. Capra. I had received multiple calls from him on the weekends, checking to ensure we had a robust presence of chaplains prepared to assist with the various emergency responses we had for the loss of a soldier within the division. Those calls usually concluded with the affirmation that we would indeed have a brigade's worth of chaplains at the motor pool that morning to serve soldiers in their grief. Such was and is his care for others. But on this morning, while the family received the support they needed from our headquarters battalion chaplain, Colonel Capra took the time to call and check up on everyone on staff who had teenagers. Calling as a leader, husband, and father, he connected with his soldiers on their concerns for their children and family and gave permission for them to take time off to spend with their families. He made People First more than a concept as he prioritized the well-being of those for whom he was responsible. It was a marking leadership moment that made us feel seen, known, and cared for, one that I'll never forget and always work to live up to."

Chapter 6

THE GREAT CULTURE SHIFT

AS I'VE WORKED WITH FAMILIES over the years, I've often thought of the quote from J.R.R. Tolkien's *The Lord of the Rings* in which Bilbo Baggins says, "I feel thin, sort of stretched, like butter scraped over too much bread."

Around 2010, there was a weariness in the air, but not so much that people were willing to complain about it. Up to that point, military families held the military accountable for issues like childcare and housing, but that is very different from questioning the institution and the profession itself.

The cultural expectation of order is canonized in regulation but more deeply rooted in the shared unwritten definitions of loyalty. On the most basic level, uniformity, hierarchy, and order ensure systemic success from small training exercises to large-scale joint operations. However, when trying to maintain order in a culture of more than two million people, rules and expectations can be misconstrued and even abused.

Insubordination is the act of refusing a direct order from a superior, but most military families believe that questioning a leader on almost any level can come across as insubordination. When combined with a long history of cultural traditions, mentoring, and "elbowing to the top," the adopted perception has been that any questioning of strategy, leaders' decisions, or the institution can be viewed as insubordination and result in negative consequences in your career.

By 2010, the concept of being drafted into the military was at least two generations in the past. Everyone in the culture had chosen to be there. They opted into a lifestyle of uncertainty, frequent relocations, and a mission-first mentality, and, in return, they and their families were provided for by the government. Even though there was growing concern about how much longer the community could give at the rate being demanded, comments like "this is what we signed up for" or "adapt, improvise, and overcome" were common even from family members who were not employed by the government or held to its regulations.

In many ways, this was a means of survival, a coping skill, developed to face the constant uncertainty and stress of a never-ending mission. Rather than honor the voice of their own mind and body that warned them of the limits of their humanity or hear the validation in the voices of their neighbors, it was easier to silence and gaslight those voices to leverage the grit needed for the next day. Besides, with a cause as important as this, why bite the hand that feeds you? It went against the cultural norm to challenge or question leadership. On a local level, there was peer pressure to not show cracks in your armor. As long as the DoD was able to provide the resources families at home needed to support the mission, families leveraged more grit. But all of that was about to change.

MILLENNIALS (DOB 1983–1996)
Collaborative & Social Changemakers

If there is any generation today that is known for globally challenging the status quo or questioning authority, millennials are it. The silent generation and older boomers protested the Vietnam War and other generations evolved causes like feminism and racial equality; however, millennials were the first generation to be raised under the microscope of social media. Perhaps it made their rebellion louder and more visible than the generation itself. If that is true, then the generations criticizing them were equally as loud.

Millennials received a significant amount of attention and criticism for the opinions they developed, in addition to how they expressed them. As the children of young boomers and the oldest of the Gen X cohort, millennials were raised under the pendulum swing toward highly attentive parenting. Where boomers were criticized for absent parenting, this new approach, termed "helicopter parenting," offered less autonomy for children by what many labeled overprotective and overinvolved parents.

Boomers tell stories of walking six miles in the snow to go to school, while Gen X tells of riding bikes until the street lights came on. But by the time millennials were old enough to ride bikes, fewer children were even going outside. Lukianoff and Haidt, authors of *The Coddling of the American Mind*, suggest that the continued concern of safety around children (beginning with Gen X) evolved into overall less risktaking, exploration during play, and essentially the "coddling" of an entire generation.

Even though crime rates were down in the '90s, public anxiety and opinions around what parenting should look like fueled an increase in 911 calls to report neglectful parenting or unattended children. Adding to the anxiety was increased media attention around parents who were legally charged with child abuse for discipline techniques of previous generations. I distinctly remember

Jessica Beagley, mother of three boys, being sentenced in 2011 to 180 days in jail and a $2,500 fine (which was then suspended), for using hot sauce as a consequence for her son lying.[120] Just days earlier, a friend of mine suggested I consider this same approach with my son rather than the traditional "wash their mouth out with soap."

At the center of this parenting shift was a perspective of and response to stress. Experts said that overprotective parenting choices were a result of adults managing their own stress of not wanting to upset their children or be perceived as absent or unavailable parents. Older generations argued that allowing kids to experience stress gave them a chance to self-soothe and problem-solve. Research eventually showed that parents rescuing children from their problems too quickly developed high anxiety, low self-confidence, and lower self-esteem.

Persevering through healthy levels of stress, conflict, and difficulty helps children and youth develop confidence, stamina, and a sense of autonomy. Angela Duckworth, author of *Grit: The Power of Passion and Perseverance*, studied what she called "gritty" students and adults and found that the most gritty people are not necessarily more talented, but have the ability to persevere through obstacles in order to reach a long-term goal. In a world that was increasingly feeling less safe, generations debated whether we were creating too much safety.

By being the always-available adult, parents of millennials were accused of creating a generation that failed to launch well into adulthood, many of them delaying life events such as attending college, getting married, or moving out. As millennials gained their voices, they demanded change, questioned authorities, and in more extreme ways, demanded that their environments adjust to their comfort levels. With social media, news outlets pumping out daily stories, and video coverage increasing, stories of millennials protesting and debating in their schools, colleges, and workplaces, earned them the unfortunate nickname "snowflakes" to mock their "uniqueness" and "fragility" by older generations.

September 11th, of course, shaped their concept of safety on a national level. While Gen X and previous generations perceived the enemy as a country away throughout the Cold War and Gulf War, millennials' experience of 9/11 made the threat seem much closer. The US response to the threat also solidified their beliefs on military action and our country's relationship with the rest of the world. Just two months after the attacks, millennials initially favored forces going into Afghanistan. However, given that 9/11 and the Afghanistan and Iraq conflicts happened during some of their most formative years, by 2005 they were "skeptical of the use of American hard power as a tool to shape the world."[121]

By early adulthood, millennials had already experienced the availability of the internet, cell phones, and social media. Their comfort with technology stemmed from spending more time with it than without it. Their contribution to technology continues, but initially they discovered new ways to innovate and make productivity online more efficient. For example, boomers and Gen X may have introduced the computer and email, and Gen X even created Google, but millennials took those tools to a new level with digital evites, enewsletters, membership sites, click-funnels, and more. They also launched new platforms like Instagram and Snapchat.

As they watched peers and influencers like Mark Zuckerberg of Facebook monetizing content, they began to question the traditional career paths that contributed to large amounts of student loan debt. Also known as the "selfie" generation, social media and online influencers redefined "talent" and introduced new ways to learn, entertain, and bring attention to social causes on individualized platforms. As online platforms quickly changed and evolved, financial and influencer success was determined by being an early launcher on a platform. Those who, for example, launched quickly and early on YouTube were quick to succeed. In many ways, this introduced for millennials (and Gen Z after them) the importance of attentiveness to new platforms, trends, and the ability to pave your own path.

A successful career was no longer defined by one career over a lifetime. Millennials wanted to travel and see the world they were seeing online. Talent could be monetized on your own as an entrepreneur, or sold to the highest bidder, especially if you were able to navigate the quickly changing tech world. Millennials also expected to be rewarded and compensated for that talent. They were more likely to leave a job that did not provide the climate they desired or align with their values, so retirement benefits were less of an incentive.

Big-idea companies like Google, Apple, and Tesla attracted millennials by providing a more collaborative workplace environment, diversity, and flexible schedules. In order to compete, the military needed to focus recruiting strategies on helping parents feel more comfortable with the military career as an option. The military focused on highlighting the career as a way to pay for education, providing job security, quick entry into meaningful work, and an opportunity to travel, in addition to the benefits.[122]

What would prove to become more difficult for millennials was their desire to be valued and respected much earlier than older generations were comfortable with. Growing up with parents heavily involved, they had become accustomed to direct feedback and two-way conversation. Boomers and Gen X, on the other hand, saw respect as something that was hierarchical and earned over time. They still valued showing your work by keeping your head down and letting loyalty lead to promotability. Millennials' more conversational approach was, and still is, often seen by older generations as disrespectful and entitled.

A millennial Marine shared with me, "Please tell older generations that we aren't trying to be disrespectful. We actually value the relationship so much, we want to provide feedback to superiors and mentors on ways they can improve as well. It is actually out of great respect of the relationship." His comment was useful insight into how relational millennials see workplace dynamics, even in a culture as hierarchical as the military.

By 2010, boomers were reaching military retirement age, which meant that Gen X was moving into command and other leadership roles. Gen X was already starting to test the waters on work-family balance in order to address the community's exhaustion (and their own). Considering there were superiors of another generation above them who still felt the pressure of an active war, there was pushback on 9-5 office work hours. Meanwhile, as millennials entered the workplace, the desire for flexibility and their version of work-life balance joined the conversation.

THE CATALYST OF THE GREAT CULTURE SHIFT

As the first decade of GWOT came to a close, the second decade began with budgetary "turmoil"[123] that would imprint a very different picture of military culture on young, vulnerable, and impressionable millennials. The first of the millennial cohort turned eighteen before or around 9/11 and had a similar experience as Gen X, with an overwhelming amount of support and funding. A majority, however, came into the force during the second decade.

In 2010, Congress delayed a large supplemental funding bill for the war efforts still going on in Afghanistan and Iraq. This delay resulted in the threat of a government shutdown. Thankfully, the crisis was averted but it introduced the idea that the government actually could shut down the military community, and it did several times after that.

In 2011, federal spending (including the defense budget, Social Security, education, Medicare, transportation, and more) was soaring, and revenues were continuing to decline with the country recovering from the 2007 to 2009 recession. The deficit was predicted to be around $1.5 trillion in 2011 and Republicans and Democrats in the House refused to agree on where there should be cuts in spending. On April 8, 2011, another government shutdown was avoided with only an hour to spare. Often there are smaller topics embedded throughout the federal budget that can slow down the entire process. The linchpin this time centered around funding for Planned Parenthood. The final agreement between congressional leaders and President Obama was a plan to slash $38 billion in spending in the coming year.

After continued disagreements on how to resolve the 2011 debt ceiling, the Budget Control Act (BCA) of 2011 was signed into law in August 2011. The intent was to create a "super committee" to focus on the spending cuts and an automatic budget sequestration process to begin in 2013 "to encourage the [supercommittee] to agree on deficit reduction legislation or, in the event that such agreement was not reached, to automatically reduce spending so that an equivalent budgetary goal would be achieved."[124]

If your family is in debt and continues to overspend, you are likely to take a look at your budget and make decisions about where to best cut overspending. Most families would cut the nonessentials first. Perhaps you'd eat out less, not splurge on as many lattes, or agree to not take an extravagant vacation this year. This is similar to what the BCA asked of the government. Budget caps for most categories, with exceptions, were created to set the level of the budget for a ten-year period to end in fiscal year 2021. Considering the defense budget makes up a significant portion of federal spending, general budget caps expected a reduction of around $1 trillion over the ten years.

One of the most noted and debated BCA exemptions is the Overseas Contingency Operations (OCO) account discussed earlier. This fund supported the active global conflicts going on and continues to be criticized for being used as a loophole for Congress and the DoD to avoid budget caps. As we've covered, presidents since World War II have leveraged their ability to use military force, even without a congressional declaration of war. The OCO account supports military operations that are war-related, considered deterrents, or support homeland security. During the early years of the Afghanistan and Iraq conflicts, the OCO fund had very little oversight, was not subjected to sequestration cuts, and was often used as a place to stash non-war funding in its operation and maintenance accounts. It has frequently come into question. However, thanks to it being out of reach, a couple of F-35s, for example, were able to be funded in the OCO during that period. Eventually, the OCO fund was authorized item by item, just like the rest of the budget request.

Also, an exception to the cuts was the military personnel funding account (MILPERS) which is responsible for pay, allowances, and some of the benefits for service members. Areas of the defense budget that were impacted included new hires, pay increases, increases in healthcare costs, acquisitions for new submarines and airframes, current or future training, a halt on new construction, and funding for installation maintenance, just to name a few.[125]

If the BCA sets the budgetary plan for reducing spending, sequestration is the enforcer. The two are not the same thing. Going back to your household budget example, if you and your family fail to cut spending in the nonessential areas, sequestration rules would force you to cut a certain percentage of spending across every area of your budget, even essential categories such as insurance, groceries, and electricity.

With the BCA already in effect, as long as the caps were not exceeded, sequestration could have been avoided. The BCA wasn't entirely focused on the defense budget, but since the defense budget was a significant amount of the federal budget, it received equal attention. In the fall of 2012, the DoD "chose to not plan for sequestration or to slow spending in anticipation of cuts" partially as "a calculated effort to maintain the pressure for a budget deal." In other words, they were really hoping Congress would figure it out. The DoD also knew that slowing spending would have "disrupted ongoing activities and frightened employees."[126] As sequestration loomed, the chiefs of staff of each branch begged Congress to find a resolution, warning of how across-the-board cuts would impact families and force readiness.

Marine Corps General James Amos testified: "The most troubling and immediate risks are those that sequestration imposes on our people. Sequestration does not hurt things, it hurts our people. The qualitative edge that the American servicemember takes to the battlefield is the fundamental advantage that differentiates our forces from our enemies. This qualitative combat edge will be severely eroded by the impacts of sequestration, leaving America's men and women with inadequate training, degraded equipment, and reduced survivability.

"While military pay and allowances have been exempted in this round of sequester, the quality of life for the all-volunteer force and their families will suffer as we reduce family programs and installation maintenance. Our civilian Marines will likewise be impacted. Ninety-five percent of our civilian workforce is employed outside the Washington, DC, national capital area. They're the guards at our gates; our financial experts who manage our budgets; our acquisition specialists; the therapists who treat our wounded; and the teachers who teach our children. The economic impact to these families and the local communities are put at risk by short-term furlough or a long-term termination. Protecting our ability to keep faith with our families and our wounded warriors is a top priority in my Marine Corps. But even this, the most sacred of responsibilities will be increasingly put at risk under sequestration."[127]

General Martin Dempsey, chairman of the Joint Chiefs of Staff testified: "There is a magnitude issue here, too. We built a strategy last year that we said we could execute and absorb $487 billion. I can't sit here today and guarantee you that if you take another $175 billion that strategy remains solvent.

"And if you're wondering why this is so hard, let me just use the Army— you know, people say, 'Well, hell, you did it after World War II; you did it after

Vietnam.' After World War II, it went from a million-man Army to 781,000—after Vietnam ... In the '90s, we went from 781,000 to 495,000. We grew it for Desert Storm—for OEF and OIF—to 570,000. It's on the way to 490,000 because of the Budget Control Act. The question I would ask this committee: What do you want your military to do? If you want it to be doing what it's doing today, then we can't give you another dollar. If you want us to do something less than that, we're all there with you and we'll figure it out."[128]

SEQUESTRATION

Sequestration was designed to be draconian, ensuring that the super committee came up with a plan to cut both entitlement and discretionary spending under expedited procedures in Congress. The idea was to include something each side cared about to force a compromise. It didn't work. In March 2013, sequestration was triggered after Congress failed to agree on how the $1.5 trillion would be cut over ten years. Much of federal spending was exempt from cuts. Defense was therefore forced to shoulder half of the cuts, although defense is only about 15 percent of federal spending, adding to the sense of betrayal.

Forced to go into across-the-board cuts, the DoD stopped some of the branch training exercises, canceled deployments (which was seen as a failure to meet a mission), and by summer, canceled installation maintenance and furloughed government employees. The goal was to halt or cut whatever possible except those services that protect life and property.

Everyone connected to or part of the DoD experienced the impact of sequestration in some way, even if they didn't know it. There is no way to measure who had it worst, nor should we. From the top down, senior military leaders were working long hours, managing multiple budgets, and making incredibly difficult decisions that would impact millions of lives.

Service members were personally impacted. By December, the force started to downsize even more. In addition to a hiring freeze, the Army announced that 20 percent of captains commissioned between 2006 and 2008 would receive pink slips in the spring, while others would be denied promotions. Enlisted would not be permitted to reenlist. Numbers, training, and equipment were reduced for all branches, some to less than what they had been prior to 9/11. With training and deployments canceled, it impacted service member readiness and would create compounded readiness issues for the future. Although the evidence is not entirely conclusive, the cuts in training and maintenance have been connected to a rise in accidents and training deaths, including a 40 percent rise in accidents involving all manned fighter, bomber, helicopter, and cargo warplanes in fiscal years 2013 to 2017, according to a six-month investigation by *Military Times*.[129]

Maintenance for vehicles and airframes was forced to get creative. One individual shared that parts were being taken from one airframe to fix another rather than waiting for new parts to arrive. Another Marine leader shared, "The situation is so dire that, incredibly, Marine aviation units have been reduced to salvaging aircraft parts from museums to keep planes flying."[130] Holds were placed on any new builds or acquisitions. In fact, the Air Force announced in 2014 their plans to eliminate five hundred planes from inventory.

There was also concern that the message being sent to our allies and adversaries was that America couldn't "get their house in order." Army Chief of Staff General Raymond Odierno remarked in 2013, "Throughout our nation's history, the United States has drawn down military forces at the close of every war. This time, however, we are drawing down our Army not only before a war is over, but at a time where unprecedented uncertainty remains in the international security environment."[131]

Also from the top down, civilian employees were put through the chaos of possible and real furloughs and layoffs. While military members' pay was threatened, it was eventually protected. The possibility of government shutdowns, according to Robert Hale, who was DoD's chief financial officer during those years, required extensive planning. Teams of people would have to figure out what government activities could continue during a shutdown and then distribute that information to DoD installations, commands, and civilian personnel around the world. The disruption would create chaos in the entire organization as employees were concerned about being furloughed without pay. Hale wrote that the cost of the turmoil of those years was more than just dollars, it created wounds and a 12 percent decline in morale among civilian government employees that continues today.[132]

THE APEX OF POSITIVE MORALE

Prior to 2013, the impact of budget cuts was mostly felt on the military side and with service members. After 2013, sequestration began to disrupt the everyday lives of military families. While family programs were not the first on the chopping block, families were not likely to know that unless they had been watching the news and listening to congressional briefings. This is what the general's wife was trying to tell me when she said, "Follow the money." In doing so, we see just how widespread this event was in shifting the trajectory of the culture and the future of our force.

In an effort to maintain and provide for an all-volunteer force, the government had created a complex web of operations. There were privatized contracts to honor and manage, families to support, and, most importantly, service members to train and deploy. Any interruption was going to create a similar web of complications. Some consider programs like marriage retreats and childcare "budget dust" compared to acquisitions and operations for the DoD. However, as civilian employees were furloughed or laid off, families experienced a confusing transition of support that was there one day and not the next. As an example, I remember Army FRSAs (the paid position that served as a liaison between command and families) suddenly gone one day. At first, I thought it was something different about our new assignment at Fort Stewart. (I was definitely not tracking Congress or defense budgets at the time.) But as I spoke with other families, they also saw their support staff disappear. Without context or understanding, the confusion sowed seeds of anxiety and distrust. Fewer than four years earlier, General Casey had stated that FRSAs would not be a "quick fix" but a long-term investment.

While not all branches had this position, it was a civilian position that impacted many families at once. These positions kept families organized, planned morale-boosting family events and information briefings, and arranged childcare to support those events. With childcare hours also being reduced and workers being laid off, opportunities for many families to get respite also became more limited. The third-order effects of childcare no longer being available deeply affected the community. If sleep-deprived, lonely spouses did attend briefings and resilience-building events, they had to try to retain important information while juggling babies and toddlers. Others just didn't show up.

Other civilian positions that were furloughed created additional ramifications in resources and programming. Commissaries (installation grocery stores) reduced their hours. Programs that taught resilience and coping skills suddenly had no funding. One spouse said, "I was a volunteer AFTB [Army Family Team Building] instructor and suddenly there were no funds for the programs—no money to print the class materials, no funding to provide childcare vouchers which were critical in facilitating spouse attendance. It was really sad. ACS [Army Community Services] staffing was cut back and they were dual and triple hatting various roles and the roles/programs all suffered as a result."[133]

Families would continue to feel the effects of the BCA and sequestration. Summer 2014 brought more layoffs. Rumors were spreading throughout the culture that service members were given pink slips while on deployment. Anxiety grew and resentment was not too far behind. Families who had no backup plans and expected to retire at twenty years nervously awaited their fate. The following is an excerpt from a *New York Times* article by Amy Bushatz, an Army spouse and eventual editor for Military.com. Back then in 2014, she was a military spouse of a service member freshly spared from the first round of cuts.

"How does it feel when the country for which you've lost youth, relationships, and health wants to get rid of you? It feels like betrayal. If the Army is our family, then this is divorce. ...

"For the entirety of our Army service the enemy has been 'over there,' far away in Afghanistan or Iraq. We grew comfortable with the danger of deployments, reassuring ourselves that in a down economy where factory-working or cubicle-dwelling civilian friends faced layoffs, we had job security. With a steady paycheck on the first and fifteenth of the month, military service felt safe compared to the nightmare of unemployment. But it doesn't anymore.

"Officer separations have been used throughout American history as a means of controlling the size of the force. ... It happened after World War II, Korea, and Vietnam. Downsizing is a normal way for the military to recover from war.

"But none of us were around then. We don't care that it's all been done before. For us it feels fresh, terrifying and even insulting. ... But now I feel a sense of survivor's guilt that my spouse has a job while so many others soon won't. Did they really escape death on the battlefield only to be a casualty of force reductions? And while my family is safe for now, this won't be the only time the Army makes cuts. Over our heads hangs the cloud of 'maybe we're next.'

"When a job is a way of life and your coworkers are your family, pink slips feel like betrayal. After years of combat deployments, training separations, child births while alone and cross-country moves, we have given so much in the service of our country. The military has become the closest family many of us have ever had. And now, just like that, some of us are orphans."[134]

There are a few incredible points this article brings up that are worth reflection. Amy's words serve as a case study of significant cultural shifts happening during this moment in 2014.

First, Amy is considered an older millennial. Her husband was commissioned into the Army in 2006. Amy's experience and imprinting of the military culture were similar to what Gen X experienced. While boomers had lived through pink slips and downsizing, those who entered after 9/11 had no personal context except war, intense demands on the family, and a level of provision that had become the norm. She is pointing out that although the military may have downsized in similar ways before, it is not part of her story or her experience, and it, therefore, requires more communication.

Another difference from pre-9/11 drawdowns was that this was happening in the middle of war, when families were most vulnerable. To those who had met the intense demands of the military, some for more than thirteen years, it felt like a betrayal. While anyone being laid off might feel betrayed, Amy was not speaking from the perspective of an employee, but as a military family member.

Millennials like Amy were motivated by their circumstances to actively bring about positive change to their environment. Pew Research Center found that millennials were very different from generations before them. They were the most racially diverse generation and "relatively unattached to organized politics and religion, linked by social media, burdened by debt, distrustful of people, in no rush to marry—and optimistic about the future."[135] They brought their vision of positive change to the military as well.[136]

Millennials' signature question is "Why?"

"Why does the military not accept women in combat roles?"

"Why does the gender of a service member matter?"

"Why can't we question a leader if we disagree or want to collaborate?"

This has no doubt made older generations bristle at times, especially as millennials demand a better answer than, "Because that's the way we've always done it."

In a culture where the response to a question or order has always been "yes, sir" or "no, ma'am" the millennial question of "why" was not only shocking but initially seen as disrespectful. It is easier to blame a generation for their upbringing and call their behavior disrespectful than to leverage curiosity and build a relationship of mentoring. Yet, the latter is what millennials were asking for. To them, respect was built through a relationship that was strong enough to handle a two-way conversation, even if the answer was that they must grow.

The betrayal Amy pointed to was that the military was not holding up its end of the bargain. Like many others who enlisted or commissioned, millennials were looking for job security and safety from a downward economy. In a world that was increasingly becoming unsafe, with homeland attacks by terrorists and mass shootings, safety became an important value for millennials. Not just physical safety, but emotional safety as well. They had grown up in homes that made room for their emotions, and now they wanted them to be valued rather than repressed or feared. Amy was not only expressing her feelings about how these changes were affecting her family and community, she was demanding dialogue.

For boomers and Gen X, emotions were (and are) a vulnerability. For millennials, emotions represent an opportunity for discussion that opens opportunities to hear another person's truth. That truth could lead to change. Hiding feelings or the truth of your experience was no longer an option for millennials.

Amy's words mark another cultural shift taking place in the military family culture: she voiced her opinion to and about the institution publicly. Now, she certainly wasn't the first, and as a writer she had a platform few spouses enjoyed; however, at the time it was still very much part of the spouse culture to be submissive to the mission. Not doing so was seen to have a potentially negative impact on the service member's career, especially if you were an officer's spouse.

Simultaneous radical shifts in technology added to the intensity of budget cuts and sequestration. Social media was booming, news outlets were competing with opinion-based blogs for clicks. Digital courage (expressing your opinion behind the safety of a screen) made it easier for anyone of every generation to share thoughts and opinions that would have otherwise been considered socially taboo.

AN EXAMPLE WORTH REMEMBERING
Carry the burden with courage and compassion.

After retiring as joint chief of the Army, General Ray Odierno served as a senior advisor at JPMorgan Chase & Co. In an article reflecting on his time in leadership, he wrote, "The foundation of any organization is trust. Trust between peers, subordinates and your leaders. Establishing and communicating right and left limits. Empowering subordinates and decentralizing decision-making within those limits. Treating everyone within the organization with dignity and respect. All of this contributes to an atmosphere of trust and pride.

"I've been very fortunate to work for leaders who mentored and developed me, tolerated mistakes, and allowed me to grow, all while providing me an opportunity for increased responsibility. Although the world may be changing rapidly, its challenges remain the same. Bold leaders adapt to their environment, address risk with confidence and parity, and then seize the moment when it comes along."[137]

General Odierno died of cancer in 2021, remembered for his dominating six-foot-five presence but more specifically for his selfless, steadfast leadership style. Former Secretary of Defense Ashton Carter said, "His commanding presence calmed the confused, and his courage and compassion helped carry the burden of loss and sacrifice."[138]

MORALE SHIFT

"The unfailing formula for production of morale is patriotism, self-respect, discipline, and self-confidence within a military unit, joined with fair treatment and merited appreciation from without. It cannot be produced by pampering or coddling an army, and is not necessarily destroyed by hardship, danger, or even calamity. Though it can survive and develop in adversity that comes as an inescapable incident of service, it will quickly wither and die if soldiers come to believe themselves the victims of indifference or injustice on the part of their government, or of ignorance, personal ambition, or ineptitude on the part of their military leaders."

—General Douglas MacArthur

Chapter 7

BROKEN PROMISES

THERE IS A PALPABLE TENSION between mission accomplishment and taking care of people that exists for all leaders, but especially larger institutions. Promises had been made to service families, maybe not guaranteeing a long-term career, but definitely for provision and support. Service members expected the essential resources they needed to go to war with confidence. For a variety of reasons, promises were falling to the wayside in the middle of major conflicts in Afghanistan and Iraq, leaving a large part of the culture feeling betrayed and millennials coming in right as it was happening.

The last place leaders want to find themselves is choosing between what is best for the organization and what is best for their people. Yet, it is all too common for businesses and institutions to find themselves in that predicament. In the aftermath of 9/11, Delta Air Lines threatened Chapter 11 bankruptcy and later terminated its pension plan for pilots in 2006. Similar to the military pension, the plan would have offered pilots 60 percent of their final earnings monthly after retirement. With a mandatory retirement age of sixty-five years, many boomer pilots had started flying in their twenties or thirties and were living comfortably on a stable salary and depending on their retirement plan.

With bankruptcy looming, Delta offered pilots the opportunity to retire early and take half their pension as a lump sum and then a reduced pension check after retirement. Around 1,300 pilots chose to leave their careers early and salvage what they could. Later, when Delta filed bankruptcy, they claimed the pension funds were insolvent and the court allowed them to escape the commitment of the reduced pilots' monthly pension checks. The real sting was that Delta continued to pay the pension plans of other employees and older retirees.

Angry, the pilots formed a group to fight the airlines and demand that they fulfill their promise. Many of them had been forced to live on a significantly reduced level of income compared to what they had planned. Delta stated in a letter to pilots, "Termination of the pension plan is among the most difficult decisions Delta has had to make in connection with our restructuring process, and we truly regret the impact this has had on you."[139]

The sudden loss of income and lifestyle was a shock to many of the pilots. Several committed suicide. The retired group of pilots took the issue all the way to the Supreme Court on the premise that Delta dealt a "bait and switch." Ultimately, they lost the case. What Delta did was legal despite the injury it caused to the people they once depended on.

In 2007, in the midst of fighting two seemingly neverending wars, Army officers began to leave service. In an effort to retain especially junior officers and fill gaps in specialties, the military offered incentives to reenlist, to transition from enlisted to the officer corps, or to move into a designated specialty. In return, bonuses ranging from a few thousand to $150,000 for senior special forces were offered for up to a six-year commitment. More service members took the deal than anticipated, some who would have stayed in without the bonus.

Just seven years later in 2014, and in an eerily similar fashion to the Delta pilots, Army soldiers felt the sting of layoffs with budget cuts. Different from the 1990s downsizing that let go of lower-ranking enlisted soldiers, the Army began laying off officers as well. One in five of those officers were prior-enlisted soldiers who had jumped into the officer corps during the 2006-2007 build-up with incentives.[140] Many were well into the eight years it took to become captain and suddenly those years meant nothing. Partial-retirement offers would reflect their last highest rank achieved. Service members and families were angry that those being let go were loyal, enduring much of the pressures and demands of war to that point.

A *New York Times* article at the time interviewed Nathan Allen, an intelligence officer who had served fourteen years. He, like many, had reenlisted and hoped for retirement benefits after a twenty-year career. Nathan stated, "Iraq, Afghanistan, jumping out of airplanes, doing all the training, leaving for work so early and coming home so late that I wouldn't even see my family during the week, and I get nothing."

Beth Bailey, a professor at Temple University, was also quoted in the article saying, "They recruit with all kinds of promises, whether it's career benefits or something more amorphous like being part of something bigger. They support families in a way that makes it a whole lifestyle. People become part of an insulated Army culture. For that to suddenly be taken away, I'm not surprised they feel betrayed."[141]

There is that word again, betrayed.

STRENGTH OF OUR FORCE UNHEARD & DISMISSED

When I read the *New York Times* article about service members facing cuts before they made captain, I saw comments left by readers like, "Welcome to the real world," "Civilians aren't promised forever jobs or guaranteed pensions," and even comments that suggested that service members were whining when they should be grateful. It is true that no business or institution can guarantee an employee's career for life. There are risks associated with working for an employer, including the business going under or changing its values, in addition to employees being fired, laid off, or not promoted. The military includes those risks as well. Promotion boards have "up or out" policies that can be career enders, moral or ethical failings can lead to discharge, and even unexpected injuries might lead to being medically discharged from service.

Similarly, many will read the story of the Delta pilots and completely dismiss their experience as coming from a privileged class and reacting with an entitled response. The fact that most commercial airline pilots earned annual salaries that put them in an upper-class bracket does not diminish the betrayal they felt and the overhaul of lifestyle they had to face, enough that several were unsure how to move forward and chose to end their lives. Rather than dismissing their experience as privileged, the situation invites us to connect on a human level to the disappointment, shock, and the stress that come with resetting a lifetime's worth of dreaming.

There is a temptation to see and hear the stories of the military culture and emotionally invalidate them simply because it is not an experience that you share. Emotionally invalidating another person is when we reject, minimize, or dismiss the emotions and/or internal experience of another person as insignificant, wrong, or unacceptable. It is incredibly discouraging and damaging to the psyche of the one sharing their thoughts and feelings. Examples could be telling someone, "It could be worse," nonverbal eye rolling, or implying they are being overly sensitive or too emotional. We all will run into various forms of cultural unconscious bias—whether that be military, civilian, social class, gender, race, etc. However, emotions offer us the opportunity to connect with people on the common ground of a shared human experience.

Many people invalidate the feelings of others subconsciously, saying they are trying to make the other person feel better, enlighten them with an alternative perspective, or evolve their emotions into a more comfortable place. In the absence of knowing what to do or how to help, it is easier to dismiss feelings.

In researching the military culture, it was enlightening to see the reactions of the civilian population to the concerns military families were expressing. The common phrases "you signed up for this" or "why can't you be grateful; you have more benefits than most civilians do," reveal a cultural bias and an unwillingness to hear the emotions of someone with a different experience, and they further divide the military and civilian community. The military culture does it as well when we minimize a civilian's work trip away from family as nothing compared to the yearlong deployment we just endured.

Yet, this is also something that commonly happens within the culture and in the ranks of leadership. As spouses like Amy began to express the frustration and impact of budget cuts on their lives, the temptation was to label them as needy, ungrateful, entitled, and whiny, especially if they were millennials. Service members like Nathan were told they should be grateful to "have a pension at all" and to "move on with their life." In response to toxic work environments and moral or ethical leadership failures, service members are reminded that instead of complaining, they should be "grateful you still have a job" and that it would be much harder "out there."

It is a bias we must be aware of and learn to address quickly if we want to be leaders who are known for compassion as much as strength. Invalidating a person or group happens quickly if we are not careful and has the power to rob us of our ability to see the story unfolding in front of us. It also may be happening over the course of reading this book. Especially when hearing another person's story from the past, hindsight has the power to diminish the true experience of the other person. We are not expected to sit with someone else's experience of difficulty if they are no longer in it, right?

Empathy, according to Gallup's CliftonStrengths, is not just the ability to put yourself in someone's shoes. I admit this is a difficult "ask" of some of our most lethal warriors, who must set aside the rights of another human, much less their emotions, to do their jobs. Empathy is actually the skill of valuing emotions as part of the equation. In this case, validating the emotions of another person does not mean we have to feel them ourselves, it just means we value them as being important to the person feeling them and therefore important to us. When tasked to come up with solutions, strategize the future, or build cohesiveness in the team, valuing the emotions and perspectives of others is a strength.

Valuing emotions, however, does not always mean we must lead with them. Empathy, alone, can steer us wrong if we abandon productive forward momentum out of fear, anxiety about pleasing others, or counterproductive efforts to make sure everyone agrees. However, as we said earlier, emotions need not be threatening or intimidating either. Whether it is an individual, group, culture, or generation, emotions are often expressed out of a desire to connect with another on a human level. Minimizing or diminishing emotions only communicates the opposite, that I do not wish to connect with you, and as a human superior to you, I've determined there is something wrong with your reaction.

Solid, healthy leadership allows room for both personal perspective and emotions. If we are too quick to offer our perspective, we risk emotionally invalidating the other person, often disqualifying ourselves as leaders in their eyes. However, listening and being able to hear the emotions of others is the gateway to earning the trust necessary to lead with a possible new perspective.

As the military culture began to express its frustration at what felt like betrayal or broken promises, the DoD's response (albeit in its own crisis of sequestration) was to emotionally invalidate its own people by not responding at all. While there were certainly key leaders here or there who tried to address the emotional state of the culture, the institution did not communicate an awareness of what it was asking of the force and their families nor the impact of sequestration on the sense of security within the culture. When faced with the difficult decision of choosing the organization over its people, failure to communicate and specifically dismissing how that decision will impact the culture is surely the way to lose the people's loyalty altogether.

THE CULTURE BEGINS TO FRACTURE

As families began to express their thoughts, feelings, and confusion, the fresh, unregulated, platforms of social media created the perfect place to direct that energy. The hateful cyberbullying trend toward military spouses became so commonplace that the derogatory nickname *dependa* was added to *Urban Dictionary* to define spouses who were dependent on their service member for financial support and benefits. Ironically, spouses *are* dependent on their service members for benefits and are largely underemployed. However, the nickname was specifically targeted at spouses who "wore their spouse's rank" and acted entitled around others. Internet trolls, now believed to be young single soldiers, would target spouses online for complaining.

Shortly after, spouses began to pick up on the trend as well, this time, targeting each other. Some spouses reached out to senior leaders asking for help. They hoped some kind of statement would reduce the cyberbullying going on in the community. The advice given was similar to the larger American culture's approach to bullying. Ignore the bully or get offline.

Senior leaders were still leading from the imprinting of the culture they knew, one that thrived in person. Simply "getting offline" was (to them) a way to take control of your circumstances and get back to "real relationships." Unfortunately, it came across as invalidating the emotions and concerns of the community, and the community had no intentions of moving backward.

Military families were now accessing support, information, and a sense of community online more than ever. Older generations tried to encourage in-person events as best as they could, but the turnout was decreasing. Boomers still desperately hoped that Gen X would pick up the baton, however, they were less available for in-person social events, leadership roles, and mentoring.

Gen X was so busy surviving an intense operations tempo perhaps they didn't realize it was their turn. What Gen X did do, though, was pick up the digital baton that was far different from prior social customs. With technology now offering flexible, remote online work, creative employment was a real possibility for spouses who had been desperate to work. While all generations were eager to test the waters of online entrepreneurship, Gen X spouses, in particular, started new businesses, military-specific online support initiatives, and nonprofits that could support families regardless of where they were located. Working from home gave spouses the opportunity to contribute financially without neglecting their core value of staying home with their children. While some did the best they could to take on personal leadership roles, online platforms seemed to fill the gaps left by cut programming.

Unfortunately, with cyberbullying and individually-run Facebook spouse groups on the rise, even in-person social events quickly gained a reputation for being a place of negativity. This first impression of the military community discouraged young millennial families from getting involved. It was not uncommon to hear new spouses saying they wanted nothing to do with the community. Many wanted their life and work to be very separate from their service member's work. Many did not want to be referenced as a military spouse or see the value in a role that seemed like a cross between *The Stepford Wives* and the dramatic *Army Wives* TV series.

The family culture was fragmenting under the pressure of war, in a desperate state of need, and dispersed in online platforms that functioned very differently from the traditions passed down by previous generations. The imprinting of those entering the services after the BCA and sequestration was completely different from those who came in before it. Millennials entered with the same expectation as other generations, looking for a culture that promised provision and the safety of a patriotic community. Instead, they were met with an entire community that was cynical, angry, anxious, and disconnected. The culture was increasingly less of a community of support and, in many ways, was years behind in diversity, inclusion, and the modern family model.

There is no doubt the decisions of Congress and DoD military leaders were hard decisions. Faced with the very important need to reduce federal spending, everyone was going to be affected. Most of the military culture was genuinely not tracking how spending cuts were impacting their daily life. Sure, obvious signs of layoffs, loss of programming, and childcare shortages made it more real, but often the slow, subtly growing consequences create the most damage.

In 2015, sequestration continued, and so did war. Millennials were peaking in numbers coming into the military, boomers were nearing or past retirement age, and Gen X served in command roles. The community had been in at least four years of budget cuts, restraints, and limited funding. As the DoD advocated for the fiscal year 2016 budget, the Joint Chiefs continued to fight for what they believed our force and families needed. General Odierno stated that "readiness has been degraded to its lowest levels in twenty years." The Army had already reduced end-strength and eliminated thirteen active-component brigade combat teams, three active aviation brigades, and more than eight hundred rotary-wing aircraft as they planned to cut ten to twelve additional combat brigades.[142]

Trying to meet the forced demands of sequestration, he said, "We will be forced to further reduce modernization and readiness levels over the next five years because we simply can't drawdown end-strength any quicker to generate the required savings. In the Army civilian workforce, we have reduced ... from the wartime high levels of 285,000 and will continue to reduce appropriately over the coming years. While necessary, these reductions in the civilian workforce have and will continue to adversely impact capabilities such as medical treatment, training, depot and range maintenance, installation emergency services, physical security and select intelligence functions."[143]

The original goal of the BCA in 2011 was to reduce the $1.5 trillion deficit over ten years. By 2015, the deficit fell to $438 billion. In a sense, sequestration was doing what it was created to do. In 2016, however, the deficit was creeping back up, reaching $584 billion. Congress passed a BCA modification raising the budget caps for two years (FY2016 and FY2017). Congress had done this two other times before to account for inflation. This time, it added an additional $8 billion for defense and $8 billion for nondefense in the OCO fund.

In 2016, President Donald Trump came into office during a politically divisive election and pledged to rebuild the armed forces. All branches of service had been impacted since the beginning of BCA 2011:

- In 2011, there were 333,370 active-duty Air Force members. By 2017, that number had fallen to 310,000. Commanders reported they would likely run out of money to pay pilots to fly the last six weeks of FY2017.
- By 2016, the Army's active-duty end-strength had fallen from 566,000 to 476,000. The Army's active-duty force was 480,000 before 9/11.
- Before sequestration, the Marine Corps had 202,100 active-duty

personnel, which fell to 184,000. Only 41 percent of Marine aircraft were able to fly.
▷ At the beginning of the surge of the '80s, the Navy had 594 ships. In 2017, the Navy had only 277 active deployable ships. The Navy needs fourteen aircraft carriers to be fully operational, yet it only had ten, as well as only 84 of the needed 160 cruisers and destroyers and 52 of the 72 attack submarines needed.[144]

President Trump's proposal for the defense budget included plans to increase pay and benefits for service members and civilians, more funding for training and maintenance, and plans for modernization in intelligence, counterterrorism, systems, and equipment to address the global conflicts and threats in Afghanistan, Asia-Pacific, Iran, Russia, and the growing number of cyber threats. The military also committed to promoting diversity, increasing maternity and paternity leave, and increasing the availability of childcare.[145] The budget was figuratively and actually looking up. But within the military culture, there was something else growing underneath the surface.

"SOMETHING IS MAKING US SICK"

In 2018, Reuters published an investigative series titled "Ambushed at Home" covering Balfour Beatty, one of the privatized housing contractors, and the neglected housing of military families. After careful review, it was discovered that Balfour Beatty had falsified maintenance records, making it appear that it had been responsive to tenant complaints when it had not. The company secured "millions in 'performance incentive fees' for good service that it otherwise often would not have qualified for."[146] Incentives for high marks were worth around $800 million over the life of a fifty-year contract. This wasn't the first time. Another article from 2011 reported a local military spouse was violently sick from mold exposure and had to pay for her own remediation crew when the housing company refused to help.[147]

An Air Force couple stationed at Tinker Air Force Base, Oklahoma, in 2018 requested maintenance on their warped flooring, concerned that their eighteen-month-old baby might put a piece in her mouth. After a water leak, it was reported that the floor tiles and adhesive contained asbestos. The maintenance logs showed that the leak was repaired in a timely manner when, in fact, it was not. This opened up an investigation revealing fake entries in one log book and, thankfully, a handwritten set that was accurate.

The manager, pressured by his superiors, admitted to doctoring records that kept military leaders from finding out. Reuters documented "sixty-five instances in 2016 and 2017 in which Balfour Beatty employees backdated repair requests, filed paperwork claiming false exemptions from response-time requirements, or closed out unfinished maintenance requests."[148] The Air Force Civil Engineering Center had dismissed concerns from employees and families and gave the company high marks anyway.

Reuters also discovered that at Tinker alone, more than four hundred new homes built in 2018 had "gushing leaks, raw sewage backups, rotten wood, and severe mold."[149] Despite not having manpower to address maintenance requests, records showed prompt responses, with one email showing staff directing employees to close "119 resident maintenance requests in four hours."[150] Reuters later discovered doctored records at San Antonio's Lackland Air Force Base as well.

By May 2019, a survey conducted by the Military Family Advisory Network revealed 55 percent of military families were dissatisfied with housing conditions at more than one hundred military bases across the country. In all, the survey found a staggering "6,629 reports of housing-related health problems, 3,342 of mold, 1,564 of pest infestations, and 46 of carbon monoxide leaks."[151] By the end of 2019, several families joined together, filing a federal lawsuit against private companies Hunt Military Communities and Corvias for fraud, mold, pest infestations, water leaks, and raw sewage. One lawsuit out of MacDill Air Force Base, Florida, stated, "In at least one instance, moldy conditions went untreated ... for so long that mushrooms grew out of the floor and carpet."[152]

HOW DID WE GET HERE?

As a reminder, there was concern in 2005 over how housing programs would be supervised, as well as how families would communicate feedback. Thirteen years later, the DoD expressed it was, "confident that privatizing housing was the right thing to do, however, we also recognize there has been a lapse in overseeing the implementation of DoD's housing privatization program."[153]

When sequestration was triggered in 2013, Odierno testified before Congress with a plea for resources, "We will reduce our base sustainment funds by 70 percent. This means even minimum maintenance cannot be sustained, which will place the Army on a slippery slope where our buildings will fail faster than we can fix them. There will be over 500,000 work orders that we will not be able to execute."[154] In its report *2013 Sequestration: Agencies Reduced Some Services and Investments, While Taking Certain Actions to Mitigate Effects*, the Government Accountability Office stated, "Officials from the military services stated that delaying and reducing installation support services in fiscal year 2013 will likely lead to higher future costs for these services due to facility degradation."[155]

While supervision of privatized contracts is different from maintenance and supervision of military-owned buildings on an installation, military leaders have since accepted responsibility for the failure to provide oversight on all of it. In 2015, Odierno testified, "We will have to make decisions at every Army installation that will impact the quality of life, morale and readiness of our soldiers... Although we've not yet seen the breaking point, I worry about when that will occur in the future."[156]

Families had gone through their chain of command, but were redirected to the contracted privatized companies. When the privatized companies failed to respond, families returned to the DoD, their ultimate provider, and around the circle they went. In 2019, *Military Times* posted online the article "Raw sewage, mold, vermin: Military families asked court to withhold rent until all houses on these two bases are certified as safe," receiving comments like these:

"I pushed these issues for my soldiers and families at Fort Bliss and was threatened with losing rank and was even moved units after using all open door policies up to the base commander. I brought in the pictures and bags of roaches and it still was not enough."

"Had a gas main leaking under our home on Fort Polk, two of our three children hospitalized for unknown reasons prior to this being discovered... had been told it was just 'stinky drainage pipes' for over a year..."

"We had mold that was painted over. You could see it on the walls. This was senior enlisted housing at JBLM."

"Balfour Beatty... had me, my wife and two small children and many other families living in condemned buildings. Not buildings that were going to be condemned, not buildings that should have been condemned, but buildings that were actually condemned."[157]

Congress started to listen, upgrading military housing through the NDAA, and committing to crack down on privatized landlords. The Senate, in a two-hour hearing in December 2019, told lawmakers and top military leaders that they continued to get complaints from military families. Senator Tammy Duckworth, a veteran, stated, "I just don't buy this argument that the chain of command can't really be held accountable in the past because we've empowered them to enforce these contracts. By nature of being in command... you're responsible.... I don't understand why not a single garrison commander to my knowledge, has yet been fired over a failure to maintain the standards."[158]

The DoD has since started to make concerted efforts to increase the frequency of inspections of homes, created a Tenant Bill of Rights, maintain an inspection checklist during changes in occupancy, and ensure residents are satisfied with move-in. They have also implemented better communication about the difference between private companies and military housing offices. Maintenance reports now reflect the quality of work according to residents rather than the response time, and the Office of the Secretary of Defense has increased its project oversight.[159]

Strauss and Howe, in *The Fourth Turning*, describe millennials as a hero generation, one usually coming of age during a time of crisis. Hero generations, like the greatest generation, go on to resolve that crisis and are defined by that for the rest of their lives. Despite the heavy criticism millennials have received, most are no longer considered "snowflakes" or "failures to launch." In fact, most are well into adulthood with taxes to pay and families to support. While it may be unclear what their exact generational crisis is to resolve, millennials have successfully advocated for racial and gender inclusivity and diversity in society, the workplace, and the military, and have expanded their reach globally on many social and environmental issues.

Shannon Razsadin, a millennial military spouse, was one of the cofounders of the nonprofit Military Family Advisory Network (MFAN), which organized the survey that brought qualitative and quantitative data to the housing crisis in 2019. Validating the emotions and concerns of thousands, she, along with many others, made it safe for families to finally feel seen after decades of housing concerns. She paved the way for spouses like Crystal Cornwall, a Marine spouse of eleven years, who testified before Congress on behalf of military families facing housing issues. Cyrstal stated that military families felt powerless and unheard, dismissed in town halls, and fearful of retaliation as they expressed concerns to installation commands and through the chain of command.

In addition, MFAN released new data revealing that one in eight military families faced food insecurity in 2019, rising to one in five in 2021. This data ignited initiatives reaching all the way to the White House to better serve military families. It seems millennials were finding ways around the bureaucracy and going straight to the source if needed. In some ways, they were also willing to become the source.

In the midst of this, Space Force launched as the newest military branch in 2019, driving a mix of uncertainty and excitement about its future.

AN EXAMPLE WORTH REMEMBERING
Be willing to go to the front line to listen.

In January 2023, Brigadier General Jason E. Kelly, commander of the Army Training Center and Fort Jackson, provided a town hall for military families to discuss housing concerns. He had already walked through the housing area and spoken to families beforehand. He expressed to families that although the post had to partner with Balfour Beatty, he was not satisfied with the housing situation and was committed to making it better.

Kelly said, "I want to listen, and I want to learn. ... I want you to leave without a doubt that you have been heard. Some of you might say you've been heard before. Not by me."[160]

Chapter 8

CONSTANT CHANGE

ALTHOUGH MILITARY FAMILIES ADOPT the cultural understanding that change is inevitable in the military lifestyle, it doesn't mean they don't need some level of predictability to thrive and be well. Many situations require the military culture to be adaptable, providing a lens for how we can improve our leadership of people through times of change.

Broad sweeping changes to the logistical movements of our military and operations tempo significantly impacted family wellness and morale over the course of GWOT. Yet leaders passively assumed that service members and their families understood these changes and underestimated how these changes would deeply impact deployment, reintegration timelines, and the community cohesiveness so integral to readiness and resilience. In order to understand the current state of our culture, especially retention, we must look at the long-term effects of constant change on the culture and learn to proactively communicate upcoming changes. We must also be willing to evaluate our own decisions by introspectively looking inward for cognitive bias, blind spots, and tendencies to avoid the messy task of leading people through the aftermath of our decisions.

One thing we can count on is that any decision for people or about people is likely to be complicated, requiring a different pace of thinking and decision-making if we do not want it to get messier. Remember, a simple solution to a wicked problem often creates additional problems. Our top military leaders making difficult decisions during the Budget Control Act and sequestration were an example. Design theorists Rittel and Webber stated that those who present solutions to a wicked problem are likely to end up "liable for the consequences of the solutions they generate"[161] due to the impact those decisions have on people. While that sounds daunting, if we can learn to regularly empathize with those we are leading, and actively challenge or examine our assumptions and knowledge about the problem before developing solutions, we are more likely to help people successfully navigate change. The more complex the situation that involves people, the more intentional we should be to dig deeper, listen longer, think slower, and seek the perspectives of others who will be affected by the decisions we are about to make.

Our culture prides itself on mission analysis, presenting various courses of action (COAs), and evaluating second- and third-order effects. Yet again and again, we see leaders omit or compromise large and often critical areas of concern. Why is that? As leaders who must balance the mission with taking care of people, our goal should be to do our due diligence and consider that the analysis of our decision's impact on people's lives is just as important as mission analysis. The people you lead are worth that level of respect. Even more, history has shown that if we are unwilling to honor that truth, people will eventually find ways to ensure their voice is louder than our intentions ever were.

How can we as leaders learn to identify what it is within us that seeks to rush a decision, especially when that decision impacts others? When making difficult decisions, especially large-scale organizational change, what we value often governs the direction of the decision. How will you measure which values are more important? What story do we, as individuals and leaders, tell ourselves to validate the truth we believe is right? Do we surround ourselves with those who are more likely to agree than to question or challenge?[162]

CHANGE IS INEVITABLE

Change management, according to Gallup, is a vital and core demand of leadership. The ability to successfully shift an entire team, business, or culture through transition and change is studied in almost every military leadership school. In fact, there is a whole *Army Leadership Transitions Handbook* dedicated to helping leaders consider the important steps of successful role change. In it, leaders are encouraged to slow down and assess the internal culture, stakeholders, and other factors to ensure a successful transition of responsibility.

To successfully lead an organization through large changes in vision, operations, and culture, Deloitte recommends an enterprise-wide approach to ensure an effective transformation. Institutions, especially those whose success is dependent on people, must be willing to cast vision, communicate clearly what will be asked of the people, walk with them through the messiness of transition, and finally lead the way toward the new paradigm.

ARFORGEN

In 2004, as the DoD was building up the force, the Army decided that a new model was needed to train and ready forces for Afghanistan and Iraq. Until that point, smaller units were deployed to fight fewer and shorter conflicts.

Now, the demand was eight to ten times what the Army had been preparing for. Army Force Generation Cycle (ARFORGEN) was developed and approved by the Army and chief of staff in 2006 as a way to "tap into the total strength of the Army, leveraging all active and reserve units, while sustaining the process by employing a rotational, more predictable plan for deployments."[163] It included three stages: reset, train/ready, and available.

This new model allowed the Army to systematically rest, train, and deploy units, rotating them through the phases and giving families more dwell time and predictable timelines of deployment. In the reset phase, soldiers returned from deployment with a goal of one year deployed, two years home, a 1:2 ratio for active duty, and a 1:4 ratio for reserve and guard members. During the train/ready phase, units would train more intensely in preparation for deployment. In the available stage, soldiers were ready to deploy.

General Charles Campbell, commander of US Army Forces Command in 2009, stated, "To the soldier on the ground and to the soldier's family, I think the one feature of ARFORGEN that particularly resonates is the fact that there is more predictability in terms of understanding when a unit will deploy, for what period of time, and to execute what mission."[164]

Interestingly, there are differing opinions about ARFORGEN from service members and families. Service members who lost valuable opportunities for experience or training look back on it with frustration.[165]

In order to move large-scale operations and a consistent supply of troops in and out of theater, ARFORGEN deployed entire brigades at once. Before, battalion leaders were offered an opportunity to learn the complex responsibility of managing training and essential tasks through the standard mission-essential task list (METL). In the new ARFORGEN model, officers relinquished that opportunity to superiors on the division level who now managed brigade training gates. As years passed and leaders were promoted to positions that required that competency, many found themselves scrambling to learn what they had missed.

The military counted on service members to relay timelines and ARFORGEN's impact on families. Families may not have known the name, but they were very aware of how it affected their homes and calendars, and those who experienced it recall the predictable patterns of deployment and dwell time as positive. The culture, as a whole, was greatly impacted by not only the global conflicts themselves but also the military's approach to fighting them.

According to one RAND study, "most installations saw as many as 60 percent of the assigned soldiers deploy concurrently, emptying out the installation" during the years of ARFORGEN. Large units deploying all at once from an installation could cause the at-home soldier population to decrease by as much as

40 percent.[166] It was not uncommon for families of soldiers to leave installations to find support from external family members during a deployment, but most families stayed in or around the installation.[167]

In response to the intense collective demand of war, the community adapted. Having a large portion of the community going through the same experience at the same time aided that transformation. Research shows that painful experiences, while subjective, are actually the social glue that bonds people together when they go through it together. In fact, one study found that even when strangers shared painful experiences, it promoted trust and increased their ability to cooperate.[168] This is, of course, not a new concept. Peer support groups have long been an effective way to enhance and accelerate treatment goals and are widely used in substance abuse prevention, grief groups, and more. The intense shared experience of the military culture has been the number one reason why so many military families develop strong, lasting bonds with each other, calling each other "family" and "the family you choose."

Over the years, I've had the opportunity to work with military spouses who chose to leave a base for external support during deployment. While many of them report feeling grateful for the support of their family members, they largely felt disconnected from the military community. Especially when they returned at the end of the deployment, some reflected that they would have made a different decision had they known. One spouse wrote, "I finally felt part of this community then I went to stay with my family in New Jersey and realized that my civilian friends didn't truly understand what I was going through. John had deployed to Afghanistan, and I realized I felt more comfortable in the military world, where there is a shared sense of purpose. I watched the news carefully, but my friends and family were disconnected from it."[169]

This is a similar frustration I hear from reserve and guard families who, while benefiting from a stable civilian community, feel disconnected from those who understand their military experience.

From 2006 to 2016, ARFORGEN aimed to train and send ready troops into Afghanistan and Iraq while giving families predictable patterns of dwell time, training, and deployment. When we lived on an installation that supported one of the largest divisions, it seemed like our entire neighborhood was deployed at once. Looking back, it is more accurate to say the neighborhood was in various stages of ARFORGEN. If a service member wasn't deployed, they were on their way home or about to leave. It didn't matter if our timing was in sync. The families on our street intimately understood their neighbors' needs. Our shared experience created immediate understanding and cooperation in the shared mission of survival.

ARFORGEN was the Army's model, however, the other branches each had similar models that came out at various times around and in support of the model. The Air Force's Air Expeditionary Force Next system (AEF Next) was approved in 2013 but never fully took root. It aimed to deploy larger teams from the same unit and to standardize dwell ratios (1:2).[170] Before that, the Air Force was not set up for longer, permanently manned deployments and therefore had a difficult time organizing a successful model. The Marine Corps implemented the Force Generation Process (FGP) in 2013, which included five phases (synchronize the force, generate the force, ready the force, deploy the force, and redeploy the force). The Navy had the thirty-six-month Optimized Fleet Response Plan (O-FRP) beginning in 2014, which had four rotational phases (maintenance, basic, integrated, and sustainment). It started much later due to manning issues and failed ship inspections. Carrying a similar theme of predictability, training as a team, and keeping the team together, the Navy's plan struggled to succeed due to continued manning shortages and maintenance issues.[171]

Together, all the branches attempted to support a large-scale, rotational cycle of force generation with the added goal of providing families with predictability. Although the branches may have been on different timelines, much of the families' experiences reflect the intense operations tempo and similar strains on families and the community.

One RAND study claimed that very few have studied the factors associated with the transition out of Iraq from a military perspective,[172] even less has been covered about how the many large-scale transitions impacted the military family culture over the two decades. So for this discussion, we will focus on the major culture-wide changes these models asked of the force rather than their effectiveness (or lack thereof) for sustainable readiness and/or deterrence for global conflicts. To do so, we will look deeper at ARFORGEN since the Army is the largest DoD branch and therefore impacted significantly more families.

Impact of ARFORGEN on Family & Marriage

The Army's goal of providing families with a 1:2 year ratio was noble; however, many describe their experience more as a 1:1 ratio, and in many cases, no ratio at all. In 2007, the secretary of defense extended deployments to Iraq to fifteen months, shortening dwell time at home, leading to "increased feelings of loneliness, anxiety, depression" and an "increased negative assessment" of the very community support spouses had come to value, according to researchers.[173]

In 2008, President Bush directed that deployments should be no longer than twelve months with another twelve months for dwell time. By 2012, deployments were reduced again to nine months for those going to Afghanistan.

After a few years at this pace, with few families understanding the changes, many began to express that the quick turnaround of deployments and dwell time was not sustainable. One year of dwell time may sound sufficient, but taking leave (vacation or time off), seeing external family members, and training up for the next deployment quickly filled the family calendar. This tempo contributed to a significant amount of public grumbling. Many active-duty couples who served during that time say they were apart as many years as they were together.

If your family has not experienced it personally, it is hard to imagine what it is like to spend half your marriage apart. Yet, so many military families, especially in the years of ARFORGEN, lived that reality (and some still do). Deployment is a challenging time for families, especially couples, to stay connected. In addition to the obvious reasons of physical distance and communication, the needs of the mission and service members' safety are prioritized. Well before, but especially during the increased tempo of GWOT, spouses were mentored on the importance of being cautious to not upset their service members or distract them from staying focused safely on the mission. Service members, likewise, make every effort not to overshare details that might heighten anxiety at home.

My husband and I learned this lesson ourselves. Christmas day, my husband joined us over video to watch our boys open presents. It wasn't until a year or more later that he shared that just hours before our call, he was running back to a helicopter with bullets whizzing through the air. On a separate occasion, I wrecked the car on the highway after hitting black ice. Thankfully we were all okay, but after we were home and recovering, I sent him the pictures of the totaled car. Visually seeing the picture made him feel far more helpless from the other side of the world. Right or wrong, after some mentoring from friends, I learned to reach out to the support around me first when he was gone before venting or sharing my own state of helplessness or discouragement.

For seasoned couples who have learned these lessons, connection during deployments can become a glossed-over, less authentic version of the relationship, almost like putting part of the relationship "on hold" until they are face to face again. In many ways, couples end up sharing more of a highlight reel rather than the vulnerable challenges or personal stretching that would normally bond them as a couple when they see each other day in and day out.[174] Every couple is different. In fact, some couples have shared their agreement that especially the service member will not share details about the deployment that are upsetting or raise the spouse's anxiety levels. My husband and I decided the opposite. While we were careful not to put the other in a helpless state, we were both clinicians and we felt we could handle listening to each other process difficult experiences. Yet, this should not be confused with the same shared experience that bonds individuals in shared pain. Although a couple may share the painful experience of being separated by deployment, they are more like two parallel train tracks that are going in the same direction but not intersecting. The world of deployment and the world at home are very different. The longer the deployment, the harder it can be to relate to or imagine your spouse's world.

The challenge of connecting despite living very different lives and the tendency to portray this more positive, independent, "everything's fine" persona during separations makes reintegration much harder for families. I often tell couples to anticipate at least half of the length of the separation for the relationship to settle down into what will feel normal. Most of us are sick of the phrase "new normal," likely because we had to establish so many new normals over the years. But it is true that most couples don't return to what the relationship was before the deployment. They find a new battle rhythm, which takes time.

There are new preferences, personal growth, new household routines, and much more. With the roles somewhat, but not quite, reversed, service members are told to "not jump into the household routine too quickly" to "sit back and ease back into the patterns of marriage and parenting." Meanwhile, spouses were, and still are, told not to fully let go of all the responsibilities so as to not overwhelm their service members. One study found that spouses "pushed away their partners in order to maintain their own emotional balance."[175] Service members, especially parents, described the loss of missing key development milestones in their children's lives while also feeling the conflict of trying to re-enter the routine of the home.[176] The result is that couples spend at least three to six months of their year back together tip-toeing around each other, again, not fully being open and vulnerable to their personal needs and likely on their best behavior.

In reality, no couple can live in that kind of inauthentic state for long. Eventually, both will individually crave control in response to everything in their life that feels out of their control. That may be an argument, the silence of disconnection, or overtly taking back control of their place in the home.

This discouraging moment of reintegration[177] can actually be healthy for the relationship (unless there is abuse or relational destruction). It is the moment when the walls are coming down and vulnerability is edging its way through the cracks of "playing it safe." On the other side, the released tension is the perfect moment for couples to redefine themselves and their new relationship. However, during ARFORGEN, leave was over by then, and training for the next deployment was beginning. Faced with another looming separation, couples were unwilling to go into a deployment with their relationship disconnected and struggling. It was far easier to put the masks back on and not "need" too much from the other person.

When couples say they spent half their marriage apart during the intense years of GWOT, many of them actually spent more than that. Many couples I talked to over the years continued the separate tracks they started. Service members bonded further with their battle buddies, and spouses did the same with their community of spouses who understood enduring the homefront. Especially for those in special operations who have the added issue of clearance that limits communication, the temptation is to choose indefinite disconnection rather than navigating creative ways to fight for intimacy.

I have seen plenty of couples find their ways back to each other, but it is not an easy path in an already rough terrain. I contend that the temptation for many couples was (and is) to choose a more "peaceful" version of their relationship where some areas stay compartmentalized from each other rather than doing the hard work of reuniting. What I saw from many couples during that time was that the fourteen-month mark of reintegration was actually where the relationship became the most authentic, where the couple began to settle into life together. Most were not able to reach this marker during ARFORGEN.

A massive literature review of military family wellness between 2007 and 2017 revealed some enlightening statistics.[178] These were some of the most intense years of force-wide tempos that cycled large groups through deployments to mostly Afghanistan and Iraq. Here are just a few of the many statistics found during those years:

▷ A study in 2011 found the clinical PTSD diagnosis rate to be as high as 30 percent for service members.
▷ In a 2008 study of military spouses' mental health, 22 percent reported that their current levels of stress or emotional difficulty had a negative impact on their quality of life.
▷ Soldiers were more likely to report intent to divorce/separate in the late 2000s, a trend coinciding with increased Army deployment tempo.[179]
▷ Spouses who feared for their service member during deployment and who filtered their communication from their service member were more likely to report negative health symptoms.[180]
▷ "Military children were admitted to hospitals more frequently for attempted suicide than civilian children," and "across several types of physical violence and harassment, military adolescents were approximately one and a half times more likely to report having engaged in physical violence or harassment."[181]

One interesting point to note is that all of the children and adolescents who would have been included in the research during the years of ARFORGEN would have been adolescent millennials and most of Gen Z. In 2022, 65 percent of Gen Z military adolescents whose parents served in the post-9/11 wartime deployments wanted to serve in the military themselves; 42 percent of those were showing signs of emotional distress, a much higher rate than civilian adolescents; 64 percent had moved up to five times in their childhood; and 11 percent had experienced domestic violence in the home.[182]

TIME FOR A CHANGE, AGAIN

In 2006, Gen X was peaking and older millennials were just coming into the military. Having never experienced the culture before ARFORGEN, incoming families imprinted on the idea that this was the lifestyle, tempo, and standard military experience. Those who experienced the beginning of ARFORGEN may have benefited from some of the initial predictability. Those who came in later may have experienced no predictability at all. Since people naturally look for patterns and predictability, many may not have realized ARFORGEN was just a temporary model for GWOT versus a sustainable model indefinitely.

By 2014, with sequestration at its peak of harm, the community breaking apart under the trauma and pressure of demand, new technology redefining community and communication, and the other branches struggling to support a similar model of predictable cycles, it was time for another change. Chairman of the Joint Chiefs Martin Dempsey testified to Congress in 2015: "We owe them and their families clarity—and importantly, predictability—on everything from policy to compensation, health care, equipment, training, and readiness. Settling down uncertainty in our decision-making processes will help keep the right people—our decisive edge—in our all-volunteer force and maintain the military that the American people deserve and expect."[183]

This was also about the time Odierno expressed, "Although we've not yet seen the breaking point, I worry about when that will occur in the future."

With troop levels decreasing in Afghanistan and Iraq, and new threats from Russia, China, North Korea, and Iran,[184] the Army knew that ARFORGEN was causing a "degradation of the Army's readiness to rapidly respond to a large-scale wartime contingency with ready and responsive Army forces."[185] The new goal was to make the force more agile, dynamic, and responsive. The Sustainable Readiness Model (SRM) was launched in 2017 to offer more flexibility, move away from fixed cycles, and avoid further readiness declines by deploying smaller numbers for the foreseeable future.[186]

With forces in more than 140 countries across the globe, SRM prioritized a steady and sustained level of readiness while also modernizing the force. What that looked like for the culture of families was varied, but for the most part, there were no large-scale deployments. Smaller groups of service members and even single members were trained and deployed to replace members coming home.

While it is normal for intact military families to somewhat isolate themselves from the community when they are in less need of support, fewer shared experiences of difficulty further eroded the common bond within the community. Going through a "solo" deployment meant fewer community members to lean on and actively "suffer with."

Much like the launch of ARFORGEN, the rollout of SRM was not exactly on the list of information distributed to families. On the contrary, unless service members educated their spouses on changes to the deployment cycle, families were likely tracking issues that were more relevant and current to their family and home. With relocations, housing issues, and the community changes impacting them directly, a briefing on force generation cycles was likely not dinner table conversation. When I asked spouses of various branches over social media if they had ever heard of SRM, those who knew about it said there was not enough coverage of it and many others commented they had never heard of it.

Going back to the power of imprinting when people are most vulnerable, consider that a majority of Gen X and boomer service members and families experienced a majority of ARFORGEN, while millennials experienced the switch to less predictable schedules, an online community, and a disgruntled, anxious force. Millennials also were less likely to experience deployments amid a larger community and may have felt, at times, less supported.

In a Pennsylvania State University literature review in 2021 (during SRM), only "56 percent of active duty military spouses [were] satisfied with the military lifestyle, and 59 percent support[ed] their spouse continuing to stay on active duty." Most of these came from the enlisted community, O1-O3 officer spouses, and dual service couples,[187] which would have been mostly millennials. The Blue Star Families 2021 survey reported 23 percent of active duty military families had lived apart intentionally, or "geo-bached" since 2016; 41 percent reported their decision was due to the spouse's desire for a career; and 49 percent reported children's educational concerns.[188] About this time, Gen Z started to peak in their entry into the force. The fumbled withdrawal from Afghanistan live-streamed across opinionated news platforms. Again, millennials and Gen Z did not walk into the same military culture previous generations experienced.

In 2022, a new model, the Regionally Aligned Readiness and Modernization Model (ReARMM) rolled out with clear goals to train units and modernize equipment for specific regions, and become a ready and dynamic force to quickly deploy in smaller tactical units "ready for competition, crisis, and conflict."[189]

The model tasked the military to become a more competitive, agile, flexible, responsive, and specialized force. All while promising predictability.

WHAT ARE YOU ENSLAVED TO?

The Army War College Strategy Conference conducted a panel discussion in 2012 on the future of the force, looking ahead to 2020. At the time, the defense was preparing to shrink the force in size and budget and, interestingly, forecasted a hopeful withdrawal from Afghanistan by 2014. Thomas Mahnken, PhD, a visiting scholar from the Philip Merrill Center for Strategic Studies at Johns Hopkins University, pointed out that our past can serve as an anchor in the way we view our future. He proposed that the best way to figure out how, when, why, and where the military is likely to leverage force is to look at how, when, why, and where it did so in the past.

In his closing remarks, Dr. Mahnken challenged the room of leaders with what he called "last war-itis." Former Defense Secretary Robert Gates coined the term "next war-itis" in 2008, stating that the DoD needed to focus on the current needs of the force rather than the temptation to devote attention and funding toward future threats. At that point in 2008, the force was still ramping up and Gates had already championed "meeting the war-fighting needs of the troops now and taking care of them properly when they get home" as priority.[190]

In his talk, Dr. Mahnken pointed out that "last war-itis" is its twin concern. His view was that even in the perceived uncertainty of peacetime, it is and will be a temptation for leaders to be driven by the last war rather than focusing on more present concerns. Challenging the most recent generation of officers who served repeated tours in Afghanistan and Iraq, he said, "There are consequences of not getting past that. The officers that led the French and British militaries in the beginning of World War II were officers whose lives, whose careers, had been shaped in very tangible ways by the experiences of World War I. There is an argument that many of them could never get beyond that searing personal experience. I think we need to be able to incorporate the lessons of our recent experience, but we need to be sure to not be enslaved by that experience."[191]

Enslaved. This powerful word had me ruminating on what Dr. Mahnken was trying to warn us about. The actual definition is to "cause (someone) to lose their freedom of choice or action."[192] What experiences or cognitive biases do leaders believe and are therefore enslaved by, preventing them from seeing a deeper, possibly more complicated truth?

Research has shown that when people work together on a project, "individuals tend to take on the core values of the group" and "compromise their own values in favor of those held by the group."[193] There are several thoughts on why this happens. On a smaller but still destructive scale, service members avoid challenging the decisions of their peers and superiors in order to manage their reputations, believing that "toeing the line" will more likely lead to promotion. Likewise, quick but cheap wins that show well on the next Officer Evaluation Report (OER) are more tempting than the deliberate and complicated critical thinking that often results in a "no" to our good ideas.

On a larger scale, in addition to the influence of groupthink that feeds our confirmation bias, the availability heuristic (or availability bias) is a sort of mental shortcut when we feel pressed to make a decision quickly. In the example of both next war-itis and last war-itis, this cognitive bias recalls the most recent information that supports our view rather than taking the time to slow down and thoroughly investigate the decision.

There are more than 150 different types of cognitive biases that can skew our perspective. Several of the comfort zone biases, which include the status quo bias, sunk cost reasoning, confirmation bias, and framing bias, prevent us from seeing an alternative perspective that takes us away or complicates our current circumstances we feel the need to protect.

Service members are taught in various schools of leadership to consider the second- and third-order effects or the downstream implications of decisions that impact military operations and strategy. However, for large-scale decisions that impact an entire culture, second- and third-order effects are more complicated and take more time to consider. The higher (and bigger) the decision, the more widespread the ramifications throughout the culture and into the future.

When faced with a complex decision that could include the ethics of harm to others (which is likely any decision as a leader), leaders should consider three rules of management from Indiana University's William Hojnacki:

▷ Rule of private gain: If you are the only one personally gaining from the situation, is it at the expense of another? If so, you may benefit from questioning your ethics in advance of the decision.
▷ If everyone does it: Who would be hurt? What would the world be like? These questions can help identify unethical behavior.
▷ Benefits versus burden: If benefits do result, do they outweigh the burden?[194]

Deloitte's second principle of "Seven Principles for Effective Change Management" recommends that organizations "Understand the institution's culture: It's critical that leaders take time to understand the institution's existing culture before embarking on a change initiative. Any undertaking that doesn't align with, act on, or uphold the institution's values will likely encounter resistance."[195]

Nowhere in the *Army Leadership Transitions Handbook* does it suggest assessing the culture of families you are preparing to lead, nor does it consider them as external stakeholders.

I recently asked my son, an eighteen-year-old Gen Z, "How do you lead well when your job is to make decisions that are in the best interest of an organization, knowing it will negatively impact people? Yet, those very people may have a vote if you will remain their leader." His response was, "Well, I guess you are going to have to communicate a whole lot more."

With concern for organizational change, the military failed twofold. One, its leaders mistakenly assumed the culture would blindly follow and adapt to large-scale change indefinitely with a never-ending supply of bandwidth and loyalty. Two, they believed they could skip the important step of understanding their own culture, which would have led to the communication of vision and value to the culture it professed to depend on.

Again and again, we have heard that military families (including service members) are the backbone of the force, strength of the force, and foundation of readiness, and yet families are compromised at the military decision table. Whether it is seen as too complex, too time-consuming, or not as valuable as the mission, this consistent choice to compromise people overwhelmingly reveals our value of being war-ready as more important than people. We are now seeing that the ramifications of that choice present serious and real risks to readiness, recruitment, and retention.

War, the threat of war, and the deterrence of war will always win when we play that card at the table against almost anything else. It is easier to sacrifice the complicated and messy responsibility of taking care of your culture when you can more easily win at the thing you are good at—war.

Author K.J. Ramsey says, "Silence is the arbiter of scarcity, the force of coercion and control that those who hold the most power wield to maintain the status quo. If power can be held only in the hands of a few, then pleasing them is what buys us belonging."

AN EXAMPLE WORTH REMEMBERING
Communicate, communicate, and communicate it again.

General Eric Ken Shinseki was the thirty-fourth chief of staff of the Army from 1999 to 2003. General Shinseki's *The Army Vision, Objective Force White Paper* and the *Transformation Campaign Plan* contained his written strategy for how to help the Army adapt to a current threat before another crisis erupted. His vision initially faced heavy resistance from military, business, and civilian leaders; ultimately 9/11 was the crisis that validated its sense of urgency.

Knowing that this kind of large-scale perspective shift would be difficult, Shinseki built a coalition of retired military and civilian leaders to help communicate to Congress and the civilian sector, while he spent a significant amount of time communicating his vision everywhere he went with Army's key leader development institutions (e.g., Army War College; Command and General Staff College). He empowered units and agencies to lead the development and training of new equipment, changed the official headgear to the black beret in order to communicate visually to the force that something was changing, and even removed "change-resistant leaders"[196] who risked sabotaging a smooth transition.

He was quoted as saying, "If you are going to make a change, make it big and bold. Walk up to the biggest guy on the block, stand in his face, and get it started. Then go around, brigade by brigade, making it make sense." And, "You must love those you lead before you can be an effective leader. You can certainly command without that sense of commitment, but you cannot lead without it. And without leadership, command is a hollow experience, a vacuum often filled with mistrust and arrogance."[197]

SPOUSE CULTURE

"Those that support us, no matter what, who are always there for us, who are there to do whatever is necessary because they love their soldiers and are willing to support them no matter what we ask them to do—that is what makes us so strong. That is what enables us to do the things that we are asked to do."
—General Ray Odierno, chief of staff of the US Army, 2015

Chapter 9

TWO FOR ONE MODEL

HOW CAN A CULTURE PUSH THROUGH so much uncertainty, war, and increased demand for support and not give up? To understand, we have to take a deeper look at the military spouse culture and its relationship with the institution. There are some deeply ingrained cultural norms, complicated relational dynamics, and taboo topics that spouses don't openly discuss with each other. The spouse culture is like an iceberg in that what you see above the surface is often safe and acceptable, yet there is much more beneath the surface. As new generations enter, spouses are opening up and in some cases strongly using their voices to live more authentically within the culture. Understanding these dynamics will help us understand their frustration and incredible grit and resilience, and why they would stay in a culture they both love and feel betrayed by.

In the clinical world, it is well known that some issues, such as addiction and trauma, can unknowingly be passed down to future generations. Especially when a family is unwilling to talk about traumatic experiences, negative generational patterns can snowball. "Instead, these families keep the traumas a secret or continue to convey them in indirect or maladaptive ways."[198]

Successfully changing our family tree for the next generation can only happen if we know our family story and the beliefs and thought patterns strongly tied to that story, and if we are willing to do something different. Cultural traditions and rules provide a similar structure for positive or negative beliefs to go unchanged or unchallenged. For decades, the spouse community advocated for changes only to have the same issues resurface in the next generation. While the relationship between the DoD and families has been publicized to be one of great pride and gratefulness, like any family or culture, it's complicated.

Trying to define the relationship between the military and spouses is a bit like nailing Jello to the wall. I spent more than a decade trying to decipher how these two entities work together, given that the military has no real control over spouses and yet considers them a key component of force readiness. Spouses are civilians, not service members, but exist within a culture that has very clear traditions, norms, and regulated expectations. It is not an employer-employee relationship and it is not a professional partnership. In fact, if you were to go to each of them separately and ask them to define the relationship, you are likely to get different answers.

Military leaders make statements like "the strength of our soldiers is our families" but also send indirect messages of detachment. Odierno's quote in 2015 about military families being the strength behind our force has become almost a mission statement for program development, encouraging weary spouses whose resilience is required. Since then, social media and direct digital access to leaders have allowed military spouses to hold the military accountable for helping them be the backbone of the force.

As of 2021, 90.5 percent of active duty spouses were female, 76.1 percent were married to an enlisted service member, and 86 percent were twenty-five to forty years old (Gen Z).[199] We also know there are more total dependents than there are military personnel and that around half of the force is married. The military's involvement in service members' personal lives vacillates between being completely detached (we can't tell you who to marry or how to run your home) to policies and regulations that govern their bodies, their voices, their professional relationships, and some aspects of their family relationships.

President Clinton's Don't Ask, Don't Tell policy of 1994 is an example of a policy that many described as archaic and overreaching compared to civilian culture. The policy "allowed gay and lesbian citizens to serve in the military as long as they did not make their sexual orientation public."[200] It deeply affected the intimate personal lives of service members and wasn't repealed until 2010. Thanks to the millennial generation's passion for social change, the overturning of Don't Ask, Don't Tell led to additional changes in and awareness of diversity and marriage equality for the military.

Military families are more diverse than ever, including blended families and same-sex, interracial, and dual-service couples. With Gen Z now entering the force, embracing a diverse and inclusive military culture, including roles within the home and workplace, is critical to recruitment and retention.

With a mission-first mentality, the needs of the military can include large movements like relocations and deployments or smaller adjustments like long unexpected work hours. This always-on mission puts families in a constant state of readiness with an ever-increasing demand for resilience. They are frequently, if not daily, asked to make sacrifices of their own to do so. For example, while American women have made considerable advancements in employment and nontraditional roles in the home, military spouses have held a consistent 22 percent unemployment rate for almost a decade, increasing to 38 percent during the pandemic.[201] This has made it extremely difficult for the military family culture to break out of traditional, cultural, and gender-specific norms that have long supported the military establishment.

TEAMMATE OR SECOND FIDDLE?

As a military couple, there is simply no easy way to jump into this lifestyle without a team mindset. When I finally agreed to support my husband's commissioning into the Army, I said, "Ok, we'll do it, but if we are in, we are all in. And we are all in as a team." Many other couples join with the same mentality, taking on family names like "Team Weathers" to encourage each other during difficult times. The demands are high and the work culture of order and regulation rubs off on the family. Drive anywhere on an installation around 5:00 pm when retreat is played over the loudspeaker and you will see mothers and children get out of the car to stand at attention or salute the flag alongside the service member, even though only service members are regulated to do so.

Another example is the deeply ingrained concept of the battle buddy. Taught early on in basic training, service members are assigned a battle buddy who is usually different from them in many ways. They do almost everything together as a way to move past differences, establish a common bond, and become more team-minded at home or at war. Spouses, likewise, call other supportive spouses their battle buddies or even use that phrase within their marriage.

It is a wonderful community we are proud to have introduced to our children. While the country remains politically divided and often prematurely opinionated, the military culture has provided a community of order, respect, selfless sacrifice, work ethic, character, and duty. These cultural traditions carry into the home and into military marriages. In fact, these shared values serve as a foundation when the military's lifestyle of demands become overwhelming.

For most of America's history, the values of patriotism, honor, and duty were the key reasons behind both the service member entering and the spouse embracing military service. To this day, it is not uncommon to hear Gen X and baby boomer spouses talk about their love of country and service. I recently spoke with a boomer couple who have been mentors to my husband and me over the years. He recently retired after more than twenty-six years of Army service. As his wife reflected on his career, she spoke of what an honor it was to "serve this great country."

For many spouses, serving alongside their loved one is an extension of their service member's military commitment and oath. When my husband and I made our agreement that day to enter as a team, I devoted my career to serving military and first responder families. Perhaps it was out of a desire to connect with my husband's strong sense of calling, or maybe it was simply because I identified with the community I was in. Regardless, it has given me a great sense of "doing my part" for those who are willing to give so much more. While not every spouse devotes their career to the military community, many volunteer.

This buy-in or willingness of military spouses to be all in proved to secure long-term commitments from service members during World War II drafts. Research starting in the '80s revealed family and spouse resiliency and wellness shared a direct correlation to a service member's mentality during the mission.[202]

During the recent invasion of Ukraine, Russians used electronic cell phone data to locate Ukraine forces by texting Ukraine families that their soldiers had been killed in action. The fear of this being true prompted families to immediately call their soldiers on their mobile devices. Russian forces then called an artillery strike down on the location where everyone's digital footprint was detected.[203] This form of electronic and psychological warfare using the family as a point of vulnerability is not a new strategy of warfare.

Propaganda like "loose lips sink ships" during World War II was distributed by the war department, and operational security continues to be emphasized and adopted by families. Especially now that cybersecurity concerns are making operational security increasingly more difficult, proper education or training could have saved the Ukrainian forces from a tragic Russian strike. Yet as this example shows, families also hold the key in the psychology of the warrior's mindset.

When I traveled overseas with the secretary of defense, troops told me again and again that they couldn't do their jobs without the support of their spouses holding down the homefront. One service member shared how difficult it was to stay focused during his previous deployment when his first wife was at home struggling with mental illness. She tragically committed suicide after his return. Now, happily remarried, he and his second wife have a healthy arrangement for how to handle stress separately and together, especially during deployment.

In a sense, these relationship dynamics between military spouses and service members are no different from any other couple as they balance family and work. However, jobs that require intense focus and include a risk of safety can have devastating consequences if the stress of family or work spills over to affect the other. Yet there is something unique about service-connected professions. It was not uncommon in my work with first responder couples to see a shared sense of service and duty. First responder spouses similarly saw their role as managing the children and home and not distracting their spouse while on duty.

Spouses of surgeons similarly do not call or distract their spouse during work hours but are far less likely to share in the purpose and commitment to the job.

Perhaps one of the biggest differences between the first responder and military lifestyle is the structure and support the military provides. While first responder couples agree that when the first responder leaves the house they are on duty and when they come home they are off duty, the military is in a constant state of readiness. In order to support that mission, programmatic support and structure have been developed to make it possible for families to be available to provide reliable stability at home. Unfortunately, this creates a mission-first mentality with no distinction between an actual threat and scheduled training. Anything the job requires eventually becomes part of the mission, and spouses are conditioned to take a passive, subordinate role to that mission.

As I assimilated into the military culture, this symbiotic relationship families had with the military was not hidden. The first message I heard from older military spouses was that the mission comes first. Likely mentored by the generations before them, they gave steady reminders to "not upset your soldier" during deployment and to use programming, social groups, and other spouses as needed. Coming in with a clinical background, I'm a bit surprised I didn't question the mentoring I received. When you live in a community that is constantly preparing for or going to places that could put your loved one in harm's way, you are willing to do whatever it takes for everyone to come home. It was also an incredibly inspiring introduction to a culture that truly took care of each other. It offered a unique and exclusive community that met almost every need, so of course, genuine loyalty develops easily toward its success.

Twelve years later, when I found myself navigating the stress of relocating during a pandemic with two teen boys, a debilitating ice storm, and a medical scare during a deployment, I reached my own breaking point. I called some of my close military spouse friends and asked their opinion on whether or not I should tell my husband about the medical scare. Each spouse I called suggested that I wait until I had more information. "Why share the worry? Wait until you know if there is a real problem before you tell him." Several shared they had been through the same thing and chose to carry the burden alone. So I did the same.

Perhaps it was out of a sense to do my part or maybe it was just what we were all taught to do. However, when my husband found out, he was hurt and angry. "We don't hide things from each other," he rightfully said. "We are a team." It bothered me that I made that choice. My husband was in Poland at the time, restricted to one building due to the pandemic. He was not anywhere close to combat, and yet my years within the culture had taught me to put the mission first, even when it was administrative deskwork.

I was also bothered by the fact that I didn't reach out to my parents. I didn't even realize until my father found out secondhand months later and asked why I didn't say anything. I chose other military spouse friends who had been through difficulty during deployment, and I felt they would be able to speak to my unique situation. I don't fault my fellow military spouse friends for their advice. I had given the same advice to others over the years. Yet this time, something about it seemed off and maybe backward from how healthy relationships function if they expect to go the distance. Surely a better balance could be achieved.

Cultural norms, messages, and traditions of the military spouse have been especially slow to evolve. Messages within the spouse culture have been around for decades. Some seem outdated yet are passed around in a way that makes you wonder if there are still traces of original expectations. "Hurry up and wait," "adapt, improvise, and overcome," and "home is where the military sends you" are casually used as tribal or social bonding to remind the family to stay adaptable to the needs of the military. "Pull up your big-kid panties" is often said to new spouses. "Embrace the suck," something service members say to each other during extreme deployment conditions, is used to motivate weary spouses.

I will never forget the young spouse who wept in my office, asking why it was considered taboo to show emotion when saying goodbye to her husband for a year. It was a good question and I didn't have a good answer. It made me wonder about some of the other messages we pick up and pass down, where they come from, and whether they still apply in today's military culture. It also made me look a bit closer into the relationship spouses had with the military. After all, we were also taught that "there is a third person in every military marriage, and the military is the mistress."

TWO FOR ONE

From the earliest days, leaders' wives, including Martha Washington, openly shared the positive influence they had on their military husbands' morale as well as their own patriotism being subservient to the military. Mrs. Washington ran the risk of being captured by British forces while traveling to various encampments and took charge of social meetings at the camps. She also served as Washington's secretary and intermediary to the point that Washington submitted her travel expenses to Congress as a "service to colonists' cause."[204]

As a reminder, before World War I, there was concern that families would become dependent if married men were allowed to enlist. Those who did marry mostly did so after enlistment. During World War II, however, a larger force was needed and married men (and those with children) were drafted and wages increased to compensate families. After the war, as the national military grew and restructured, wives were beginning to be seen as more of an asset than a risk.

The military lifestyle, especially overseas, made it difficult for wives who wanted to work. Women in the civilian community were advocating for equality in the workforce, even though the gender and cultural norms of that time still carried the heavy expectation of women in the home. With this still being the standard, investing in the home, raising children, and later volunteering in the military community allowed military wives to contribute with a sense of purpose and a way to serve their country.

Until 1941, much of the ideal behavior of a military wife was implicit. Officer wives took the lead in mentoring the next generation on the topics of marriage, homemaking, child rearing, and following rules and regulations of military protocol even though they were not military personnel. However, the surge of new military wives during and after World War II made it difficult for officer wives to mentor effectively. To help fill the educational gap, an Army commander asked officer wife Nancy Shea to write *The Army Wife* as a semi-official guide for preparing wives for deployment, benefits, and social etiquette. Shea collected information from other officer wives on social behavior (such as greeting various ranking service members), hosting coffees and teas, and even proper dress codes for each occasion. Shea also discouraged wives from working, saying that the most important place for them was in service to the military, the home, and the community through volunteering.

In 1956, the military saw the value of providing better access to health care in the benefits package for families. On a space-available basis, active-duty dependents, retirees, retiree dependents, and survivors could receive medical care in military treatment facilities with priority given to active-duty members. Benefits continued to be a successful recruiting tool and soon, the number of dependents outnumbered the number of personnel.

Throughout the '50s and '60s, military wives continued to grow in recognition as unofficial ambassadors during overseas assignments. Germany and Japan were now allies and military wives softened the American reputation of the military and marriage. Many silent generation and baby boomer families remember that officials became much more direct in telling military wives that they could do their part for their country by supporting their soldiers and the military community. As military programs grew and military wives organized social community events, mentored, and volunteered, the military became more dependent on the "two for one" model of a service member and his wife.

Shea's *The Army Wife* would be printed six times between 1941 and 1966 and distributed to not just officer wives, but all wives stating, "whether she is the wife of an officer, an NCO, or enlisted man this knowledge is of vital importance, for her behavior and attitude will have a direct effect on her husband's efficiency and chances of promotion. It can also mean the difference between an awkward, unhappy hitch in the service, or a fulfilled, exciting way of life."[205] By 1966, the manual would be revised by one and two-star generals with a statement of "full cooperation of Pentagon Authorities" printed on the inside.

In 1965, the Army created the Army Community Services (ACS) program, which provided training on parenting, financial counseling, childcare services, domestic violence and child abuse awareness, and more. With limited paid staff to run the program, the military depended on spouses to volunteer. Meanwhile, spouses were encouraged to continue tending to their social lives with social events that were quickly becoming a cultural tradition.

Shortly before the end of the Vietnam War, Congress moved from the draft to an all-volunteer force. Recruitment numbers dwindled. The military-civilian divide grew. No longer able to force citizens into service, the military began to see families as central to the success of recruitment. Benefits and support programs were no longer seen or used as a reward for service but as an appealing recruitment tool for families looking for job security and health care during the economic crisis of the '70s.

While civilian social welfare programming faced major cuts, questions were raised on whether the military was creating its own military welfare state. General George S. Brown, the chairman of the Joint Chiefs, argued that military service was "a uniquely deserving vocation—more like a calling—and entitled to special benefits." From the top down, leaders implied that civilian jobs were easier and "less than" compared to those "called to serve their country."[206] Referring to service as a calling implied there was something special and unique about the caliber of person who would answer that call. It also justified the protection of the special privileges that came with it. Rather than cuts to the DoD budget, funding increased for additional military benefits, including more programs for military families. In essence, it was the military that first pointed out the separateness of the military lifestyle from civilian life.

IT'S NOT PERSONAL, IT'S BUSINESS

In almost every romantic comedy, there is a moment when one person wants the relationship to move forward and the other is not quite ready to commit. Seeking to end the confusion and tension, the burning question is finally asked, "Where is this relationship going?"

It is a question meant to define the relationship and make the implicit explicit: "What are we really doing here?" As part of the audience, you share the desperate hope that everything leading up to that moment was not in vain.

Nothing is worse than being all in only to find out the other isn't. And yet, no one wants to move forward in a relationship you can't fully commit to.

Comparing the DoD's relationship with military spouses to any kind of romantic relationship may seem odd. Yet, a lens of courtship may make it easier to understand the complexity of a relationship that is dynamic, confusing, and tumultuous at times, while still loving.

Up until the mid-'80s, military spouses enthusiastically and with great patriotism volunteered their talents and resources to the military culture without much complaint. Whether out of a sense of purpose, filling an unmet need for a career, contributing to the development of the community they relied on for support, or simply following the cultural standard, military spouses largely built and ran the family culture of support at home. The military benefited from the investment of military spouses in that it contributed to higher morale in service members and families, created a community of support for families during deployments, and helped recruit new families. The volunteer model sustained itself without requiring significant funding. It was a mutually beneficial relationship.

You might say that this relationship sounds more like a business partnership, and perhaps it is, except that there was never an actual agreement or expressed expectation between the DoD and military spouses. An oath was never taken, and no documents were signed. Even today, military spouse benefits are provided through the service member's social security number or DoD identification number. It is an arrangement based on unspoken agreements and expectations. From a business and organizational consultant perspective, any partnership that lacks clear boundaries or definitions is headed for trouble.

This brings us back to our analogy of a courting relationship. Every relationship, even business or friendship, begins with some kind of courtship. Two entities meet, discover what they have in common, and deepen their relationship through shared values. For example, in a business partnership, two entities decide they will be more successful working together than independently. They agree on what they will each offer, leading toward achieving their shared goal. As they work together, they establish trust in the relationship and deepen their bond even though their brand, internal culture, and working style differ and are not governed by the other. Even business arrangements where separate entities overlap and influence each other still hold autonomy. In an acquisition, two independent cultures merge to execute the difficult task of defining a combined culture. All of these require intense, frequent, and clearly communicated expectations, policies, and agreements from all parties.

To offer one more comparison, the first responder culture is one where no expectations exist between the first responder profession and the family. Even though families would agree it is necessary to be available and flexible to provide stability at home, it is entirely up to each family to make those arrangements. Couples are also expected to navigate how to manage their lifestyles without the support of programming, social events, or childcare.

What makes the relationship between the DoD and military spouses so unique is that over time, the language and dynamics shifted. During the world wars, the needs of the military were supported by the country and its citizens. The private sector contributed by hiring more employees, including women, to support war-related production. Women were paid for their skills like any other employee. There was a clear agreement on the exchange of dollars for skill.

Most people recognize when something shifts in a business or friendship into something more. There is a moment when you realize that at least one side is investing more than what was originally established in the relationship. The party that is more invested starts to pay attention to the other's needs and realizes that the other can meet some of theirs. And while some of those characteristics might appear in even the healthiest of business relationships, there must be clear boundaries to keep the relationship professional.

After World War II, when military wives became soft assets in military efforts, they were rewarded for their character. During a time when women longed for a sense of purpose and impact outside the home, the military encouraged in both subtle and direct ways that you had influence not only in your husband's career but in the lifestyle you would have as a family. Those needs were met and validated through recognition, awards, and the promotion of your spouse, giving you even more opportunities to serve. The higher rank your spouse achieved, the more responsibility a military wife carried for mentoring, being available to host and attend social events, and more. Military spouses were willing to give their time, talent, and future to the military, which also invested a lot in them.

In all fairness to the military, wives did contribute heavily to some of the expectations and responsibilities based on their husband's rank and leadership. They thrived and enjoyed the community they helped build. Yet, the fact that the military encouraged their efforts and even tied professional advancement opportunities to it likely muddied the boundaries. The difference here is that the military was and is a profession for the service member, but it became a lifestyle and identity for the military spouse. The military's response to the spouse's willingness to volunteer talent validated a much deeper need.

As the operations tempo and culture of the military lifestyle increased, spouses had a reasonable argument that they needed additional support from the military because of what the military requested of them. Often, the DoD positively responded with more support and benefits, recognizing that family wellness contributed to the organization's success. A relationship of trust developed as families began to view the military as a listening and caring provider that responds to the needs of its people.

Reaching back to their original intention of not creating dependent families, however, there were other times when it was very clear the military could not meet all of the needs of military families, especially when it would have to be justified to Congress.

It wouldn't be until the '80s that boomer military spouses would begin to ask the DoD, "Where is this relationship going?" While women were already years ahead in the civilian workplace, military spouses organized annual worldwide symposiums to voice their concerns. In their first meeting held in Washington, DC, they identified thirteen areas for improvement. More than anything, they wanted the lines of communication between families and the military to be clear and reciprocal. They also highlighted that they were a free, voluntary asset that maintained a culture, becoming another recruitment and retention benefit.

This wasn't just about spouses volunteering, though. Inflation, the oil crisis, and stagflation (a combination of inflation and economic stagnation) of the '70s left the nation in a decline. In 1980, military pay for soldiers had dropped 15 percent and families struggled with housing shortages, inadequate base housing allowance for off-base housing, and more. Enlisted families, especially those overseas, struggled to buy their families the food they needed. Some were moonlighting and others applied for food stamps. Officer wives' clubs opened food and household pantries and some commanders opened dining facilities on the installation to help feed families. Some concluded that the Army in particular had not been ready for the number of soldiers who enlisted with families. Spouses argued it was families that were saving the day.

Another top concern was relocation. While spouses understood moving was in some ways essential to the job, they asked that families have at least six months' notice before moving and an increase in financial compensation for moving expenses. They argued that relocation forced many wives to leave jobs, disrupted education for kids, and disconnected them from established support. They suggested employment education and counseling to help families reestablish their two-income homes faster after the move.

Spouses also promoted that they "constituted neither passive nor dependent extensions of their husbands or the [military] itself" and that "the decades-old Nancy Shea 'two-for-one' model of the army wife" was outdated and unsustainable.[207] They wanted recognition for their investments and the opportunity to gain outside employment, even if it meant less time to volunteer. Spouses no longer wanted a relationship that suggested "friends with benefits."

Shortly after the symposium, spouses continued their grassroots advocacy for reform, knowing it would be easier to change the system from the bottom up than the top down. Some changes were made, including a hotline families could call for support and a liaison who would report concerns directly to the Army's deputy chief of staff. They also removed the term "dependent" from policy papers, publications, and orders and replaced it with "family member" or "spouse." Other changes would come more slowly. Spouses continued to organize the symposium annually to express their concerns and continued to educate spouses with workshops like "How to Develop a Successful Grassroots Family Program."[208] By the third symposium, more than five hundred attended from all over the country.

In 1983, Army Chief of Staff General John A. Wickham Jr. wrote his white paper, *The Army Family*, after he and his wife experienced limited access to childcare facilities, inadequate buildings and centers, and concern for service member suicides. His white paper would shape the views of Congress and President Reagan as he implored that the Army had a "moral and ethical obligation" to take better care of families.[209] He also pointed out the relationship between families and the Army for what it had become: "Once a private matter, [the family] is now an organizational concern. Geographic mobility, changing family structures and the recognition that competition between family and organizational needs can be destructive to both parties ..."[210]

In response, Congress passed the Military Family Act in 1985, establishing the Office of Family Policy within the Office of the Secretary of Defense. The office was responsible for "coordinating programs and activities of military departments to the extent they relate to military families" and gave authority to survey families on the effectiveness of family support programs. In 1988, a congressional inquiry over allegations of child abuse at a number of installations led to the Military Child Care Act of 1989 to improve the quality and safety of military childcare.

The military, in essence, affirmed its partnership with families and invested more. Spouses had felt taken advantage of and wanted to know that the military saw them, valued them, and was willing to reciprocate. The military responded with an investment of support and an incredible campaign to "take better care of families." What started as a mutually beneficial arrangement that fit within the cultural norms of the day evolved into one based on need. Not just need in the sense that a business partner is dependent on the other for a successful product, but a codependent relationship that involves emotion and perceived and real obligation. Cynthia Enloe, a feminist scholar who studies military families, argues that the Army white paper only "strengthened its control over its members through dependence."[211]

With additional family programming and support came the need to run programming with limited manpower. The relationship between families and the DoD remained complicated as military spouse mentoring and volunteering had become a deeply cultural, unquestioned part of the community experience. Especially following the grassroots movement of the '80s, spouses weren't directly asked to volunteer; they were reminded that as civilians they could make their own choice to volunteer or not. Yet, programs specifically designed to increase readiness through support and information distribution for families, like FRGs and ombudsman groups, were mostly run by military spouse volunteers.

Although spouses could work if they wished, it was not a popular choice in the spouse community, and doing so often limited opportunities to connect with other spouses. Some programming and groups would be scheduled during work hours, preventing opportunities to work outside the home and vice versa. In many ways, programs that families needed were dependent on them to exist, facing budget cuts if not utilized. With no real proof of whether a spouse's lack of participation could negatively affect her husband's career, underlying anxiety continued to grow and the family tree narrative was unchanged.

With Vietnam becoming a distant memory and the Cold War ending around 1991, wartime felt more than a generation away. With no major war to help recruit, ads spoke of the military being a "family" for those with a special calling. In return for serving your country, the military would provide for your family almost like a parent.

The country by this time was examining its own social welfare programs and concerns were building that the military was a military welfare state. Instead of attracting the skillset needed to build the force, it instead attracted families looking for benefits like childcare, health care, education, and housing. There was also now vocalized concern that families were becoming more dependent on the military, and spouses, specifically, were becoming entitled and demanding more support that was already expensive.

Motivated to address the growing dependency and protect itself, the military changed its messaging away from family and instead toward families becoming more self-reliant. The Army even changed its official motto from a promise to "take care of its own" to "an awkwardly worded, half-hearted pledge to 'take care of its own so that they can learn to take care of themselves.'"[212] As another way to distance the military as a sole source of provision, family support programs such as housing, childcare, and healthcare were further privatized and handed over to external contractors and businesses for oversight and management.

Within a decade, the military successfully rerouted and outsourced as much "military welfare" programming as possible while still touting the message of providing all the enticing benefits a family could need. In other words, the military still recruited with the promise to take care of its own, yet much of its side of the relationship was contracted out to the private sector. This is an incredible shift in the arrangement between the military and families, one that was not clearly communicated. In essence, it not only shifted responsibility away from the military, but it distanced the military from very real stressors the institution's needs created. Complaints around childcare would be directed through the chain of command but handled by the privatized childcare contractor. Housing was now maintained by privatized housing agencies. Traumatic brain injuries from missions the military organized were now addressed through TRICARE's privatized insurance companies. And stressors of the lifestyle were handled eventually by DoD contracts to address family mental health services.

While there is nothing innately wrong with the military's shift toward privatization, it was never fully communicated to families that the relationship was changing. For most families who were not tracking the subtle shifts of messaging around responsibility, it seemed like the military was offering more support and investment than generations before ever had. As privatized benefits grew, the military became an enticing lifestyle that created a way for both service members to serve their country and spouses the privilege of staying home to raise their children. Even into the '90s, boomer spouses who did not want to work saw the lifestyle as an opportunity to raise their children. The military culture, in-kind, continued to reward volunteer service with recognition for dedication.

For spouses who dedicated so much, the DoD seemed fully committed to the relationship. When there were problems with the military's privatized programs, families took the problems to military leaders expecting leadership to take responsibility and respond with equal or more investment. However, the blame shifted to the entities actually in charge. It seemed the answer to "Where is this relationship going?" was met with a pause and then, "It's complicated."

I recently read an article titled "Emotionally Unavailable Partners and The Highly Sensitive Person" that describes the characteristics of these two personalities in a relationship. It described the emotionally unavailable partner as one who can be "charming, engaging, and make you feel like they are committed. They may even be physically available at all times. But as time goes by and as the relationship deepens, something in you feels lonely, dissatisfied, and you are not sure why." Highly sensitive people in long-term relationships with emotionally unavailable partners can feel like they are somehow "too dramatic," "immature," "needy," or "too much." Under this description was a quiz outlining ways to know if you are in a relationship with an emotionally unavailable partner. Here are some of the statements:

- When you bring up a disagreement or raise a potential conflict, an emotionally unavailable partner may distance themselves, withdraw, or counter-attack, rather than connect with you to resolve the conflict.
- They make jokes about how "crazy" or "too sensitive" you are. They may even suggest you should seek professional help for being "too emotional."
- An emotionally unavailable partner may intellectualize a lot. When you talk about something intimate or express a deep feeling, they do not give a personal response but a quote from a theory, a book, or a famous saying from someone else.
- They may try to make you feel guilty for wanting more emotional connection than they are willing or able to give.
- They are passive and withdrawn. When you seek more reactions, such as asking them how they feel about what you had said, they withdraw further and refuse to communicate any further.
- When you share something in more depth, they seem to check out or have to distract themselves. You have to ask yourself: are they there with you, or are they waiting for you to finish?
- They try to make up for the lack of emotional intimacy by showering you with physical attention (e.g., elaborate dates, expensive gifts). When you express feeling emotionally alone, they may blame you for being demanding or ungrateful.
- Instead of joining with or matching your emotional intensity and excitement, they try to tone it down. For example, they may ask you to "chill out" when you are sad or anxious, or ask you to "calm down" when you are excited.
- They hardly ever respond enthusiastically to your ideas and ventures. When you bring a new idea to them, they act as a critic, albeit a well-meaning one, rather than an equal who joins your enthusiasm as a partner.[213]

By the '90s, military spouses were being labeled as needy, overly sensitive, and entitled. They were criticized for constantly asking for more support or programs. On top of that, they existed within a culture that continued to teach them to behave in ways that reflected their service member's rank. From the spouse's perspective, the military would seem incredibly available and willing to provide for their every need, only to detach quickly when held accountable.

In the 1990s and 2000s, much of the change toward self-reliance and resilience stemmed from leaders sharing the growing burdens of spouse complaints and deployed service members requesting more support for their families. The military didn't change its willingness to offer support for the challenges. Even employment became a new focus for the DoD during President Obama's White House Joining Forces Initiative in 2011, which advocated for the hiring of more veterans and military spouses. There was an acknowledgment that the military lifestyle contributed complexity to military spouse rights, so the DoD invested more to make that more attainable. Yet employment is still perhaps the biggest challenge for military spouses and the unemployment rate remains high.[214]

While the military has invested a considerable amount of support, it keeps its distance through privatized management. The DoD has a continuing need for the strong backing of families, which keeps it from setting more firm, clear boundaries on the limits of its provision. The answer to "Where is this relationship going?" became a message of, "It's not personal, it's business."

In the words of Meg Ryan's character, Kathleen, in the romantic comedy, *You've Got Mail*, "All that means is that it wasn't personal to you. But it was personal to me. It's personal to a lot of people. And what's so wrong with being personal, anyway? Whatever else anything is, it ought to begin by being personal."

AN EXAMPLE WORTH REMEMBERING
"If they're important to you, they're important to us."

Military spouse Lauren Hope, executive director of Second Service Foundation and a millennial "was a military girlfriend in the height of lengthy deployments (2006-2009). I moved to a new state to spend time with my boyfriend before his first fifteen-month deployment. [I was] new, scared, and unsettled; he did everything he could to make sure I was taken care of while he was gone. Well aware of the rules around family readiness groups being exclusively for military families enrolled in DEERS, he asked the battalion commander's wife, 'May my girlfriend participate in the FRG in any capacity?'

Her response was short, but every bit of the kindness I needed in my life at that moment. 'If she is important to you, she is important to us.' That olive branch built a magnificent foundation for my future as a military spouse. Mindy Newsome offered me a seat at the table—to learn, to understand, and to empower. Such a small interaction that has had such a lasting impact on my life, and guided me as I encounter others. Be kind and be welcoming, you never know what ripple effect your actions may have."

Chapter 10

TRAUMA BOND

ANY SYSTEM THAT WANTS TO CHANGE must take a hard look at the rules and beliefs that support its dysfunctional or destructive pattern. It also helps if everyone in the system agrees it is dysfunctional and destructive. Military spouses often perpetuate expectations of themselves more than the military does.

The military has its own established reasons for keeping officers separate from enlisted. Maintaining a sense of leadership and order is crucial to its success in a variety of settings. On the installation, there are very clear physical examples of this separation. Housing areas for families are organized into separate neighborhoods and, in some locations, separate sides of the installation. Usually, there are neighborhoods set apart for senior officers, filled with nicely maintained and sometimes historic homes that overlook golf courses. Most would be comparable to the middle- to upper-class housing in a civilian community. General officer homes are often the nicest and biggest homes on the installation. Depending on location and the responsibilities of the general, their homes may be maintained by the local military historical society, with the couple given enlisted help to support social and catering events in the home.

While it sounds lovely, according to a study by Henrietta McGowan, officer spouses are more prone to isolation, depression, and anxiety, due to the real or perceived pressures and expectations of the role.[215] As a service member is promoted to higher ranks, this isolation continues to grow because families under the officer's authority are less likely to allow themselves to be vulnerable to the boss. Considering officer spouses have more traditionally taken on the roles of mentor and spouse leader, this separation and isolation from enlisted wives creates a skewed, privileged perspective on the experience of the military culture.

For example, when my husband was commissioned into the Army, he started as a captain. Even at that rank, our family qualified for the new housing being built for officers and NCOs. When we moved to our second assignment, he was a captain, and again, we waited on brand-new housing. While I'm sure the military was also putting in new housing for enlisted families somewhere during that time, I have not seen this to be true in the seven locations where we have lived.

While my experience as a new spouse during the force build-up of the 2000s was impressive, and I considered the military overwhelmingly benevolent, that may not have been the case for enlisted families who were receiving poverty-level income and living in housing that was falling apart. My experience was and continues to be privileged in that our family has comfortably been able to live on one income, and without facing food insecurity, mold-related health issues, lack of childcare, or lack of follow through on government provisions. From the officer spouse mentor perspective, it is really easy to continue messages of "be grateful" and "suck it up" when the life stressors are different and you see the next promotion's housing on your morning walk in the golf course every day.

There are other ways this separation makes it into the spouse culture. Although not regulated, spouses are encouraged to respect rules that protect lower-ranking service members from misuse of power. For example, while the community of spouses is well known to support each other throughout a deployment, it is not appropriate for the spouse of a higher-ranking service member to ask the family of a lower-ranking service member under their leadership to pet sit or perform any other task or favor of similar kind. The idea is that the lower-ranking service member and their family are more likely to agree to the task because it is coming from a superior, and it is harder to say no or differentiate it from a professional order.

A senior spouse who has decided to become an entrepreneur making and selling products, cannot sell that product officially on the installation or to the community without considering regulation or how it may impact their service member's career. According to the current Joint Ethics Regulation (JER): "Personal commercial solicitations by the spouse or other household member of a DoD employee to those who are junior in rank, grade, or position to the DoD employee, may give rise to the appearance that the DoD employee himself is using his public office for private gain. In such circumstances, the DoD employee's supervisor must consult with an ethics counselor and counsel the employee that such activity must be avoided where it may cause actual or perceived partiality or unfairness, involve the actual or apparent use of rank or position for personal gain, or otherwise undermine discipline, morale, or authority."

In other words, if a spouse sells their product and it comes across to someone as solicitation and misuse of their service member's rank of influence, there will be a meeting to evaluate the situation and their service member will be counseled to come home and discourage the activity from continuing. Today, spouses have become creative as entrepreneurs, opening businesses online and reaching customers all over the world. However, there are still very real consequences to spouses' behavior and character in the community.

These are just a few of the many ways the service member's profession can become enmeshed with the personal lives of families living and working in the same community. This is also why spouses tend to easily adopt the regulations and behaviors of the profession even if they are not themselves regulated to do so. Additionally, the expectation that spouses of service members in command should become the senior volunteer spouse who leads support for families has resulted in much negative attention toward some officer spouses "wearing their spouse's rank."

No one likes change, even if they are miserable. Many will choose to stay in the current system rather than face the disruption and dysregulation that will come from trying to change it. Even though the boomer spouses' grassroots efforts of the '80s led the way to some level of change, remember what they were fighting against. They were a younger generation asking for new policies but also challenging an older generation that did not easily see the problems being pointed out and had very much benefited from the system. They were a small but mighty group of change agents trying to shift an institution that had long settled in a comfortable pattern of codependency.

In family systems work, the goal is often to understand the dynamics of a relationship, the healthy and unhealthy patterns of behavior, and then help the individuals within the system develop new and healthier patterns. Trying to control or force the other person to change is aggressive and unproductive. But we do have control over ourselves.

Couples, families, and even large cultural groups can establish unhealthy patterns that can be difficult to change, especially when that pattern provides stability and certainty. For example, a military spouse (we'll call her Laura) came to me for counseling because she wanted to improve communication and connection in her marriage. Her husband was in special operations and he loved to hunt. On top of an already stressful lifestyle of unpredictable deployments, he would come home and want to go hunting. Sometimes, he went straight to the woods rather than coming home first. Laura knew, deep inside, that it was reasonable to ask that he come home to spend family time first before doing what he enjoyed. However, she was taught that he needed time to decompress from whatever he experienced while on deployment. He also expressed that hunting was his happy place. Whenever she would bring it up, an argument on whose needs were more important would ensue. They would hit a stressful stalemate, which only validated his need to go out into the woods.

The pattern was destructive enough to cause anxiety in their marriage, and Laura and her husband wanted it to change. Yet, the pattern also provided a level of certainty. They could count on it like clockwork. When she asked for more investment from him, he got frustrated. The more he was frustrated, the more he wanted to be alone in the woods.

The two went round and round until, eventually, they found themselves trapped in a pattern they created themselves. Although both of them wanted something different, it was much easier to stay in the certainty they had created rather than face the conflict and chaos that a new pattern would cause. In this case, a new pattern would have him sacrifice some of his transition time to serve her and Laura would support his love of hunting with an agreed-upon plan.

Any new pattern will feel like chaos at first. When the boomer spouses of the '80s advocated for change, there was no doubt chaos: uncertainty of whether the military would follow through; the struggle to build, fund, and test new programs; and surely underground push-back from silent generation spouses who saw no problem with the structure and expectations of the military culture.

Three things can happen when faced with the chaos of possible change:

- The chaos is so uncomfortable, so unbearable, that it convinces you to abandon your efforts and return to the old pattern,
- The other person abandons the relationship altogether (this is what most people fear and therefore choose the first path), or
- You persevere through the chaos and the other person adopts a new pattern, one you have started or one you create together.

Working on a new pattern with an individual, couple, or family relationship is one thing. Changing an entire culture is another, especially when only part of the population wants it to change.

In 2004, as my husband was going through Basic Officer Course, I was invited to a conference to join the other new military spouses from his class. The conference's goal was to introduce us to the new world we were about to enter and to learn the acronyms and traditions of the Army. At each of our seats was a binder full of content, organized with colorful tagged categories such as military ball etiquette, ranks, and acronyms.

Neatly placed on top of my binder was an army-green book with a gold eagle on the cover. *The Army Wife Handbook, Updated and Expanded, Second Edition* by Ann Crossley and Carol Keller. Originally printed in 1990 titled *The Army Wife Handbook: A Complete Social Guide*, it was now in its seventh printing. There was an awkward giggle in the room as the two male spouses in the room asked if this book was for them, too.

In fairness and transparency, I appreciated the mentoring and assimilation into what feels like a very foreign culture to a new military spouse. A conference like this is invaluable, especially when you desperately want to respect the deep heritage the military has created and don't want to risk embarrassing yourself with social customs other senior leaders have long adopted. In fact, in 2018, I was asked to lead that same conference for the next generation of spouses coming in and I accepted the challenge immediately as my opportunity to give back.

I also quite liked *The Army Wife Handbook* and still have it today. It has been my go-to for etiquette (albeit vintage now), how to address various ranks of soldiers, and how to host social events, complete with proper invitation announcements, appropriate dress codes, and so much more. It is a gem for long-lost etiquette and social behaviors that still prove valuable. Yet in the preface, Crossley shares the many changes that her generation saw in military family structure, roles of women in combat, and divorce rates leading to second marriages. She states, "I found that these women are caught up in the dichotomy of wanting to reject the traditional Army-wife roles as their 'responsibility,' yet wanting to support the unit and their husbands' struggle to remain competitive in the shrinking Army," which led her to work on the second edition. I thought it was interesting that this message was still being encourage.

The Army Wife Handbook has since been updated again and renamed *The Army Spouse Handbook* adapted by a good friend and mentor Ginger Perkins, wife of retired General David Perkins. The cover is now blue. I recently polled the spouse community on social media and asked them to go on a scavenger hunt for either of the books or any others they knew about. The goal was not to discredit the writing or mentoring of spouses but to evaluate the current messages we are passing down and ask if those messages still fit within the changing culture around us. It was also to check on the subtle influence of the military's endorsement by selling it in the base exchanges. The scavenger hunt results showed Air Force, Marine, and National Guard versions, and almost all spouses reported having their book close by on their bookshelf or coffee table. Some had recently seen them sold on the installations.

Old patterns are hard to change, especially when they are most familiar. Since the pandemic, Gen Z and millennial spouses have voted with their feet. In-person events like coffees, teas, and other social events are minimally attended; most who do attend are Gen X and baby boomers. Military spouse clubs are low in attendance unless they are supporting larger causes that also impact the civilian population. The newest generation is starting a new pattern and is seemingly willing to persevere through the chaos. They will not tolerate a pattern that does not match the progressiveness of the civilian culture they came from.

LOOPING IN SURVIVAL MODE

Psychologist Abraham Maslow's hierarchy of needs, or Maslow's ladder or pyramid, teaches that there are five categories of needs that must be met to reach a person's fullest potential. At this peak state of self-actualization, one has a sense of purpose, knows what brings them peace, and more importantly, has all their other needs met, allowing them to fully embrace that state.

The first four categories are: basic needs (sleep, shelter, food, sex, clothing, air, and water), safety (physical and emotional security, and structure), love and belonging (friendship, intimacy, and sense of connection), and esteem (respect, achievement, recognition, self-esteem, and status). There have been many variations of Maslow's ladder since he published it in 1943, but one of my favorites was one I found years ago when I was preparing for a military spouse event.

Over the year, I had been asked to speak on the same three topics to spouse groups: purpose, identity, and resentment. I wasn't completely surprised that spouses were struggling with these topics. It was around 2015 and the population was weary. Families thought surely by now the global conflicts would be over, but with no end in sight and many (knowingly or unknowingly) struggling with the impact of budget cuts, they had been repressing their own needs for a long time in deference to the mission.

As the internet and social media boomed with spouse group pages and the new wave of dependa cyberbullying, I was beginning to see the resentment leaking through the cracks. In marriages, in forums, and at social events, spouses were desperate for a sense of purpose and wanted to break out of the constraints they had felt for so long. Like a quiet sickness, resentment spread as an undercurrent in a culture that seemed to thrive on the surface.

Their craving for purpose immediately made me think of Maslow's ladder. I thought perhaps it would be a good tool to help them discover what kind of purpose they sought. I came across one in particular that had a phrase that jumped off the page. Under the second rung of the ladder, safety, which included shelter, routine, and structure, there was the phrase "fear of the unknown."

What spouses were asking for decades ago, and still today, was a chance to achieve some sort of actualization. It was a longing as innate as the physiological need for water and air. Their service member was getting the chance to climb the ladder. Although somewhat starting over with every move, once the basic needs of food and shelter were met, service members quickly jumped back into a routine of physical training every morning, meeting their colleagues who all wore the same uniform and shared a mission. Despite shifts in responsibility with each assignment, promotions continued to build confidence and self-esteem.

Spouses, on the other hand, were in a constant insanity loop of the first three levels. With every relocation, they would continue to settle the household, help children establish their routine at home and school, and maybe by the six-month mark consider finding a spouse group to join on the installation. They lived in a constant state of instability and lack of structure. Once they were settled, they quickly began to anticipate disruption again. The fear of uncertainty is where they lived most of the time.

Meeting their need for love and belonging was often the category they were encouraged to jump into first, rather than waiting until the other needs were met. Often, opportunities to volunteer quickly were offered as the answer to thriving in the community. In some cases, depending on the rank and responsibilities of their service member, they walked into spouse leadership roles almost immediately. For those wanting to work, it was nearly impossible to consider a need for purpose outside the home until everyone else was farther up the ladder. When or if they did find a job, it was soon time to move again.

Those who struggled with childcare, overpriced or inflated housing markets, mold in housing, or even food insecurity, rarely ever made it to the second level of Maslow's ladder. Meanwhile, most spouse leaders were well within reach of actualization themselves. Spouses confidently and fiercely advocated for issues that affected their children or the whole family unit. However, as I traveled around and presented Maslow's ladder to spouse groups and leaders, I found that spouses could not identify their own needs beyond sleep and a break from their children. When challenged to think past that, they would offer blank stares of "I have no idea what I need" or shrug their shoulders and say, "I guess a spa day?"

When it came to couples issues, I noticed that many spouses were fighting resentment toward their service members. They saw them leave for work every day doing something they loved and growing personally and professionally. As military spouses reached burnout and limited opportunities, service members seemed to get away, have time with colleagues or friends, alone time without children, have food prepared for them, and ample time to go to the gym during deployments, especially unaccompanied tours like Korea. This, of course, wasn't an entirely realistic view of deployments considering the service members felt deeply conflicted about their time away from family and some deployments were quite dangerous or traumatic. However, the military spouses' perspective was not entirely untrue. What caused even more division was that they rarely articulated any of this to their service members or external family members because of how selfish it sounded.

At home, the cultural message of "don't upset your service member" was hard to shake when the service member returned home. Many spouses felt torn between wanting to give their service members time to rest and reintegrate and their own desperate need of respite. Their resentment and jealousy only grew if their spouse played long hours of video games, trained extra hours in the gym, or took up hobbies that pulled them away from the family. Many counseling sessions were spent teaching communication, healthy boundaries, and identifying one's own needs for rest, achievement, and purpose.

Knowing that spouses rarely shared honest feedback on how they were doing publicly, I conducted an anonymous survey in 2018 to ask some pointed questions about how spouses felt they were doing and where they felt a sense of purpose. Spouses were invited from a range of social media groups and represented all branches. Almost eight hundred spouses took the survey and some of the results were shocking. A majority of spouses stated that their favorite part of the culture was the camaraderie of the community and the places they were able to see and travel to as part of the military. The biggest challenges contradicted their positive remarks, with a majority expressing deep loneliness and frequent relocations disrupting relationships and jobs.

I also asked, "Is there anything you wish you could say about the military lifestyle, but feel you can't because of how it might sound or be interpreted by your spouse or others?" A few comments stood out:

- This isn't what I imagined our life would be like.
- There's a lot of expectation for me to bend to accommodate my husband's schedule and professional goals and barely any room at all to make my dreams a reality. Any professional progress for me has to be made in consideration of his career first.
- I wish he realized and acknowledged that our PCS experiences are very different. A move gives him an instant peer group, an instant job (that someone else sets up for him to step into), a sponsor to tell him all the things he needs to know locally, an instant doctor … and on and on.
- It's fucking hard.
- It is both stressful and rewarding; it can very easily be an emotional roller coaster.
- Being completely honest, I wish my husband wasn't in the military but I'm not sure we would be together if his life took another path.

Also interesting was the question of whether they felt they had a sense of purpose and identity outside their role as a supporting spouse. Almost 64 percent said yes, however when I asked them to describe their purpose, almost everyone listed their role as a spouse or parent. The only exceptions were those who had a career or were giving back to the community as a volunteer.

What concerned me the most about this was that they defined themselves first by the role they played for others. For those who were parents, this was going to be challenged when their children left the home as adults. Finally, and most disturbing, was the following question I hoped would get past the trained culturally positive answers: "Answer this question with the first thing that comes to your mind: How are you really doing?"

While many said "okay" and "pretty good" an alarming number responded with a negative answer. Here are just a few of the 751 answers that stood out:

- Swimming in the deep end without a life preserver.
- I just feel alone and worn down. After almost thirteen years of this lifestyle, I feel alone and disconnected from the outside world more than ever. I'm getting tired. I want to be a normal family with friends and family nearby, gatherings, support, fun, etc. and instead I'm just here ... gearing up for my husband to go on another deployment after only being home for four months after those two years.
- I no longer have an identity.
- Lonely. I am so very lonely at this duty station. We have only been here a few months and I just haven't been able to settle in here. Fear of making good roots here because I know we are leaving in a few months for another assignment.

After several years of helping spouses find value in themselves and identify where they were on Maslow's ladder, I noticed another issue. Spouses, especially those who had been in the military lifestyle for some time, lacked not only the ability to identify what they needed for self-care, but they could not actually ask for what they needed.

TRAINED PASSIVITY

I decided to shift my teaching to focus on healthy communication and assertiveness. My favorite definition of assertiveness is this: I have the right to my thoughts, feelings, and perspective. I also have the right to ask someone else to change their behavior if it is destructive to me. I can ask for what I need by honoring my rights and yours in kindness.

Every human has these basic rights. If you met a child who was bullied at school, you wouldn't blame them for having strong feelings about being bullied. You would also likely want to encourage them to detach or stand up to those who were harming them. Yet as we develop strong attachments in relationships, those basic rights and boundaries seem to get blurred.

For most children and adults, unhealthy approaches are reinforced: passive, passive-aggressive, and aggressive communication. Consider these definitions as a point of reference:

▷ *Passive*: Allowing your rights to be violated by failing to express honest feelings, thoughts, and beliefs, or apologetically expressing your thoughts and feelings in a way that others can easily disregard.
▷ *Passive-aggressive*: Indirectly expressing aggressive feelings through passive resistance, rather than openly confronting an issue.
▷ *Aggressive*: Communicating in a demanding, abrasive, or hostile way. It is insensitive to others' rights, feelings, and beliefs. The usual goal of aggression is domination, forcing the other person to lose.

Most people can identify which form of unhealthy communication they lean toward, especially when hurt, angry, or defensive. Interestingly, almost every group of spouses I worked with, regardless of age, branch, or generation, leaned toward passivity. When we covered assertiveness as a healthier way to communicate, most saw it as a form of aggression. But it's not.

▷ *Assertive* communication is the ability to communicate your thoughts, feelings, opinions, and in some cases, request for change directly with kindness. It does not violate the other person's right to their feelings or opinions or force them to change. Assertiveness speaks for yourself while valuing the other person.

Spouses, though, when offered an opportunity to practice identifying what they needed, even when it was a need that only they could provide for themselves, thought it was asking too much. If they needed someone else's support (like their spouse) to help them meet a need, unless it was a basic need for their children or family as a whole, they struggled to do so without expressing it in passive or passive-aggressive ways.

I discovered that the spouse community had been habitually trained over time to become more passive, regardless of how they entered the culture. Spouses who were passive by nature were moldable, slipping easily into the cultural norms of putting the needs of the military first. Spouses who were aggressive by nature, perhaps young and immature, or even bold and confident as a working spouse, were likely mentored by senior spouses in a more submissive direction or ostracized from the spouse community until showing otherwise.

Consider the following phrases current military spouses shared when asked about messages they remember most from mentors and spouse leaders:

"You are 'voluntold' to be there."

"Semper Gumby (always flexible)."

"The military always comes first; your marriage will always come second."

"Forget your career, you have a soldier to support."

"The one thing we have in common in this room is that we all love and are loved by a selfless courageous human being."

That last one is hard to read, as it brings up another topic of hero worship and the assumption that the service member's oath to service is always a reflection of their values and character. However, on top of the messages that remind spouses to anticipate their human needs will be unmet or secondary to the military's needs, there are the overt messages of expected strength and resilience.

"Military spouses are some of the most resilient humans on the planet."

"Suck it up!"

"You knew what you were getting into."

"Our military families are strong and resilient."

"Oh, you think this is hard? Wait until [example of a difficult situation]."

"Bloom where you are planted."

"Just another opportunity for resilience!"

Resilience is a word that has become demeaning to spouses after being overused as the military's top description of families. The word resilient was adopted by the military to continue the self-reliant message of the '90s. In 1996, Richard Tedeschi and Lawrence Calhoun coined the term post-traumatic growth, to describe "the experience of positive changes that occur as the result of the struggle with major life crises."[216] Over the next three decades, the military would use the word resilience in its policy and programming as a goal for families. RAND, a research think tank often contracted by the military, published works on how to build and enhance resilience within military families.

The intention in sharing these statements from the spouse community is not to criticize the well-meaning, overworked, volunteer senior spouses who made those statements. The military lifestyle, especially during the last twenty years, has been challenging enough to burn out almost every generation of military spouses. Without encouragement, coaching, a positive community that understands, and even tough love, spouses would crumble under the weight of the responsibility that has been placed on them.

Yet, even I am aware as I write this that I have a twinge of fear, a senior spouse voice in the back of my mind reminding me that "we should not point out problems without solutions." My good friend and mentor would say, "Don't complain about it unless you are willing to do something to fix it." However, the clinician and advocate in me sees a perpetual pattern of behavior and messages from both the military and our community that are clearly not helping families fully thrive.

IS THE MILITARY CULTURE A WELFARE STATE?

Many scholars and experts in policy have made the point that the military system, in an effort to protect itself and maintain a state of readiness, created a military welfare state. The argument is that civilian social welfare programs are designed to be available only when the normal supply of the family, market and economic resources breaks down. The military, on the other hand, provides significantly more (housing, education, retirement, healthcare, childcare, etc.) as an expected right of being a part of the profession and community. In addition, a substantial percentage of the defense budget goes into these benefits, especially healthcare and retirement, pulling funding from federal spending that would support civilian welfare.

Other than concerns around policy and funding, experts in economics and sociology point out similarities between social welfare and the military system. They contend that the military's benefits of healthcare and education attract low-income individuals and families, offering more than the civilian sector can provide, and jobs that rarely translate back. For spouses, any attempt at self-actualization (employment, outside interests, etc.) requires more support than the military is able or willing to offer, and potential problems are created for the family. For example, the frequently changing schedule of an active-duty service member requires a level of flexibility that few employers are willing to consider. Service members often do not have the same flexibility in their jobs, even when home. Add to that the likelihood of relocation, and many businesses do not see military spouses as a positive investment, regardless of skill or talent.

After years of working with Laura and her service member, more issues surfaced. The stress and conflict in their marriage were not only about hunting, but the husband's use of alcohol, questionable relationships with other women, isolating his wife from the support of other military spouses, and suicidal threats during arguments. The husband abandoned therapy, and Laura and I switched to focusing on her well-being and safety. It took more than a year to work with her on the plan to move with her child to her parent's house.

The issue was not whether she felt it was the right thing to do. She was ready for change even though she deeply hoped her husband would pivot toward fighting to save his family. The biggest concern she and I both had was her starting over. The special operations community provides a bit more stability from the frequent moves of the conventional forces, but she had no work experience and no other medical benefits.

Even though the DoD made significant progress in advocating for businesses to commit to hiring more military spouses, very few spouses sought jobs or used the government's career coaching and resume writing programs. Why? When I polled the general population of military spouses, many feared the stigma of employers being hesitant to hire or feared employers could not offer the flexibility they needed to solo-parent. It was one thing to ask for grace during a deployment, but the constant need for them to be available for frequent interruptions during the day for child-related issues on top of the everyday responsibilities of running a household seemed unfair to the employer. Another concern was that the government-provided childcare centers were backed up with extensive waitlists and no relief in sight.

As an active-duty military spouse, trying to build and maintain a career has been incredibly challenging and exhausting. Only two moves in, I had already fought two states to obtain licenses to practice counseling. I argued that my volunteer counseling hours mattered just as much as paid hours. Considering their board rules did not mention payment as a factor, loopholes like this helped my advocacy. I would have likely quit had the board not finally relented. If I had known then that I would fight three more states and several insurance companies (including TRICARE, the military's insurance provider) over the next decade, I would have given up on my career and chosen something quiet like watering plants at Home Depot. Credentialing with an insurance company and transitioning my license would have been needed even if I had worked for an employer, but this was my attempt at a more flexible entrepreneur track.

For many, it is simply too discouraging to start over again and again, ever looping on the bottom three rungs of Maslow's ladder. Eventually, spouses begin to passively resolve themselves to do well with what they've been given rather than want something different. Many are overqualified for the jobs they end up working, and some opt to give up on the idea of a career. While the military would say they are being self-reliant and making an individualized choice, others would argue they have little choice at all.

One spouse shared her frustration with her lack of employment opportunities even after she had used all the DoD spouse employment programming available to her: "Optimism is necessary, but there is a ... rose-colored glasses level of delusion when it comes to military spouses. In the same five-minute discussion about an average $200,000 loss of lifetime earnings in twenty years, constant uncertainty, 20-25 percent unemployment rate, and lack of opportunities to build savings/retirement, they will say it's an honor to be a milspouse. I'm always confused as to what's great about being in those statistics or how moving to Killeen, Texas, was a rare travel opportunity I should feel blessed to experience. It's valid to say this experience sucks. Not everyone is content with staying home, not utilizing their skills to the fullest, and living in undesirable areas. ...

"A similar level of unchecked optimism goes for [service organizations and DoD programs] that say they had a successful year ... yet nothing has changed for military spouses when it comes to employment for the last decade. Are you successful if nothing is changing? ...

"Each support resource I went to tried to instill in me that I needed to volunteer in order to be capable of having a job, as if I didn't work before. They don't take no for an answer. Community service should never be about personal gain; it should be about a passion and desire to help without expecting something in return. ... I have never seen veterans with a career gap as long or longer than mine be told they need to volunteer to receive access to job placement in group programs. There's just something about telling women to do free labor in order to access jobs that rubs me the wrong way."[217]

According to experts like Jennifer Mittelstadt, a political historian and author of *The Rise of the Military Welfare State*, what differentiates the military from civilian social welfare to an alarming degree are the norms and values of a culture that expects an almost unattainable level of loyalty and order to maintain these benevolent resources. Right, wrong, or indifferent, the military continues to reward and promote the mentally fit military couple who are team-minded in the use of their talents to benefit the goals of the military establishment. Cracks in wellness and loyalty to the cause create a vulnerability to the service member's career. If a spouse struggles at home to provide stability or comes across as too needy, it is not uncommon to hear the message from leadership, "Get your household in order" or "Your marriage is not the military's problem." It has been so ingrained in the culture that the directions to the promotion board now specify that the board will not consider the spouse's volunteer service when assessing the service member for promotion.

Stephanie and Darren are a high-functioning military couple who have worked hard to put their marriage first when possible. They regularly include exercise and spirituality in their life as positive coping tools and shared interests. Darren switched to special operations a few years ago to stabilize his family, although deployments were likely to be more frequent. Just weeks after Stephanie had their fourth child, Darren deployed. Stephanie, despite employing her coping strategies, was having a difficult time sleeping. The adrenaline needed to function during the day constantly flooded her system, making her nauseated during the day and sleepless at night. She tried working out so she would be more tired, eating right, and even avoiding naps to try to get a good night's sleep. After weeks of not sleeping well, she asked for help from the community and was not able to get the help she needed. In many ways, what she needed was practical support around the house so she could reduce her responsibilities. After no luck finding help or support, she went to counseling.

Feeling trapped in her circumstances, Stephanie told her counselor that she did not feel well and was worried about damage to her brain from the lack of sleep. The clinician encouraged her to seek medical help. While Stephanie was able to find a family member to travel in the next day to help, she had to find her own childcare that day before she could take herself to the medical center.

When Darren heard what was going on back at home, his leadership's response was, "You can come home but you will have to turn around in five months and do another six-month deployment." He was also told, "Your wife doesn't have a DoD number, you do. So that's not my problem, that's your problem."

Hearing this story shook me. Not only because I know this couple well, but because this was a couple who knew all the right things to do to be self-reliant and resilient. Yet the help the family needed to support the mission wasn't there. Thankfully, Stephanie took the courageous step to put her health first.

Starting in 2018, the DoD started to track and publicize military spouse suicides. In 2019, the rate was around 8 percent. In 2021, a clinical group set out to assess the military spouse perspective on suicide and the military mental health resources available to assist spouses. The results found that spouses experienced a "loss of control and loss of identity" within the culture and struggled with the fear and stigma of getting help. They also had difficulty finding culturally competent providers. One participant shared, "We've had situations where wives were struggling, but ... he couldn't get off that day, he had to report in because she's not at the hospital ... it's not serious."[218]

What does this do, then, to a group of people who depend on an institution for their needs (including basic survival, safety, love, belonging, and esteem) only to have that same institution implicitly detach when they need help the most?

In 2011, I was working with a military spouse who was in a marriage that had turned physically abusive. Sarah (not her real name) was not in denial that the relationship was abusive, but like other victims, found herself on a loop about actually leaving. We had extensively covered the importance of a "go bag," shelter, and a separate checking account that would provide financial security, but the emotional and sometimes spiritual struggle of leaving a relationship she had invested so much in was the hardest to work through.

Like Laura and others I had worked with, Sarah wanted her relationship to survive. She loved her husband and he "would always come back around" to the sweet, gift-giving, apologetic version of himself she had fallen in love with. Also adding to the complexity were their children. They were in a good school on the installation and had just gone through a deployment with their dad gone for nine months. How terrible would it be to take them from their dad now?

For weeks, Sarah and I went over her emergency plan while wrestling through her feelings of guilt and grief that her marriage had an extremely slim chance of survival. One day, she said something that has haunted me since, "Where am I going to go?" Sarah started, "I have no job, will have no benefits, no work history for a resume. I would have to start all over from scratch. Even worse, no one will understand. He is a hero to everyone else. Leaving a soldier, especially one who has gone to war, is literally the most 'unpatriotic' thing you could ever do. I feel completely and totally trapped."

Sarah left therapy and the last I heard she stayed in the marriage.

Trapped is a word commonly heard in relationships that involve power, control, and cyclical patterns. But Sarah wasn't just talking about her marriage, she was asking me to hear it in the context of a military marriage. I had also never thought about how the civilian community would respond to a military spouse leaving her service member and lifestyle, seeing her as "unpatriotic" or "less than" due to hero worship.

Please hear me say that the military and DoD are not abusive. While service members may feel "trapped" into a commitment if they are not happy, spouses are civilians and are not obligated or forced to get involved. This is likely why many younger generations are choosing to set boundaries by living away from the lifestyle and considering it a job like any other.

What Sarah was pointing out was that there was a complicated relationship between spouses and the military that makes it very difficult to do anything but stay. We hear of families approaching retirement with goals for a second career, only to struggle in every way to transition out. Recently a good friend whose husband retired as a colonel said it felt like leaving a cult with the amount of 'rewiring" they've had to do in retirement. Another veteran told me he struggled to translate his military job into the civilian world even after getting a master's degree in criminal justice. Additionally, I've worked with many military spouses who grieved leaving the lifestyle more than their veterans did because it was the only life they knew.

Part of the difficulty that some face in leaving involves the stark difference between the tempo and demand of military life and the civilian world. While many service members were lost in those two decades, many more came home with trauma. Yet, how do we measure the trauma spouses have endured with minds, hearts, and bodies after living in a constant state of stress for so long?

Arin Yoon documented her experience as a military spouse through photography. In her *National Geographic* article, "A military spouse reflects on life over two decades of war—and what comes next" she writes: "The military does offer some classes to prepare spouses for military life. There's even a guide called the 'blue book,' but it's more about how to throw a luncheon than how to handle secondary trauma. Military families shoulder so many burdens—from the ever-looming threat of a loved one being killed or injured to substance abuse and mental health crises. It can be incredibly hard. ... For the military community, the fact that the war is over doesn't mean they get to return to civilian life. They're still doing war fighting exercises and the training that can save service members' lives. They—and their family members—are always in a state of preparation for war. We can never say, 'Ok, it's over.' For us, the question is: 'What's next?'"[219]

I don't know if anyone could say this tribe hasn't been traumatized by the longest war in US history. And I would argue that Arin's (and any other spouses') trauma isn't secondary. It is their trauma, their experience over however many years of war-connected stress from home. Theirs is also a primary trauma.

For spouses, the trauma is connected to war, but more to the demands of war on their life and home. It involves their relationship with the DoD, which holds incredible power and influence in and over their lives even though it is also a benevolent provider. It is not to be blamed, and yet, it can give and take away, ask and not provide in return. It holds the power in the relationship.

Trauma bonding can be explained with the following metaphor. You are in the desert and the person you are with is dragging you along through the sand and heat. You are parched, starving, and hurt from being dragged along the sand. If that same person who has been starving you and dragging you along the sand then hands you water and nurses your injuries, they become your salvation. You look at them differently. You now need them even though their behavior is hurting you. It is a trauma bond that is difficult to break.

What I found in my search for the authentic truth beneath the surface of the military spouse iceberg is a community largely grateful for the services the military does not have to provide but does. From bowling alleys to theaters, the Morale, Welfare, and Recreation (MWR) services make sure families have access to affordable events and activities. There are at least four different ways families have access to counseling or coaching services for every area of life. There are programs created to support special needs issues as well as sports teams and piano lessons for kids. Just recently, the secretary of defense announced initiatives to make food more affordable in commissaries, raise housing allowances to help with inflation, and pilot fellowships for spouses to address the employment problem. There are so many resources and programs to address every area of life that DoD leaders often wonder why they are underutilized or why families are still unaware of them.

It can be truly overwhelming (in the best way) to follow your spouse into a career that also offers an entire culture designed for your success. Everything you could possibly need on your worst day of deployment and solo parenting is available to you within a short drive and practically at no cost to your family. Even better, you don't have to earn any of it. In fact, you are regularly reminded that as a spouse you serve your country too and the military has an obligation to take care of you. After a while, it is difficult to not adopt those messages that remind families how great they have it and that there are proper ways to point out any problems.

But as I saw with Stephanie and many other spouses of service members from all ranks, concern for a spouse's well-being is not limited only to relationships with dysfunctional patterns or abuse. Even spouses with healthy marriages find it difficult to thrive, even with access to DoD resources.

Further, most advocacy stems from deep concerns around subpar entitlements spouses were promised and depend on. As I look back throughout the generations, I am inspired by and grateful for spouses and military leaders who listened. However, I also feel disheartened as I read about issues brought up in family forums of the '80s because they read exactly like today's advocacy.

Up until 2020, the primary advocacy topics concerning military spouses and families centered on spouse employment and childcare. In 2021, food insecurity emerged as a top concern. While not a new topic, we have now data on just how many are struggling. According to MFAN's 2021 research:

▷ Nearly 59 percent of families reported "moderate to poor health"
▷ 23 percent of enlisted families indicated food insecurity, roughly five times the number of officer families
▷ 70 percent of enlisted families that identified food insecurity found federal assistance like SNAP and WIC helpful[220]

Over the last few years, I have personally heard stories of spouses who have eaten ice rather than dinner so they could afford to feed their children. The economic market is inflated, and housing prices off installations are so high that families cannot afford housing. There is insufficient housing on the installations, and many still have mold. The recommendations of most nonprofit organizations and even First Lady Jill Biden sound similar to what was requested almost forty years ago:

▷ Increase the availability of childcare
▷ Address BAH for inflated housing costs
▷ Raise service members' pay
▷ Address the low availability of health and mental health providers
▷ Reconsider how frequently military families are relocated

The last two decades have inarguably been tumultuous for military families. Many families who joined since 9/11 realize they joined an institution that supports a never-ending war-deterrent timeline rather than one that fluctuates between war and peace. They were initially offered benefits, bonuses, and social programs better than the civilian sector. Their families gained an all-inclusive community, complete with housing, childcare, schools, and even future education they could gift their children. What about the worry and disruption of relocation? Most recruiters answered by saying the average move was every three years and that the community the military provides helps ease the stressors of those moves for a family. In reality, unless you are in a specialized field like special operations, medicine, or another profession that is based in a limited number of locations, the average move cycle begins at two and a half years and increases in frequency as the service member is promoted.

Families, especially enlisted, quickly discovered that housing and income were subpar, many qualifying for food stamps. Mold and water contamination on the installations were also shocking experiences. Imagine the frustration that new millennial and Gen Z military spouses experienced when encountering weary and resentful Gen X and boomer spouses quitting volunteer positions to address the deterioration of wellness in their own families.

The operations tempo offered little rest, and the pandemic demanded even more from military families. As the world returned to a new normal, the support system the military created did not. Some of the privatized programs struggled to come back to full strength, and spouses reevaluated where to put their energy.

With the 2021 withdrawal from Afghanistan, spouses faced a task they had not anticipated. They had spent the better part of twenty years supporting their service members, families, and the military institution for a war everyone was determined to win. While veterans across the country revisited their traumatic experiences from Afghanistan as it fell back to the Taliban, the nation vocalized that it was our next Vietnam. Already-weary spouses faced the overwhelming thought of having to revisit combat trauma with their spouse that they thought had already been processed.

So why don't families leave? For spouses, much of that is contingent on the years their service member has committed to the military and the "up and out" progression of the military career. If the family can hold on, the next assignment usually offers additional incentives, possibly exciting locations, better housing, and, if you're one of the lucky few, you might just get a view of the golf course.

AN EXAMPLE WORTH REMEMBERING
Empathy in the most vulnerable moments is unforgettable.

Millennial Aimee Selix, 2023 AFI Space Force Spouse of the Year, shares: "In March 2019, my husband was called for short notice deployment. This was a devastating blow to our family as we had been told that chances of him deploying anytime soon were slim. Part of the previous summer was spent separated as he was in weapons school, so as you can imagine this was a hard topic to help our then eleven and seven year olds to not only understand but to accept. In a matter of weeks we went from having family vacation plans to a quarter of our household being separated once again. I know these things aren't uncommon in the military but it doesn't change the fact that they take a toll on families, especially ones with young kiddos who don't quite understand how these things work. Hearing the news, a fellow spouse and friend, now one considered to be more like family, suggested we make new summer plans and visit them. I didn't think she was serious but she insisted. So, I planned the trip from Colorado to Washington, DC, loaded up my boys and my mom and took off.

"During our stay with this family, my boys were able to connect with another military family who got it. This family ... included them in their midnight waffle tradition, which my boys started upon returning home, and they connected with them on such a level that it helped them understand they aren't alone. All of this helped me immensely. This spouse has always been there for us. She has shown us over and over that she cares about people, a true testament to a great leader. Over the course of many years, she has driven me to pay her kindness forward. Little did I know, taking her up on that offer would not only help my boys but that it would be the ignition to finding my passion and purpose. It is said that you never forget the way a person once made you feel. This person has always made me feel valued, appreciated, included, and loved. And it has made all the difference."

SOCIAL SHIFT

"The most important language in the world
is the language of the person you're trying to speak with."

—Colonel Jeffery Peterson, USMC (Ret.)

Chapter 11
SOCIAL MEDIA & COMMUNICATION

BY 2018, MILITARY CULTURE WAS FRAGMENTING under the weight of intense operations tempo and the community moving more online. Social media and technology deeply influenced the larger American culture. Changes were especially noticeable within a military community that was geographically separated from external family and friends and largely dependent on other military families for support. Examining the impact of social media on how military communication shifted, as well as how it influenced core generational motivations and work ethic, provides insight into some of the generational differences we are seeing in the military work environment and family culture.

While social media has changed the way we influence others into action, online communities, social media platforms, and personal websites have successfully reshaped our understanding of marketing and outreach and, unfortunately, become psychological weapons.

To simplify and understand how this shaped the military culture during a vulnerable time, I will make some generalizations about generations. Keep in mind, however, that true connection to a particular generation depends on many variables, not just birth year. Access to information and community shifted the way the military culture innovated, solved problems, and possibly inherited new ones.

TECHNOLOGY SHIFTS FAMILIES FURTHER INWARD

As early 1970s games and entertainment moved from brick-and-mortar businesses in the community to inside the home, families shifted their time further inside. Gen X kids who had spent hours a day in front of the television, movies, and video games were already being labeled anti-social and addicted to screens. *The Oregon Trail* video game was released in 1971, transforming the Macintosh computer into a tool for entertainment as much as it was for productivity. In 1972, *Pong* by Atari turned a television set into a personalized, quarter-free arcade. Additionally, movies were no longer chained to theaters. Families could wait for their favorite movies to go on VHS and rent them at the local video store.

Parental concerns for Gen X's safety had already made neighborhoods and strangers feel less safe, resulting in fewer kids playing outside unsupervised. With Nintendo Entertainment System (NES) licensing to third-party developers and distributors by 1983, the home video game industry took off. In 1989, Nintendo released the handheld Game Boy, making video games portable. Cassette tapes had already made music portable as well. When the Sony Walkman came out in 1980, anyone who could afford it could listen to their favorite music with the privacy of a handheld cassette player and headphones. The first portable mobile phone came out in 1984 by Motorola and was nicknamed The Brick. It had a battery life of thirty minutes and cost a whopping $4,000 (adjusted for inflation, this would be more than $11,000 today).[221]

The World Wide Web went public in 1991, and email gained popularity in the 1990s with platforms like America Online (AOL), Yahoo, and Hotmail. In 1998, President Bill Clinton's sex scandal with White House intern Monica Lewinsky changed the internet forever. News updates and privately recorded tapes involving Lewinsky were posted online, creating an internet storm of comments, shares, and what she called the "first moment of truly 'social media.'"[222] By 1999, the internet had grown to more than four hundred million users.[223]

By the late '90s, parents and experts were concerned about the long-term mental and physical health risks associated with screen time, and the oldest millennials were just beginning to hit their adolescent years. Called Gen Y at the time, it was their parents who were on the front wave of discovering what the internet could do. When planes hit the World Trade Center towers in 2001, there was no social media or YouTube live-streaming; most of the world still relied on traditional news sources. As people went to the internet to get more information, media websites like CNN and *The New York Times* removed photos and graphics to reduce loading time and strain on the servers.[224] Meanwhile, the BBC took the news to other parts of the world.

The aftermath of 9/11 shaped the way people communicated and stayed in touch. According to a Pew Research Center study conducted a year after 9/11, Americans reached out via email to family and friends they had not spoken to in years, and 83 percent had maintained those relationships a year later. Websites allowing comments provided "a way to connect emotionally with a virtual community whose ties were not geographic, but bounded by common experience."[225]

In 2002, the Blackberry launched. It was a mobile smartphone that featured newly-available 3G cellular service, wifi, calendar, email, internet browsing, and a front-facing camera.

Facebook Shifts Online Connection

In 2003, Myspace became the first platform dedicated to personalized public profiles. It allowed users to connect with the profiles of friends and family as well as long-lost friends and acquaintances. Meanwhile, Mark Zuckerberg, a millennial and Harvard student, launched Facemash in 2003, which would eventually become Facebook in 2005. With the rise of Facebook, Myspace peaked in 2005 with more than twenty-five million users before it began to decline. Also in 2005, LinkedIn launched as the internet's solution for business connections. YouTube launched as well before fading into the background to prepare for its comeback in 2009. Twitter joined the space in 2006. In these beginning years of social media, millennials were introduced to the online world during their most formative years and are considered digital immigrants in that many of them still remember a time before the internet and social media.

In 2007, Apple launched the iPhone with the first ever touch screen and apps that detached social media from the desktop, as well as an in-phone camera and the ability to text. In 2008, Facebook surged ahead of Myspace with thirty million users. By that time, 55 percent of millennial teens were on at least one social media platform.[226]

Facebook's relationship labels, such as "friending" a person or setting an "in a relationship" status, introduced new ways to communicate how relationships began, evolved, and ended. As teens (and adults) built public profiles and began to build their "friend network," the "like" feature, introduced in 2009, ultimately altered the way people, especially adolescents, saw their real likability and status.

The rise of social media introduced an entirely new way to connect families with service members who had internet access during deployment and training. With photos and videos uploaded to profiles and the ability to share posts, friends, family, and acquaintances could follow, like, and comment.

The Blackberry and iPhone introduced the ability to access almost everything the landline phone and desktop could do from home. It also opened up the ability to work remotely. With access to email, calendars, and more, it became more difficult to turn work off after hours. One Marine shared his memory of how smartphones changed the work-life balance. "Work never ends," he said. "I could be reached by colleagues and notified of even the most minor things after work hours." His wife echoed that frustration. "I remember the shift," she said, "The look on his face changed when he was home. Work was never over."

The portability and productivity of the smartphone was a significant shift for our entire country. In the military culture, it introduced new positive ways to stay connected to loved ones. However, in a culture that already did so much of life together, it also had the potential to invade family time and important opportunities for reintegration.

For spouses, Skype was previously the video platform of choice to connect families during the deployment, but then new chat features like Yahoo Messenger on smartphones made it easier for military families to stay connected on the go. I remember the first time I was able to chat using Yahoo Messenger with my deployed husband while the kids and I were at the playground. Mobile devices unchained military families from home landlines and reduced the likelihood of missing a call from a deployed service member.

This would be the beginning of what many military spouses describe as an addiction to their devices as a means of connection to their service members. Before the smartphone, if your deployed service member called and you were not home, you missed their call and would have to wait until they could call again. The smartphone made it possible to leave the house and not miss a call. Even when service members are home, safe, and reachable at the office, it is still socially accepted and encouraged by fellow spouses to pause whatever you are doing and answer when your service member calls.

By 2009, Facebook was up to three hundred million users with birthday reminders, the ability to create events, and a feature that allowed more visibility to group pages.[227] With so much of the military culture steeped in the social tradition of information distribution through in-person events, early adopters of social media (primarily Gen X and millennials) created group pages on Facebook as a way to distribute information quickly to military families. For those with mobile devices, information was in your hands before you even got home and checked your email. Printed newspaper subscriptions began to decline as online news spread more quickly and was easily updated in real time. By 2013, 61 percent of respondents in the Blue Star Families survey said their service member's unit used email to distribute information and 37 percent were using social media.[228]

Facebook groups also were effective for installations, neighborhoods, and spouse-run local groups to distribute mass information and connect families even before they arrived at their next assignment. Rather than waiting to establish community through in-person events or being limited to one circle of families, people now had access to an entire online community for local recommendations, school reviews, and even organized playground meetups.

Some of the most notable online groups were not organized by the DoD, nonprofits, or other businesses. Instead, spouse-owned groups dominated the platform. Lizann Lightfoot's "Seasoned Spouse Deployment Masterclass" Facebook group welcomed spouses of all ages and stages to share the deployment experience. Retired Colonel Steve Leonard, known on social media as @DoctrineMan,[229] began as an anonymous cartoonist with his tongue-in-cheek Doctrine Man web-comic in 1999. After moving to Facebook, his page evolved into a successful social media presence that continues to encourage dialogue on military and leadership topics mixed with a bit of humor. About this time, US Army W.T.F.! Moments also gained a viral following.

In many ways, social media made connections and relationships within the culture easier, and in some cases better. It broke down barriers between the officer and enlisted families; spouses found belonging in online groups simply out of their shared connection of being married to a service member rather than worrying about rank or etiquette. Families sent to overseas locations had a way to remain connected with supportive relationships established in previous assignments, reducing the isolation they might have otherwise experienced.

By 2013, more than 60 percent of military families were using social media to stay connected with other families,[230] with 49 percent seeking informational support, and 42 percent seeking emotional support, especially during deployments.[231] By 2014, 67 percent of millennials accessed the internet through their smartphones for information and to gain opinions from their peers.[232]

The rise of social media also took connection and community out of the hands of the DoD. Service members were still held to regulations that governed how they portrayed themselves, but lines were becoming blurred with profiles being personal and therefore not regulated. Military families found social media to be a much more efficient way to thrive. Why depend on your service member for information, go to an in-person event, wrestle with tired children, or put them in childcare for updates that could be posted online?

Meanwhile, older generations saw no replacement considering their most formative years in the military culture thrived with in-person experiences. As boomers eagerly waited for Gen X to pick up the baton, Gen X saw the new capabilities of social media and prepared to run an entirely different kind of race.

Instagram Shifts Brand Access & Self-Image

Although Facebook was the most widely used platform by military families to stay connected and find support, Instagram launched in 2010 and reached one million users within the first year. From the outset, the platform came with filters to make uploaded photos more visually appealing.

Millennials are said to have encouraged, if not pushed, Instagram's movement toward aesthetic perfection. Soon after its launch, third-party apps created even more filters, tempting users to curate content that was visually appealing, thus elevating their profile's status. Leisure time, vacations, and experiences became status symbols, creating profiles that were more of a highlight reel, a false perception of people's lives, than the authentic version. Tech reporter Sarah Frier, author of *No Filter: The Inside Story of Instagram*, stated, "A filter on Instagram was like if Twitter had a button to make you more clever."[233]

In addition, platforms like Twitter and Instagram allowed users to have access to celebrities and others not easy to friend or engage with on Facebook. The pressure to accrue likes was already having a negative impact on mental health and self-esteem. There was concern for false social comparison[234] due to following celebrities and unattainable lifestyles, leading to depression and body dysmorphia. Similar to how '90s supermodels like Kate Moss influenced a generation of eating disorders,[235] Kim Kardashian (celebrity millennial and, at one point, owner of the "seventh most popular Instagram account"[236]) is said to have influenced not only additional photo filters that could change your body shape but an entire generation of women seeking plastic surgery.

The perfection of self, home, work, and life added to the already-building anxiety of FOMO (fear of missing out). People worried they were missing out, not just on what others their age were doing, but that their own lives may not be as perfect as they could be. Jeremy Tyler, PsyD, professor of clinical psychiatry at the Center for the Treatment and Study of Anxiety, has worked with patients impacted by the stress of social media. "Social media expression is inherently biased because very few people aim to post about their flaws," he said. "There are so many ways perfectionism can drain quality of life. Often, patients think that if they try hard enough to be perfect, they'll be the best. But what we know is they just end up trying harder."[237] What we have seen in the years since is a continued rise in anxiety, depression, and burnout.

Although this sounds more like a problem for women due to Instagram's largely female audience, studies have shown a similar impact on the mental health and body image of male Instagram users. The millennial aesthetic was not only contained in Instagram, it impacted art, design, home organization, and more. Clean, perfect images boasted shades of white and "millennial pink." Even the iPhone came out with a rose-gold version.

SHIFT IN MARKETING & COMMUNICATION

As I speak with military leaders who are eager to understand, reach, and lead the youngest generation, their concerns narrow down to two topics: motivation and communication. We often think that motivating others is about finding the right words to inspire them, like a coach motivating a football team to push harder. Yet there are different approaches to motivation. In counseling and coaching, the motivational interviewing approach is designed to "strengthen personal motivation for and commitment to a specific goal by eliciting and exploring the person's own reasons for change within an atmosphere of acceptance and compassion."[238] In professions where there is a less intimate relationship and more large-scale communications, the psychology of persuasion and marketing essentially does the same thing. It is knowing the people or groups you are serving before attempting to move them into action. And as GI Joe taught early on, "Knowing is half the battle."

At a recent training of military leaders at Fort Cavazos, a discussion around an upcoming military ball started a dialogue about how to motivate more millennials and Gen Z to sign up and bring their spouses. A young millennial leader spoke up to say it was not something he wanted to attend and therefore he struggled to convince other families. This piqued the curiosity of others in the room, who asked why he would not want to attend something that has long been a fun, celebratory tradition. He answered he would attend out of obedience (since his superiors were asking him to go) but the event did not speak his "language."

When asked what he meant, he said he really couldn't answer it. The older generations stalled with how to "sell" an event or idea to someone who has no context of what it is or represents. The young leader had never had the experience of going to a military ball at the end of a difficult deployment where lives were lost. He had not experienced the celebration of life, comradery, and belonging with colleagues who had become cherished friends. He had not experienced a predeployment ball either, where he could introduce his spouse to those he was about to go to war with. This was the context many Gen X and boomer leaders wanted to preserve and share. The fact that this millennial was uninterested threatened ties to a tradition that represented a deep sense of belonging.

Thankfully, the commander of the Fort Cavazos group invited transparent conversation during the training, however many service members and families do not feel comfortable openly sharing their thoughts, feelings, and opinions in front of superiors, and fewer families are showing up to town halls. Social media is the window into their world.

In 2019, General Robert "Abe" Abrams called it "imperative" for senior leaders to be engaged on social media. In addition to adding humor and showing your humanity, he said, "It is important for us to acknowledge that the higher you go, the more insulated and distant you are from where the rubber meets the road. Social media gives us the opportunity to hear directly from our service members about what is bothering them, or the challenges facing them and their families. It is not always glamorous—not unlike reading the inside of a porta-potty at NTC—but the majority of the time, you will get unfiltered, grassroots feedback, and that is something we can all benefit from. Your subordinates are all over social media. As such, these platforms provide another venue for them to have access to you—think virtual open door policy. After establishing credibility, you might even get the chance to actually help people."[239]

Once you understand your audience, communication can be tailored to address those needs. While they could have disconnected, the older Fort Cavazos leaders had an opportunity to leverage curiosity and mentor appropriately.

Technology provides leaders a window into the lives of those they lead, a chance to learn what excites, motivates, and frustrates them. This is one of the reasons social media platforms and support groups have seen an increase in membership. For instance, the US Army W.T.F.! Moments Facebook page went viral because it gave service members an opportunity to send in funny photos and videos anonymously that depicted Army life—both the absurd and monotonous parts of it. Because of its anonymity, it also became a place to share the ugly and toxic side of leadership, starting dialogue but also raising fear of public accountability for leaders. As service members submitted toxic, mean, or absurd leadership demands (e.g., being forced into mandatory fun), they were able to voice their frustrations. Comments and shares validated frustrations and opinions on a larger scale. While leaders could ignore issues or choose to remain naive, ultimately, they would be seen as disconnected leaders. To echo the words of General Abrams, awareness gives you insight, credibility, and opportunities to make adjustments in the way you lead and communicate. That takes curiosity, but also the courage to consider additional ways to lead.

From a marketing and outreach perspective, the rise of social media offered companies and institutions instant access to the military culture as well as metrics to show the effectiveness of their reach. Faster than email, anyone who wanted to motivate an audience into action could communicate and distribute information directly, as well as instantly survey an audience. Any layperson could make a social media graphic, post it, monitor the post for questions, and wait for the audience to share it with others in the community (free word-of-mouth marketing).

For the military culture, this shift from in-person to online communication was a nice break from long email threads of information and was much more efficient than in-person events. In the absence of key positions during sequestration, the capability and speed of social platforms to connect, communicate, and advertise events was critical. Similarly, many businesses invested in young social media managers and temporarily moved away from traditional marketing efforts.

In 2007, Facebook unveiled Facebook ads, allowing users to pay to "boost" social media graphics, ads, and content as a way to target audiences and raise the visibility of their posts. In 2009, Facebook added algorithms to "bump posts with the most likes to the top of the feed" and in 2015, started "downranking pages that post too much overly promotional content."[240] Easy marketing and communication suddenly got more complicated. However, the culture had already assimilated to a different way of thinking. Considering this was occurring during the build-up of the force and the busiest and heaviest years of deployments, many military families were likely unaware of how these new features affected the actual visibility of information.

As algorithms made visibility more challenging, Instagram and YouTube showed the power of influencer reviews over traditional marketing or expert reviews. In 2009, after some identity adjustments, YouTube expanded its bandwidth and added new features for users to create personal profiles and collect subscribers. Early adopters of the platform were able to monetize their content with YouTube ads. Influencers on other platforms migrated to YouTube, expanding their presence and profitability. By 2013, YouTube hit one billion monthly users, far surpassing any other platform other than Facebook.

Influencers became a way to break through visibility challenges, even in the military community. With regulations around what service members could say or be paid to do, military spouse influencers made the most decisions around finances and day-to-day life. With employment challenges still a top concern, suddenly military spouses could be influencers for businesses and nonprofits seeking an audience with the military community. Many Gen X and millennial spouses saw an opportunity to continue building creative, remote careers.

Social media had so overtaken our attention and replaced our way of connection that the culture slowly adopted the assumption that just because a post or graphic was uploaded, it was effectively messaged. Whether the community realized it or not, they had evolved their way of communication and outreach without seeing the vulnerability of the platforms becoming a single point of failure should they shut down or change the algorithms. Just as we must be careful not to assume "email sent" is "message received," we must take the extra time to consider the complexity of social media. "Graphics and posts uploaded" does not mean "information distributed."

Although boomers and older Gen X adapted to new technology, they initially still preferred in-person meetings, conference calls, and printed memos. Millennials and younger Gen X, however, had grown up with email and smartphones. For them, unnecessarily long conversations were a waste of time, and phone calls that demanded a quick response raised anxiety. Instead, they felt communication could be faster through email, text, project management software, or message apps. New platforms not only made work more productive but allowed the user to respond when ready. It also organized information in one place. A *Forbes* article explained the rules of millennial communication like this:

▷ If it can be said in an email, send an email.
▷ Always send an email, if sending an email is possible.
▷ The only reason an email should not be sent to communicate basic information is if the conclusions, objectives, or answers are not yet decided upon, and multiple people should be present to weigh in.[241]

Note that the third rule invites collaboration, but only when there is still room for deliberation. There is often conflict between millennials and older generations about when a subject is "open for discussion" or closed, especially in the military where the default has most often been one-way direct communication from superiors. It is important for leaders to understand that two-way conversation and feedback have almost always been present for millennials and have allowed them to be change agents in their environments. On a large scale, their digital access gave them a direct connection to the world and, therefore, direct influence. Whether it was following their favorite celebrities, musicians, political figures, or brands online, voicing positive or negative feedback publicly through comments, reviews, and blogs often led to large-scale consumer changes in behavior. In the military, the use of the internet and technology to rally other voices behind social causes effectively influenced changes in sexual harassment advocacy, diversity, regulations for female service members, and even congressional attention to the housing crisis.

Communication is no longer just from the top down. Instead, it is multidirectional and not always controlled by the brand or leaders at the top. Even when presenting in front of command teams and leaders where millennials are present, I have found them to be much more communicative and willing to challenge information. I can usually feel Gen X and boomers in the room shift in their seats, bristling under the awkwardness of what feels like disrespect toward the guest speaker. If I am honest, as a Gen X, I have a similar internal quick reflex as I was brought up to listen to the expert and have opinions later. However, what I respect most about millennials is their push-back on blindly adopting information simply based on someone's personal experience or status. Especially when they have proven that listening to a diversity of perspectives leads to important changes in the way people are seen and treated.

Rather than viewing it as disrespectful, older generations can see these moments as an opportunity to either mentor the relationship when it is not a time for collaboration or honor the opportunity to hear an alternative perspective. I like to call it "educate or evolve." More important than adopting the information as their own, is the establishment of mutual trust and respect in the relationship. If you establish the latter first, safety in the relationship is established and learning becomes transformative.

ONLINE MILITARY CULTURE A VULNERABILITY

As technology made connection and communication more efficient with less need for face-to-face conversation, the lack of regulation and security had the potential of putting personal information at risk. This was especially concerning for the military community, which has long been protected by regulations that protect personal identifiable information (PII). Email became another platform to evolve with automatic enewsletters, click funnels, cloud storage, and digital marketing tools available to the average layperson to design a streamlined, marketable business. Successful in the civilian culture, pockets of the military community saw this as a solution for email distribution and storing information for continuity. This proved to be problematic with rosters and email lists containing PII of military personnel and family members protected as a matter of national security. Third-party software programs, enewsletter platforms, and even social media had the potential to put information at risk.

In 2015, ISIS (Islamic State of Iraq and Syria) posted the names, photos, and addresses of around one hundred troops online and called for attacks on them.[242] Just a month prior, a spouse's Twitter account was hacked, sending threats to a group of military spouses, stating, "We know everything about you, your husband and your children and we're much closer than you can even imagine. You'll see no mercy infidel!"[243] Examples like these made it even more challenging for military leaders to keep up with advancing strategies. With the rise of cyberbullying and cybersecurity, the DoD scrambled for solutions to keep families safe while also not over-managing personal accounts.

Also in 2015, the number of millennials using social media rose to 90 percent compared to only 12 percent in 2005.[244] Gen X users grew to 77 percent from 8 percent. One study showed that service personnel's usage of social media tripled between 2008 and 2013.[245] Divisive, polarized dialogue and "trolling" also escalated in society and military culture, thanks to anonymity and the lack of nonverbal feedback and eye contact between users.[246]

A 2014 dissertation focusing on a group of Marine spouses' use of social media revealed that although they "appreciated the ability to communicate and connect with others, they also described distressing aspects of social media use (e.g., stalking, derogatory postings, and discovering upsetting information during deployment before the news is public)." Casualty notification protocol of "blackout" communications (no one being able to communicate in or out until the family of the service member killed in action is notified) is to protect the family members and the process. Yet, nothing was stopping service members from connecting through social media platforms with their loved ones or information leaking out prematurely. Family members were hearing of their service member's death on social media before getting formal notification. Spouses asked the DoD to get involved.[247]

In 2011, Snapchat launched as a platform to "picture chat" in real time. It was quickly adopted by millennials and later Gen Z youth with filters and a feature that makes images disappear after a short period of time. Initially, users saw this feature as a safe way to share pictures that would not be forever archived online or seen by parents. However, as other users started screenshotting these images and sharing them with peers, a new level of cyberbullying grew, especially among teens. As Snapchat's daily users reached 150 million, other platforms offered to acquire Snapchat but were unsuccessful. It offered a platform away from those used by other generations.

Discord launched in 2015 to connect users through affinity interests, especially gaming. In 2023, Discord hit 150 million monthly active users, allowing voice and text while in video games. While most social media apps and platforms have a thirteen-year-old age minimum, Discord's is seventeen, with fewer monitoring and parenting controls. However, Discord's features and popularity grew among teens as it replaced other communication apps used by parents and schools, pressuring younger teens to join early to stay connected despite its access to adult content.

Social media was not only here to stay, it was actively changing communication, problem-solving, and community. During this time, the military community was also going through sequestration, layoffs, fewer in-person events, and logistically supporting two wars. By 2020, a study revealed that active-duty military members spent more time on every social media platform each day than the general population.[248] For millennials who were entering the military culture for the first time, their imprinting was not only the experience of a culture riddled with weariness, anxiety, and resentment. They viewed the military culture as more of an online culture than an in-person one. This was, after all, their experience of the life and world they had grown into.

Meanwhile, a younger generation was at home, likely playing *Fortnite* and *Minecraft* with people they had never met, and arguing with their millennial or Gen X parents that screen time was not rotting their brains. As Google Chromebooks replaced Trapper Keepers, monitoring YouTube and games on devices became more challenging in schools. Parents questioned how and when to introduce their children to the virtual world and also how to protect them from it.

AN EXAMPLE WORTH REMEMBERING
Leading with empathy bridges the generation gap.

Army intelligence officer Austin Von Letkemann launched his comedic account called @MandatoryFunDay on TikTok, Instagram, Facebook, YouTube, and Discord to "bring positivity to the armed forces through comedy."[249] Poking fun at everyday nuances of military life with short video clips, he has also used his following and influence to touch on serious topics like veteran health issues and suicide awareness. During a podcast interview, Austin shared, "Ninety percent of people ages nineteen to twenty-nine have a social media account. That is where those people are. I think one of the biggest issues is that disgruntled veterans are getting on whatever platform they are on and bashing the military. And the guys who have gained the most from it are staying quiet."[250]

When it comes to his own leadership style, Austin said, "I've learned during my military career to lead with empathy because it allows us to bridge that gap … especially with our younger generation. We as leaders are in a deficit in trust right off the bat. I think that they don't trust us because we're in positions of authority, and because of that, we have to earn their trust."[251]

Chapter 12

THE ENEMY IS NO LONGER A COUNTRY AWAY

IN RECENT DECADES, SOCIETY has experienced what some call the most drastic communication shift in five hundred years.[252] The Gutenberg Press of 1440 was the last to make such a historic impact on the world's access to information. With its ability to print and distribute text, the press was credited for starting the information revolution. By 1500, printing presses in Europe had distributed around twenty million books. In 1517, Martin Luther wrote his *95 Theses*; his thoughts and ideas were printed and circulated within days throughout Europe, leading to the Protestant Reformation,[253] shaping church history and spirituality from that point forward.

The current information age, which began in the mid-twentieth century, evolved swiftly. By 2000, there were 361 million internet users. By 2010, that number grew to two billion across the globe.[254] In 1945, the volume of knowledge was said to double every twenty-five years. In 2010, it was predicted that information on the internet would double every five years,[255] and by 2020, the volume of knowledge was doubling every twelve hours.[256] As of 2023, 93 percent of US adults access the internet, 99 percent of ages eighteen to twenty-nine.[257] Today, information is shared thousands of times faster than during the printing press era; some artificial intelligence processing systems can generate textual responses in seconds based on user prompts and referencing more than "eight million documents and ten billion words."[258]

GENERATION Z (DOB 1997–2012)
Digital Natives

Gen Z holds the key to recruitment challenges and is the topic of greatest interest and concern considering that they make up around 40 percent of the force.

While they were initially expected to continue some of the values and trends of millennials, like every young generation, they swing quite the opposite. Instead, they are being compared to their great-grandparents, the silent generation. Originally called iGen by generational researcher Jean Twenge, PhD, they were born to Gen X and older millennials and outnumber the boomer generation.

Born into a time of war, Gen Z has no personal memory of 9/11. They watched a country divide over politics and global conflicts and families split over personal opinions. They witnessed the impact of the 2008 economic crisis on their parents, leaving them highly motivated by financial security and education. Like the silent generation, there is a quietness about them, at least compared to millennials, as they wait to adopt their own opinions in a very opinionated world.

Dr. Twenge found that Gen Z, like millennials, prolong adolescence and delay adulthood. They tend to drink less and get their driver's licenses later. They spend exponentially more time online and less time in person, reading books, or with traditional media. For many, reading feels "slow." Their minds are accustomed to scanning for information rather than reading an entire publication.

New studies are showing that because of their use of technology at such a young age, the visual and multi-tasking portion of their brain is more effective than in previous generations. (This confirms, unfortunately, what my teens have been telling me; that they, indeed, are listening to me when they are on their devices.) While many adults and experts were initially concerned about the attention spans of Gen Z, it is now apparent they can scan the environment, conversation, and text to find key information within eight seconds. Key information could be hyperlinks, keywords, or discerning if content is relevant to them.

I spoke with a military leader who was giving a brief and was shocked that a Gen Z soldier left within the first five minutes. When he later asked the soldier about it, the Gen Z soldier said that the beginning of the briefing sounded irrelevant to his position and he wanted to get ahead on other things. As shocking as that is within a culture based on hierarchical order and expectations, it is another reminder for leaders to consider opportunities to mentor and adapt.

Overwhelming amounts of information have inspired Gen Z to be direct communicators, valuing efficiency a little differently than millennials. Gen Z describes efficiency as finding quick ways to do laborious tasks (valuing time and energy). Rather than being on a couple of platforms, they look for how to best use multiple platforms. For example, they may use YouTube for learning and entertainment, Facebook for its marketplace and connecting with family, Instagram for following brands and creating personal highlight reel portfolios, and Discord for interacting with peers around gaming.

Their communication is also more direct and efficient. Different from the millennial cohort that enjoys opportunities to collaborate, debate, and offer feedback, Gen Z has a phrase, "Say less," which is another way of saying, "Roger, got it" or "Say no more." However, valuing time and energy does not always translate to choosing technology over relationships. While they prefer text, chat, and email where it is most efficient, the pandemic disrupted crucial milestones when almost all human interaction went online. Gen Z experienced the technology as not always the best answer for everything. In-person conversations are valuable where dialogue is needed, especially in the form of mentoring, which they crave. In fact, reports are now showing that Gen Z is already choosing opportunities to work in the office, rather than remotely, where mentoring and the development of social and employment skills are more available.

When it comes to communicating with Gen Z, breaking through the noise is the easiest way to capture their attention. They still check their email, some almost daily, but prefer reminders via opt-in texting to break through the volume of information. One Gen Z service member told me that other generations often think he's being difficult, disrespectful, or less communicative when he is direct. "There is so much information coming in at once, I just need to know what you want me to do, and then I'll go do it," he said. He tends to get overloaded with a long list of tasks at once and prefers more manageable amounts of information. "If it is due in three months, it feels overwhelming to have all of it come at me now. We do much better with reminders on the next thing, later."[259]

The silents were named for being risk-averse and lacked strong vocal opinions on matters they hadn't formed opinions about. Gen Z is similar; however, Gen Z is passionate about social and environmental issues and is deeply impacted by the experiences of Black Lives Matter and marriage equality.

Recruiting and retention numbers of Gen Z are historically low. A Pew Research Center study in 2020 showed that only 64 percent of Gen Z trusted the government.[260] This puts incredible pressure on an institution like the military, which tends to be a decade behind some cultural advancements in the civilian world. Awareness and the visibility of global conflicts have made Gen Z more hesitant and concerned about future wars and also less trusting.

The concept of "serving your country" is also evolving. Some members of Gen Z enlist to protect and defend their own country; others find creative ways to get involved in global conflicts that impact humanity without making it their careers. In 2022, as Russia invaded Ukraine and censored its own people from the reality of war, social media activists attempted to circumvent Russian censorship by posting images and comments about the war in reviews for Russian cafes and restaurants on Google Maps. For example, this comment was attached as a review for Starbucks in Moscow: "Five thousand Russian soldiers died in Ukraine. ... Your president deceived you. There was no genocide in Ukraine. It was a pretext for attacking a sovereign and democratic Ukraine."[261]

One young Gen Z female service member I spoke with, just a couple of years into her military commitment, said she and other Gen Z are still trying to decide if America is actually where they want to call home. This is a long way off from the sense of patriotism and duty of those just a few generations before them.

MASS-SHOOTER GENERATION

When the oldest Gen Z was only two years old, two millennial high schoolers opened fire in their school in Littleton, Colorado, killing thirteen and wounding twenty before they committed suicide. It was the worst school shooting in US history at the time and the nation felt it as people sat glued to their televisions. Similar to 9/11, anyone who can remember it can likely recall where they were and what they felt.

In response, schools across America began a zero-tolerance policy for disruptive behavior, implemented school shooter drills and emergency plans, and evolved school security. While there were far more deaths in gun violence outside of school every year, mass shootings were increasing, and coverage and awareness of them were too.

While Gen Z may not remember some of the earliest mass shooting events, their parents do. Gen X and older millennials, who grew up during the increased anxiety of public safety, became concerned about taking their children to school, church, and even movie theaters. By 2021, more than forty states required active shooter drills for schools. Creating additional anxiety in the young generation, many drills were scheduled but not proactively announced to children. Depression and anxiety numbers climbed and research showed that one in five Gen Z children felt unsafe in their school.[262] While some have said this makes Gen Z more fragile, others believe they are more hardwired for difficulty and challenge.

This may seem like a dark perspective of this generation; Gen Z considers themselves quite dark as well. Their most formative years included an oversaturation of information about almost every topic imaginable. Globally, humanity was becoming more aware of people at their best and worst. Browsers and social media were a window into any global perspective, including mass shootings, war, opinions on gun laws and politics, social causes, and more.

Gen Z has furthered the mental health conversation. Online conversations and humor gave them the digital courage to talk openly about stress and its impact on their mental health. As a result, Gen Z is willing to advocate for a more transparent discussion of depression, anxiety, and even thoughts of suicide.

Sometimes it even comes out through humor and artistic expression.

Similar to the lost generation's Dadaism, Gen Z humor is often cynical, sarcastic, and dark. Swinging opposite from millennials, Gen Z's digital communication makes every effort to appear authentic, imperfect, and even flawed. Created out of the frustration of not being able to live up to standards set online, memes and images bring humor through grammatical errors, blurred images, sarcasm, and phrases that hold a cultural meaning among Gen Z.

Trying to describe Gen Z humor is a bit tricky and pointless. In fact, part of what makes their humor so unique is how quickly it changes. While Gen X and millennials circulated the same memes throughout the first decade, Gen Z quickly grew annoyed with how often the same jokes were passed around. Their contribution was to change it up or even evolve the joke further. Meme-generating apps and platforms allowed any individual to change up the image, make new ones, and keep the joke evolving.

In the early 2020s, Gen Z moved beyond memes to short video clips on TikTok, YouTube Shorts, and Instagram Reels, mocking especially millennials for what they coined the "millennial pause." This term was given to "those who have an awkward pause at the start of a video to check to see if it's recording" compared to Gen Z, who are "so confident in their recording knowledge that they don't need to check if they pressed the record button.[263] Also setting them apart are fewer selfies stored in Gen Z devices, replaced by a collection of memes, images, and videos they found and stored for sharing with friends.

As knowledge becomes more easily accessible to all, some say that creativity will become the primary creator of economic value, with an expected imagination age to follow the information age. Gen Z is testing the waters of digital art. With new creators of video games, anime, and art, this generation is asking how intellectual and creative property can be bought and sold digitally. Using digital currency as art in the virtual space is becoming a new reality.

RISE OF THE VIRTUAL WORLD

Throughout the decades, the American family culture slowly turned inward into homes while the internet offered an ever-expanding opportunity to connect with anyone, anywhere, all over the globe. The world outside became more unsafe, not just with stranger danger, but in schools and possibly even by a fellow student sitting next to them in class. During the pandemic, everyone could be considered "unsafe." The online gaming community seemed far more accessible.

YouTube continued to grow as a platform for entertainment, influencers, and do-it-yourself (DIY) tutorials. In 2011, the online video streaming platform Twitch launched as a way for video game influencers to live stream video games and connect with the gaming community. Video games became less linear and more experience-driven. As a comparison, *Super Mario Brothers* on the NES allowed for multiple players but centered on completing one level at a time, ultimately beating the boss at the end. By 2015, the Xbox Live Marketplace changed the way people play and buy video games; 54 percent of gamers were using games to connect with friends and 45 percent to connect with family.[264]

Some say we moved on from the information age to the experience age.[265] Games like *Call of Duty* were creating an experience for gamers all over the world. As one production contractor described, "What I've been told as a blanket expectation is that 90 percent of players who start your game will never see the end of it ..."[266] For younger Gen Z, worlds like *Minecraft* were a place to connect with friends, explore vast, never-ending worlds, and be creative.

For generations, scientists and engineers have been trying to master virtual reality (VR). In 2012, the Oculus Rift was invented and later bought by Facebook in 2014 for $2 billion. With Sony quickly following up with a VR headset for the PlayStation, the virtual reality world took off.[267] By 2016, hundreds of companies began creating VR products. In 2019, Steam, another gaming platform, saw one million headsets connected in one month.[268] In 2021, Facebook announced its plans to invest $10 billion in creating augmented reality (AR) and VR hardware. Companies such as Apple are actively searching for opportunities to merge reality with a host of opportunities for entertainment, productivity, and more.

Gaming isn't the only place the virtual reality world is impacting culture. Art galleries are creating exhibits where AR "wakes up" the art, making it come to life on your device. Productivity apps are being developed to allow remote teams the ability to work and collaborate in a virtual environment. Mindfulness apps create a completely immersive experience of meditation with a VR headset.

Trying to list out all the advancements in technology here would be counterproductive. As I write this, Instagram and other platforms have evolved to what many are calling social entertainment rather than social media as influencers flood platforms with creative video-based content. By the time this book is published, technology and artificial intelligence will have evolved even further, outdating whatever is written. What is important to note is that as technology advances, younger generations are often first to catch the wave and pilot new ways to communicate, work, and build community. Gen Z may be the generation most like the silents, but both would say they aren't as silent as you think.

Can't We Just Disconnect?

In 2019, service members around the world tuned into a live stream of the annual Association of the United States Army (AUSA) Conference in Washington, DC, to catch a glimpse of the anonymous creator of the US Army W.T.F!

Moments Facebook page. AUSA, which usually centers discussions around modernization, family forums, and readiness, had found a way to connect with service members and families, who would have otherwise not attended or watched the live Facebook videos, by bringing together a group of influencers for a panel discussion called "Risky Business: Leadership in the Information Age." The anonymous creator turned out to be Retired Sergeant Major Kenneth Ramos, a former psychological operations soldier and the voice of *Sergeant's Time*. Other panel guests included retired Colonel Steve Leonard (@DoctrineMan), then-Brigadier General Patrick Donahoe (@PatDonahoeArmy), and then-First Lieutenant Kelsey Cochran (@LadyLovesTaft).[269]

When asked how to incorporate social media and an online presence into personal leadership, Leonard replied, "When [you're] not there, [service members] are going to look for other people who claim to be leaders, but might not be providing the best example. If you're not out there, then you're not really engaging in the same way most of your people do."[270]

Brigadier General Donahoe was investigated for gaining national attention for comments on Twitter defending female soldiers.[271] He was defended by thousands of soldiers online. Once cleared of the charges, he retired as a two-star. He commented, "The richness of the discussion outweighs the risks," as long as you follow the "Don't be stupid" rule.[272]

I thought about Donahoe's words recently after I witnessed a small college in Kentucky trending online with what was being called the "Asbury Revival 2023." The college held its normal Wednesday chapel service, and nothing seemed out of the ordinary until a group of Gen Z students stuck around afterward to pray. Soon after, students started to return from their classrooms with a feeling that they should "go back to the chapel." The chapel service lasted sixteen days straight with worship, prayer, and testimony going on 24/7. This wasn't the first time in history something like this had happened, but it was the first time it had ever been live-streamed on social media for the world to see.

As news of this mysterious yet authentic movement went viral on almost every social media and news platform, thousands of people started traveling from all over the country to see and experience what was happening in this small town in Kentucky. They were drawn by the authenticity of an experience that included imperfect musicians, stories of suffering and community, and a genuine "come as you are" presence that was undeniable by all who attended. Soon, the same experience popped up in almost twenty other universities as other Gen Z students wanted the same on their campuses.[273] Online, people from every faith tradition were asking questions, entering discussions, and even debating about what an event like this meant for a culture (Christian) that had, up until then, been largely unattractive to Generation Z.

As my own family went to church on a Sunday morning during this time, I wondered if the church staff was going to address the event that had gone viral and attempt to provide context or encouragement on what was happening in the culture. Eagerly, I sat through the service and waited as the worship team performed their perfect set of songs and the pastor shared the scripted message that had been planned for months. I thought about how the perfectly planned service completely contrasted with what everyone was flooding Kentucky to experience. I left thinking how disconnected the leaders were to completely ignore what was happening. The best I could reason, either they weren't aware of what was happening online, weren't sure how they felt about it, or didn't want to deviate from what they had planned. Whatever the reason, they missed an opportunity to be authentic with the congregation.

What makes engaging with the online world challenging, other than the increasing number of platforms that evolve quickly, is burnout. As humans, we were never created to manage this much information and this many relationships at once. As each new generation enters adulthood, they are demanding a better work-life balance that includes the right to disconnect from the digital environment when not at work.

Disconnection is trending, with social media sabbaticals, going "off the grid," and the return of the flip phone. Young military families no longer support an all-encompassing lifestyle that supports an "all in, all the time" culture. As a way to set boundaries, service members are no longer signing up their families to get information from commands. Some are even choosing to live off the installation to allow for more of a separation from work and home.

In response, it is tempting to disconnect as a leader as well. In a way, this new generation is giving permission to healthily disconnect and is asking that more leaders model it. However, the military culture is now just as much of an online culture as it is an in-person one. Avoiding or ignoring where a large portion of the population meets, laughs, dialogues, and seeks help is to completely miss the opportunity to look out another window into perspectives you would never otherwise access.

AN EXAMPLE WORTH REMEMBERING
If something's not right, change it.

On January 13, 2023, Major General Michael McCurry, commanding general of the Army Aviation Center of Excellence and Fort Rucker (renamed Fort Novosel), Alabama, wrote his expectations for all service members under his authority on how to protect a healthy work-life balance. It was later shared on

US Army W.T.F! Moments as a positive example:

"I want to set forth my expectations for communication through text and email on holidays, weekends, and after prescribed duty hours, which may vary across the command.

"Since assuming command, I have noticed nonurgent emails and text messages over weekends, holidays, and after prescribed duty hours. I understand that we do not always complete our responsibilities during the duty day and often use weekends, holidays, and after-duty hours to catch up. Sending messages to our subordinates during nonduty hours creates an environment where leaders are tethered to their devices when they should be spending time with their families, themselves, or recharging for the week ahead. The better we are at time management and effective dissemination of information, the less we will rely on twenty-four-hour communication of important but noncritical/urgent information.

"We owe it to our soldiers, army civilians, and families to provide good leadership and predictability. Predictability helps ensure uninterrupted personal time outside of prescribed duty hours. I encourage leaders to engage in direct, timely, and effective communication during duty hours. When subordinates are in a leave/pass status, to include weekends or Day of No Scheduled Activity (DONSAs), do not text or email them unless it is an emergency circumstance outlined in our Commanders Critical Information Requirements (CCIR) reporting, including any information related to life, health, or safety. Leaders (including me) have an opportunity to model mindfulness to their subordinates and teams by thinking through the consequences of sending noncritical/urgent information outside of prescribed duty hours versus waiting until the next duty day."[274]

BREAKING POINT

"We are the people's army, and we always have been,
we come from the people and we defend the people.
That's our purpose. That's the only reason we exist.
And we have to maintain the trust of the American people.
Right now you have it. We have it. But trust is a fragile thing.
And every time somebody breaks the trust,
you're chipping away at the trust, the bond, the cohesion,
between the people, and the people's army,"

—General Mark Milley, chief of staff of the US Army, 2016[275]

Chapter 13
CULTURAL BREAKDOWN

MILITARY SERVICE DURING THE COVID-19 PANDEMIC often meant exposure to a threat that was not fully understood and could put one's own family at risk. The swift efforts of the DoD to protect and mitigate risk to service members and their families, as well as the humanitarian response to the nation, were impressive and served as a model for other countries.

I will never forget the garrison leadership at Fort Leavenworth, Kansas, who communicated daily updates during the lockdown and rationed supplies. I was in awe at how the Fort Hood Carl R. Darnall Army Medical Center systematically distributed vaccines to thousands of military personnel, families, civilians, and retirees with excellence and order. As a military spouse and civilian, it is difficult to comprehend our force's talent and loyalty until you experience it firsthand and in real time.

Is it possible for an entire culture or people to have a psychological break? What would that look like? The COVID-19 pandemic throughout 2020 and 2021 would be the closest many of us could describe to one that not only affected the country, but the entire planet. Considering everything the military culture had endured up to that point, as well as how much they were already asking to be heard on various issues, the pandemic was about to stretch them beyond anything they had experienced. While it can be tempting to look back in hindsight and say military families were resilient, the pandemic shifted the culture in some ways that paralleled the civilian culture. However, the stress of two global conflicts remained.

Gen Z experienced the pandemic during their most formative years. For those who joined the branches during the pandemic, the crisis imprinted their conclusions about government, healthcare, education, human connection, safety and security, and, of course, what they believe the military culture represents. Considering technology and educational shifts also shifted how Gen Z learns and sees authority, Gen Z's relationships in the workplace were also impacted.

Specifically, the pandemic affected both the military culture as a whole and Gen Z. Their reactions are starkly different in that the culture was nearing a breaking point while Gen Z was beginning adulthood with new excitement and ideas to offer the world. Holding both of these extremes at once is key to understanding the current recruitment and retention crisis. March 2020 is a historical marker for any generation who experienced it. Being called the "9/11 of the Gen Z generation,"[276] the pandemic lockdown is remembered as the beginning of a difficult and lengthy season, blurred by the fog of confusion and fear.

LOCKDOWN FOR THE MILITARY CULTURE

On February 24, 2020, US Forces Korea (USFK) announced its first case of the coronavirus. General Abrams, commander of USFK, was praised for his early actions, "including setting up a 24/7 operations center to monitor the situation, tracking employees who had traveled through mainland China, raising health protection conditions on post, developing communication plans, requiring self-quarantines, and undertaking additional precautions as the virus spread."[277] Italy soon followed with a large and swift outbreak. With so many military families living off the installation, US Army Southern European Task Force, Africa worked with the Italian government as Italy imposed its lockdown.

Mitigating the spread of the virus among service members was a top concern for the DoD. The Spanish Flu of 1918 had been intertwined with World War I, when "influenza and pneumonia sickened 20 percent to 40 percent of US Army and Navy personnel" and high morbidity rates "rendered hundreds of thousands of military personnel noneffective."[278] On March 11, 2020, Secretary of Defense Mark Esper sent out a memo pausing all DoD travel plans, delaying homecomings for deployed service members, and considering quarantine measures before they could go home. Any families scheduled for overseas moves had their plans paused. The only exception was for mission-essential personnel.[279]

On March 13, another memo was released, stopping domestic travel entirely, which included any approved leaves, training, and PCS movements (regardless of what stage they were in). For those who were in the middle of deployments or anticipating a homecoming, deployments were now on an extended pause with no answers.

On March 15, COVID-19 was detected on board the USS *Theodore Roosevelt* while at sea. With confusion among the ranks on how to handle the situation, those infected were evacuated to Guam. "Over 1,200 crew members eventually tested positive for COVID-19. Of these, about 20 percent were asymptomatic, twenty-three were hospitalized, four needed intensive care, and one died."[280]

Fear and anxiety escalated for especially the Navy and Marine Corps community as the DoD discussed how to address extended deployments. A spouse who was facing how to quarantine her husband if he was able to return said, "It's a huge mix of emotions. My heart wants him home, my children want him home. We have not seen him since June. At the same time ... we are a family of six with four children under the age of nine. That is a lot to take on, being indoors for two full weeks."[281] Another expressed the full weight of the pandemic on top of what she had already endured: "Added stress like this during uncertain times is one of those things military families like ours don't see coming but carry along with everything else that nearly twenty years of continuous operations bring. Ready to be a normal family again. Right or wrong, I am so over sacrificing for the greater good."[282]

For families across the country, the lockdown was filled with fear, panic, and daily check-ins from White House briefings on social media, YouTube, and TV. Military couples who were able to be together felt an initial sense of gratitude. The idea of having quality time with family away from work, even under a cloud of fear, was a strange respite for those who had revolved around the mission for years. This was especially true for couples who had spent half of their married lives apart. Now, the mission could be family.

A few weeks into the lockdown, however, couples were beginning to acknowledge mixed emotions about how difficult it was to reintegrate in a completely new way. While it was a rare opportunity to reconnect as a family, they were not accustomed to spending this much time together. Strong, independent roles and patterns had been adopted over the years making it very difficult for couples to adjust. Concerns of domestic violence rose across the nation as the pandemic only exacerbated previously dysfunctional relational dynamics.

As isolation wore on, the nation grew impatient. From June to July, the infection rate for the military "shot up from 10,462 cumulative cases to 37,824." The figure included more than 14,300 infections among active-duty troops, as well as total cases reported among civilian workers, dependents, and contractors since the pandemic began."[283] While our experience at the schoolhouse in Fort Leavenworth had mostly intact families, neighborhoods were able to lean on each other for support. Other military families across the globe, however, felt far less connected. A Blue Star Families survey found that 70 percent of military families lived off installations, and a third felt there was no one to ask for a favor.[284]

Considering extended halts on the force could produce "power global vacuums," reduce force strength, and disrupt families further, the DoD found successful ways to reopen progressively with telework, enforcing quarantines where needed and creating restrictions for social distancing to reduce the spread of the virus. By May, the DoD had systems for rapid testing that allowed for better troop movement and had established quarantine centers in and out of theater.

IMPACT ON EMPLOYMENT

Like the rest of the country, military families faced similar employment challenges throughout the pandemic lockdown, including balancing work, home, online education for kids, and other responsibilities. Those looking for employment prior to this likely felt they were on an indefinite hold, and many who started jobs lost them as the country locked down. Those whose jobs were not easily transferred to remote work suffered. According to the NMFA survey, 49 percent of respondents "reported lost or reduced access to childcare" in addition to losing jobs due to the lockdown; 49 percent of respondents experienced a reduction in family income.[285] The Blue Star Families 2020 Military Family Lifestyle Survey found that, "more than four in ten active duty spouses who were employed prepandemic stopped working at some point, mainly due to layoffs and furloughs."[286]

With the military's gold standard childcare just as affected by the pandemic, military spouses who were dependent on childcare in order to work suffered as well. The Blue Star Families survey reported that for the first time in history, spouse employment made "the top five list [of concerns] for active duty service members."[287] It also found that "83 percent of active duty families who live off-installation have housing costs that exceed their housing stipend" and "a large majority of these families (77 percent) pay more than $200 per month in out-of-pocket expenses to live in these areas." Some families choose to live off the installation in order to access better school districts for their children. Food insecurity also rose from "one in eight respondents in 2019 to one in five," and "an increase in those experiencing hunger rose from 7.7 to 10.5 percent."[288]

Three years later, the DoD childcare system continues to face challenges in staffing since the pandemic. It remains the largest obstacle to spouse employment with a waitlist that has forced the DoD to find creative methods to expand childcare into the civilian sector and into homes.

COVID-19 AND GEN Z

Meanwhile, Gen Z spanned from elementary school to college age, making the pandemic's impact on education and learning especially critical. This generation also makes up a considerable portion of the military children who experienced the intense journey of their parents in the years prior. While all generations were affected by the pandemic, it impacted each one in distinct ways.

With roughly 88 percent of the force consistently averaging forty years old and younger, a majority of military families have young children who are middle school-aged and younger. With nearly "900,000 military-connected children of all ages worldwide,"[289] the Department of Defense Education Activity (DoDEA) provides government-funded prekindergarten through twelfth grade education to military kids all over the globe. Although DoDEA provides education through high school, it is not uncommon to find multiple elementary schools on larger installations to support the majority of the population and military families with older children living off the installation for access to larger state-funded schools.

With the country in complete lockdown, schools sent students home and scrambled to evaluate how to move education completely online. Each state and district varied in how it proceeded, which meant military families also had varied experiences. Schools needed to figure out how to finish the second semester as well as innovate new strategies for keeping kids and staff safe. Many schools and higher education institutions had used on-demand courses and online curricula for years, however, it was a privileged option for families who had access to the internet and computers.

Quickly, video calling platforms like Zoom, Google, and Microsoft Teams allowed businesses and schools in America to continue remote work during lockdown. Microsoft Teams had already been adopted within the military culture. Chromebooks or other devices were distributed where possible to students in need. As the spring semester finished, schools helped kids and teachers limp across the finish line. Over the course of the year, the stimulus package, the American Rescue Plan, and President Biden's infrastructure proposal helped fund devices and internet services for families and schools.

Military kids were already at a disadvantage due to relocations and constant adjustments before the pandemic. Many schools continue to lack a clear understanding or awareness of the needs of military-connected students. Assimilating to schools with different policies and procedures, new cultures, and sometimes moving in the middle of the school year creates education gaps and behavioral problems that come with readjusting. As an example, one December move, our son's last math teacher was behind on the fourth grade math curriculum, and she aimed to catch up in the second semester. When we arrived at the new school for the second semester, the teacher was ahead. It took two months of studying math for four hours every night to close the education gap he experienced.

There are many other stories of military children missing education blocks for learning topics such as colors or months of the calendar, or missing whole high school credits necessary for graduation. These frustrations are shared among the military community, and organizations like Military Interstate Children's Compact Commission (MIC3) and Military Child Education Coalition (MCEC) have made incredible progress in helping parents and schools communicate better through laws that support military kids as they transition. As it was already the second semester of the year, for the first time, college boards waived SAT and ACT scores as safety could not be guaranteed at testing centers.

Now that schools across the nation were transitioning online, parents worried about summer move cycles, knowing it would ask even more of their military kids, who were already struggling with online school and lack of peer support. Precautionary measures, although crucial, "created some confusion, anxiousness, and even depression, at times, amongst both the adolescents and their parents as relational isolation prevented the adolescents from being able to say goodbye to friends in the manner they normally would. In addition, the techniques they normally used to make new friends or to begin forming new relationships was difficult, and even impossible in some instances."[290]

Schools also varied in their approaches to reopening for the fall, which sparked intense and volatile debates and protests from parents and schools who had differing views about the virus, masks, vaccines, and policies for reopening. While the whole nation was experiencing these changes and disruptions to children's education, military families added this on top of war-related stress, lack of respite, and a level of uncertainty that even for the military was abnormal. Their Gen Z kids were now absorbing the complexity of the pandemic's impact on their safety and security in school.

While some schools remained closed and completely online, other schools offered hybrid options or reopened with mask mandates. Masks for small children and those with special needs were especially difficult to enforce. Parents, who depended on schools for their children's education and to care for them during the workday, were angry that their children's right to education was limited by the school's policies. Schools, however, expressed a right to keep other children and staff safe as well. Meanwhile, the public was also becoming increasingly opinionated about the virus, vaccines, and various approaches to protecting students and teachers. As tensions rose, Gen Z, which had already seen the country and their families divided by politics and vaccines, saw division again regarding their own health and education.

In the decade leading up to 2020, homeschooling became a popular option for military parents who wanted more control over educational gaps caused by frequent moves. As the pandemic impacted in-person gatherings and districts debated school policy throughout 2020, homeschooling rose from 8 percent to 12 percent for military families, a rate higher than the civilian population.

GRIEF IN THE MILITARY CULTURE

The impact of grief is a topic not covered thoroughly enough within the military culture. Although the loss of a service member is honored through rich tradition and ceremony, families grieve significant losses throughout their time in the lifestyle. The military community has long provided a family away from family; however, there is a loss of closeness that comes with time away from those who have always known you. Technology has definitely made connecting easier, however returning home to mourn the loss of family and friends can be challenging and expensive, especially if the service member is deployed. The continued grief process is further complicated by having to leave family and friends and return to military life.

During my time at Fort Leavenworth (and before the pandemic), I started a spouse group that met weekly for personal development and support. The topic of grief was a top request, one I had not tracked closely enough. Spouses shared the deep emotional grief that they had carried for years surrounding the loss of family members. One shared that her intense grief of losing her dog surprised her. She, and many others, shared how they had bonded with their animals through moves, deployments, and difficult times. When it was time to say goodbye, they didn't have a "home" to bury them or a place they could return to. Not knowing what to do and not wanting to leave them behind, they kept some of their ashes until they could decide.

Only a couple of years later, I would experience this myself. Immense grief, loaded from the years of moving, flooded me when the vet asked what I wanted to do with my pet's body. Our boys and I had chosen our family dog from a shelter during our first deployment. A military spouse friend that I nicknamed "the dog whisperer" helped us pick him out since my husband was in Afghanistan. When we traveled, military neighbors would dog sit. He would ride with me to the post office to mail care packages and distract the boys during difficult days. He graciously adapted to six different homes. I trained him to be a therapy dog, inviting military clients to pet him when they needed grounding from anxiety in the counseling office. Then, he was gone. I realized answering the vet's question was to acknowledge a truth I had not said out loud before. Knowing we would be moving soon, I barely got out the words, "I don't know. I have no home to take him to."

Throughout the pandemic, other "small deaths" were occurring for families. They were less severe than the loss of life, but they were important milestones missed or altered. This was most significant to small children who struggled to understand the complex circumstances of an invisible virus. As Gen Z military teens graduated from high school, installations creatively rallied to offer graduation parades for the seniors to drive through neighborhoods and be congratulated by the community. Other important events, including weddings and birthdays, were either virtual or socially distanced. While these creative methods definitely lifted spirits, Gen Z describes missing milestones, such as prom, a first semester of college, and peer interaction key to their development, as losses they will never be able to reclaim. Blue Star Families also found that even though some employers found ways to create flexibility through remote work, only 32 percent of active-duty spouses were able to continue working in a remote or telework arrangement.[291]

There is grief associated with losing something you've worked hard to establish. Grief in the idea of starting over again. We have seen this in the spouse culture and our military kids. There is, indeed, a limit to the amount of grit we can muster before some are willing to throw in the towel. To date, America is still calculating the number of students who did not return to finish high school or college due to the pandemic.[292] The DoD Spouse Education and Career Opportunities (SECO) program helps spouses obtain employment (including remote) but remains underutilized, raising a question of whether spouses actually want to work. There are businesses willing to hire, yet spouses are not applying due to difficult obstacles.

The ever-present threat and/or experience of the death of family and friends from the virus added an additional layer of trauma to the pandemic. Families who lost family members to the virus were unable to visit the sick and dying or to attend funerals. Any kind of gathering was considered a possible "super spreader event" especially as news stories circulated on how the virus was still active on bodies seventeen days after death.[293]

The process of grief is not only crucial to well-being, it is also sacred among most cultures. Ceremonies, even in the military, honor the dead by coming together, honoring and mourning the life that was lost and receiving the comfort of others. With ceremonies canceled and families isolated from loved ones, grief was stunted and made complicated by the isolation from the dying, as well as the trauma of the treatment of the dying.

Post-traumatic stress is talked about now more than ever. Trauma is now broken down into little "t" and big "T" categories. Hindsight graciously dulls our memories, however. The pandemic was traumatic for many; the military culture may have experienced it through the lens of other acute ongoing traumas.

MENTAL HEALTH CRISIS IN THE MILITARY

In addition to the medical community being stretched beyond their personal and professional capacity during the pandemic, the mental health community carried the heavy burden of psychologically supporting individuals and families.

Prior to the pandemic, most mental health providers provided in-person services in brick-and-mortar locations. Like much of the medical community, telehealth was new for many of them. Veterans Affairs (VA) had been piloting telehealth since 2017, investing billions of dollars in its VA Video Connect. The VA's new program proved to be a successful model in that they were able to provide 2.6 million telehealth visits in 2019.[294] Within months of the lockdown, the president announced a national emergency, allowing providers and insurance companies to approve and evolve services to telehealth quickly. This included TRICARE and the DoD's other behavioral health programs.

In just the Army alone, "virtual behavioral health encounters increased from 13,000 to 68,600" from January to April 2020.[295] Although those numbers decreased by July, the Blue Star Families survey for 2020 revealed that "more than 50 percent of active duty respondents reported that their overall happiness, personal mental health, children's education and children's mental health were worse or much worse than before the pandemic."[296]

As a clinician and TRICARE provider, I was amazed at how quickly providers were set up to provide online services and able to file claims throughout the pandemic. TRICARE quickly approved and paid for in-network and out-of-network services with ease. Approving and innovating HIPAA-compliant platforms quickly was not as easy. Yet, thanks to the National Public Emergency Order, I believe families were able to get access to help faster than if it had not been in place. Although concerns for domestic violence and suicide were high, studies have since shown that suicide rates for service members and family members did not rise above the numbers recorded prepandemic.[297]

The isolation felt throughout the year impacted the entire country. A Pew Research Center study found that long before the pandemic even hit, "one in ten Americans [felt] lonely or isolated all or most of the time."[298] Now that there were quarantines, social distancing, and remote work and school, isolation was having a significant effect on people's well-being. A study out of Canada found that 29 percent of individuals who were quarantined for SARS emerged with post-traumatic stress symptoms. Those numbers were especially higher for those forced into quarantine.[299] As COVID-19 quarantine regulations came out from the DoD, the Centers for Disease Control, employers, and schools, mass confusion spread about which symptoms warranted being tested, how to obtain COVID-19 tests, how to understand different types of tests, and more.

As I was writing this book, we marked the third anniversary of America's lockdown and the initial panic and confusion around the virus. Feeling anxiety in my own body connected to the memories of that time as well as the research

I was digging into, I decided to reach out to military families on social media to see if they still felt any anxiety or body trauma connected to 2020. I received an overwhelming response. One service member shared that putting on a mask to enter a medical facility brought back difficult memories. Many spouses shared their stories of moving overseas or across the country during the lockdown. One person had moved three times during the pandemic. Another spouse shared how traumatic it was to have a baby, knowing that her spouse may not be able to be in the room with her. Another who moved in September 2020 shared, "Everything was closed where we left (no proper goodbyes) and everything was closed where we arrived (no welcome, no clubs, no friends). I initially lost some weight from the stress and uncertain food availability during lockdown in California. My husband and I ate tiny portions to make sure there was enough food for the kids. Then I gained a lot of weight after the move because food was more available, and I think my body was adjusting to the depleted hormones from all the stress. Still working to lose that weight and recover, two years and one additional move later."[300]

As I opened up my calendar throughout the fall of 2020 for military families, most were still recovering from the initial lockdown. Those with high-risk family members or who had additional life stressors on top of the pandemic stress were not functioning well. For six hours a day, thirty hours a week, and for almost a year, I was on a loop encouraging clients to implement the smallest tasks that made them feel like they had a sense of control in their life. "Walk to the mailbox," I would say, "The one thing we have control over is our breathing," Session after session after session. One of my clients has children with special needs and went to the doctor to get her allergy shot. She wasn't aware she was positive for COVID-19 and went into anaphylactic shock.

With a second wave of COVID-19 hitting the country in the fall and winter, families were no longer just afraid of the virus, but were wrestling with deep convictions and opinions about vaccines and the military's mandated vaccine policy for service members. Parents were paralyzed on how to encourage or protect their children when schools were making decisions they didn't agree with or the rules kept changing. Due to the constant circulation of news and opinions, military families were disconnecting from family members who were once supportive, but now strongly questioned their decisions. Amid decision fatigue, people isolated more. Even in my virtual events, it was like families had adapted so many times to so many things, they just couldn't anymore.

Over time, military parents watched helplessly as their children's mental health declined and then theirs as well. Reports showed 27 percent mental health deterioration of parents.[301] Experts shared a 10 percent to 20 percent rise in depression among military Gen Z teens ages fourteen to twenty-one, with 32 percent expressing anxiety. Mood disorders in girls were twice as common as those in boys and at adult rates.[302] The larger concern, however, would show up in Gen Z youth just a few years later.

Many parents put their own wellness on the back burner but sought wisdom and advice for how to help their kids survive when it seemed their whole worlds had turned upside down. There were no perfect answers for what was happening to our country and in our military culture. There was no choice but to persevere. Some advised parents to "be models of resourcefulness and mindfulness" and provide a "flexible routine that fits with school schedules." In addition, parents were encouraged to "stay positive … stay active, stay scheduled, stay connected, stay informed, and stay safe."[303]

Although this advice was not wrong, it was too much. Everyone in the nation lost access to key support resources, temporarily but in some cases indefinitely. Parents became their child's friend, sports coach, spiritual leader, teacher, accountability, and parent, and also served as a childcare provider, employee, spouse, friend, child, sibling, and any other role they had leaned on or provided before the pandemic.

Under the pressure of being all things, encouragement to be more proved too much. In exhaustion, parents allowed more screen time and connection with friends over video games and online. Video games and the online world began to innovate creative solutions to allow connectivity despite social distancing. Netflix streaming made for a great escape for adults as well.[304]

Mental health issues for Gen Z became a norm rather than an exception. How could one come out of such a season unscathed? Where the enemy was once a country away, a family member, neighbor, or classmate could make you sick or cut you off for a different opinion. Police tension and racial riots played out in real time, live streaming on YouTube and social media. Gen Z's dark humor coping skills bonded them in their experience of growing up in a dark time.

The millennial approach to perfection was not only unrealistic for Gen Z, but inauthentic and a standard they refused to live up to. Their focus would be the opposite, demanding authenticity and transparency, even if that meant vulnerably showing your flaws. In many ways, their dark humor of World War III and death was a mirror of what they saw in the world every day. Although it is hard to comprehend why anyone would joke about death or suicide, Gen Z has been sending the older generation a message that if you can't talk about it, then you aren't being honest with yourself, or them.

Although there is research that shows that military kids tend to be more resilient as they transition into adult life and college, the public conversation around mental health for Gen Z may suggest relooking at the data. In 2015, 30 percent of children experienced difficulty from deployment alone. At the time, drug and alcohol use was rising among military teens, as were anxiety, depression, and suicide.[305] As one military psychiatrist and researcher said, "That doesn't necessarily mean that military children are mentally ill, but that they're distressed."[306] This, of course, was five years before the pandemic.

Since then, the Army public health report of 2021 reported numbers that the DoD, leaders, and parents could no longer ignore: 51 percent of parents reported that "their child(ren) appeared to experience emotional, behavioral, or other difficulties since the start of the COVID-19 pandemic."[307] NMFA reported 42 percent of respondents having low mental well-being and 45 percent moderate mental well-being.[308] Other reports indicated a rise in Gen Z seeking out therapy to help with coping. With 65 percent of military teens considering entering the force, we will also need to reevaluate mental health diagnoses limiting the entry of recruits. Gen Z will be the ones to continue that conversation.

As some traumatic reactions began to dissipate, a new mental health crisis emerged: a shortage of mental health providers. Partly due to the increased demand for care, it was also because providers had burned out themselves.

Clinicians are taught to manage transference between themselves and a client. Transference refers to one's own reaction to something a client is sharing. If you recognize transference, you are taught to take it to a colleague or supervisor, or in cases where it prevents you from treating the client without bias, you must make a referral. Throughout the pandemic, professionals were going through the same trauma, right alongside our clients. There was no way to escape it, deal with it, or refer clients to someone else. We were all in it together. However, the amount of regulation it took for providers to continue to give throughout two years of a pandemic exhausted much of the profession. Some took time off, some switched to coaching, which was more portable and focused on milder issues. Thousands quit the profession entirely.

Innovation during the pandemic also changed the mental health industry, increasing options for HIPAA-compliant telehealth platforms and integrations for client management software. The pandemic forced providers who had not already adjusted to telehealth to quickly create home offices that were conducive to therapy and to learn new software. Many found that telehealth was not only cost-effective due to less overhead, but it gave them access to clients from all over the state rather than just locally. Concierge services became more popular as well. Companies like BetterHelp created ways to provide telehealth with chat features that enabled clients to connect with therapists outside the session.

With inflation and the pandemic impacting clients financially as well, some found that private pay or moving out of network with insurance companies was easier and gave them more control of their businesses. Many mental health providers who had previously contracted with TRICARE to serve military families no longer wanted to be under a contract that paid less than other insurance companies. Many military spouse mental health providers who had wanted to credential with TRICARE were weary of advocating for themselves after relocations for licensure portability. Choosing to provide private pay or out-of-network services removed that burden if they weren't part of a group practice. Joining DoD's programs like Military OneSource as a clinician paid even less than TRICARE. Even though it was considered nonmedical counseling, it required the same effort and skill from the provider. So fewer signed up, and fewer stayed in.

In the past, contracting with TRICARE helped clinicians connect to clients, as in-network providers used the TRICARE directories to refer patients to other in-network providers. But TRICARE's system for updating provider directories involved a lengthy process that left the directory inaccurate, frustrating providers and families looking for referrals. By moving out of network, providers could have more control and get paid closer to what they actually billed for services. Plus, it was easier than ever to manage and market a therapy business.

When the national public emergency ended, new technology and ways of getting help remained. As businesses reopened, some providers did not return to office locations. TRICARE stayed with the alternative ways of delivering services, but the process of filing claims went back to the old model. With medical and healthcare costs rising, referrals to out-of-network providers (which would cost the government more) were more difficult for some military families to obtain, keeping referrals in network for providers willing to sign contracts again.

To say there is a shortage of mental health providers is only partially true.[309] With government healthcare costs being a significant part of the defense and federal budget, it is an area the defense is constantly being asked to reevaluate. As mental health providers seek to be compensated appropriately for their services, the government will need to evaluate what constitutes medical care versus help with life stressors. Mental health care is quickly becoming a luxury item for families who feel they must choose to find help on their own dime.

A current shift is a growing conversation around performance psychology as a way to fill the gap. Since the beginning of the profession, mental health care has lumped together mental illness, which benefits from a diagnosis, sometimes medication, and medical oversight, with mental health or wellness that includes life issues such as mild depression, grief, anxiety, and stress. Using the medical model, mild issues would traditionally get a diagnosis, making it a medical condition insurance companies would pay for. However, many people are not dealing with mental illness but the significant stress that has come with the lifestyle.

The new conversation around performance and holistic care separates performance and life stressors from the medical model, reducing the need for a diagnosis, reducing the stigma, and normalizing ways to use coaching and what we know in sports psychology to help people function better in stressful situations and with life stressors. The DoD provides Military OneSource as a resource for nonmedical counseling or coaching, which has been underutilized. The question is whether new conversations around the DoD providing access to human performance optimization[310] could be the answer.

With Gen Z making mental health a more public topic, the discussion of how the mental health field is valued will also rise. Psychology, social work, religion, and sociology are consistently listed as majors and professions that pay the lowest when entering the workforce.[311] As individuals across the nation went back to work after the pandemic with a new mindset of what they valued, "quiet quitting" and choosing not to overwork felt like a healthier work perspective. The mental health industry did the same, they just waited until the postpandemic wave subsided before they began to consider their own self-care.

MEASURING A CULTURAL BREAKDOWN

Back when I was an undergraduate student, one of my professors handed out an inventory called the Holmes-Rahe Life Stress Inventory. It was a long list of life stressors with points assigned to each, and the intent was to mark those you had experienced within the last six months. Included were options like the loss of a family member, a medical diagnosis, a move or relocation, changing schools, marital separation from a mate, and many more. The goal was to see which of the three categories you landed in at the end: a tolerable range of stress or in two concerning zones that predicted a 50 to 80 percent chance of a major health breakdown within the next two years.[312]

I've thought about that inventory a lot since marrying my soldier and entering this lifestyle. I wondered what the inventory would say now that we were in a constant state of uncertainty that included frequent major life upheavals. I wondered what it would mean if your body was in a constant state of stress and change with no end.

A myriad of health problems are associated with stress. Stress is linked to heart attacks, strokes, weight gain, and hormone imbalances and affects every system and part of the body. What tools like the Holmes-Rahe Scale invite us to do is break our denial around the amount of stress we are carrying, remind us of our ability to reclaim control of our well-being, and inspire us to make necessary changes where possible. Obviously, no one wants to experience a major health breakdown. So people initially look for ways to reduce their workload, set better boundaries with others, start new resolutions, and focus on what really matters in life. Easier said than done, successful change requires that we know what we value most and are willing to prioritize. Even still, most of us find ways to overestimate how much we can handle, and we take on more.

It's one thing to be responsible for the added stress in our lives, but many times we are in circumstances out of our control. A good friend of mine has a son with multiple medical diagnoses, including autism and Down Syndrome. Her husband has a significant traumatic brain injury (TBI) that keeps him up all night long, vividly living out the traumatic memories of his time in service. I am always inspired by their ability to endure yet another surgery or sleep study and somehow function throughout the day.

I often wonder (and sometimes ask) if they have ever reached a breaking point or discovered a way to discern the limits of what they can endure. She has said she gets frequent comments like, "I don't know how you do it," or "God will never give you more than you can handle" and is quick to say there are days when she doesn't feel like she can handle more. She and her husband implement key self-care strategies to keep themselves afloat when times get especially difficult. She also reminds me frequently that everyone's experience of difficulty and suffering is subjective and no less significant than hers.

Especially when key relationships and responsibilities are on the line, most people will choose to persevere rather than quit. Even if they feel like they don't have a choice but to move forward, most quickly survey the cost of quitting and rule it out. Author Angela Duckworth calls the choice to persevere, grit. It is the passion to persevere through difficulty in order to reach a future goal.[313]

I have long appreciated and used Duckworth's definition of grit in place of the military's overused resilience. Whether it is getting through a day, week, month, or season, grit is a powerful descriptor for enduring momentary difficulty in the hope of reaching an end goal, and military families are the grittiest group I have ever met. Their ability to push through almost anything to reach that homecoming, next duty station, another new job for the military spouse, or next promotion, and then do it all over again is incredible.

However, what is the cost of living in a constant state of perseverance and grit? In 2022, the USO asked me to research what military families needed most following the pandemic "season." We knew the culture was survey-weary, so we decided to host virtual "sensing sessions," a term the military uses for "an open and safe forum" for discussion on any major topic participants are concerned about. We invited couples from every stage of life and generational perspective as well as experts who serve in the military culture.

I was not surprised to hear that couples were worn out, detached, worried about their kids, and needing respite, especially after the pandemic. However, an interesting pattern emerged when speaking with spouses. When I asked the simple question, "How are you?," almost all of them answered with some variation of a sarcastic, "Well, the world isn't burning, so I guess I'm good!" Now, after living through a crisis on top of an already stressful lifestyle, they had two speeds: crisis mode or managing. Anything less stressful than a full crisis was considered normal. After reflecting on what I noticed back to them, everyone agreed that when you've lived with chronic stress for so many years, your threshold for identifying what is actually stressful is, unfortunately, much higher.

For much of the military culture, the decade leading up to 2020 was like an intense roller coaster of uncertainty, deployments, and broken promises. One spouse responded to my use of that metaphor by describing it as more of a solid, uphill climb than a roller coaster. She said, "There were very few periods that felt like reprieve in those ten years." She continued, "Most years were just slogs. High operations tempo, high training, high instability, and high stress. And all while raising toddlers to early school age. If it was a roller coaster, it was one where it was dark, with no clear end and no warning for what was ahead."

While there will always be seasons and life stressors we need to simply persevere through, seasons change. The crisp fall air rescues us from the oppressive heat of summer, and just when winter has convinced us to give up, spring wakes up something new and exciting. Humans need seasons to change in their own lives in order to be well. Without it, we become ill. Humans need respite from stress or we will be less likely to persevere in another season.

When I work with people in a chronic and persistent state of stress, it is a red flag that something needs to change. If a season is never-ending, then it isn't a season. If there are stressors out of their control, then how they manage that stress must change. In the years before 2020, there was no respite, no change of season. Families were aware of their stressful state of existence but were too gritty or felt too out of control to choose to do anything else but lace up their boots and persevere. The culture, after all, rewards the resilient.

It is always easier to look back and minimize the amount of stress, fear, and uncertainty that happened during that time. Hindsight gives you the benefit of time, information, and clarity. It provides a knowing you may wish you could take back to your younger self. At some level, years after lockdown, the country and military culture will still be recovering from the collective trauma.

In March 2021, I asked military family members to take the Holmes-Rahe Life Stress Inventory I had been introduced to more than twenty years earlier.

We were not quite out of the pandemic fog, in fact, we were dealing with another wave of COVID-19, and into another deployment. Much of life was still uncertain, and in-person social events had barely resumed. A good friend of mine, a military spouse and clinician as well, let me know she had been using the same scale to collect and measure stress levels in the special operations community. Together, we tallied the numbers and were not surprised at the results.

Of the more than three hundred responses, 43.8 percent had a 50 percent chance of a major health breakdown and 41.6 percent landed in the 80 percent chance category. Together that was 85.4 percent. Even more striking, the inventory counts any total above 300 as in the highest risk category. About 6 percent of those who participated from the special operations community scored above six hundred, several were at or above one thousand.[314] In a culture that rewards the resilient, what is the cost of being in a constant state of perseverance? What does a breakdown actually look like culturally? It made me think back on Odierno's words from 2015: "Although we've not yet seen the breaking point yet, I worry about when that will occur in the future."

First, military families got a taste of what it looked like to be a family again during lockdown. The culture has shifted since then and how families want to spend their time has shifted as well. Most generations emerged with an unwillingness to sacrifice their personal time and/or family for the cause, in the name of patriotism, or for a mission that had no end. Gen X had always wanted a healthy work-life balance, millennials were willing to start the conversation, and now Gen Z is and will continue to expect a career that respects healthy balance, promotes wellness, and respects family and work boundaries.

While everyone's experience and opinion may have been unique, how stress was stored in our bodies, as a result, was equally unique. It remains a trauma we have still not recovered from. It takes more energy to make arrangements to meet in person than it did prior to the pandemic, and families are more aware of the costs associated with leaving their homes. Evaluating whether you will attend a military family event, for example, costs physical and relational energy, and may even cost your evening if the kids are overtired by the time you get home. Events that used to be a stress relief, such as a military ball, are now not only financially costly for young families but may require childcare and cost family time they no longer want to give up.

Millennials and Gen Z are choosing to find ways to separate their family time from their job. They are less likely to show up for events outside of work hours. They are more protective of their family and their family's time than older generations were at that age. Geo-baching has steadily been on the rise for military families who choose to voluntarily live apart rather than relocate with the service member. The Blue Star Families report from 2020 revealed that "23 percent of survey respondents had geo-baching arrangements during the previous five years. Of these respondents, about 49 percent reported that children's education was one of the reasons for voluntary separation. However, if families had a child with identified special needs, that percentage rose to 65 percent."[315]

Geo-baching is especially hard on families and couples, and I have not recommended it unless there is a medical, financial, or familial crisis. In some cases, I have seen geo-baching used as a solution for the exhaustion of military life where the family no longer wants to serve alongside their service member. However, as the pandemic added stress to young Gen Z's education and mental well-being, more families made decisions to "choose the kids" over the needs of the military by creating their own stability.

It may sound like a contradiction, but due to Gen Z's experience of having their entire lives become virtual, they have a new appreciation for in-person gatherings. Considering that they see online platforms and the virtual space as tools to be used more efficiently, they look at in-person opportunities (especially education) as another tool to choose from. Does the event require an in-person meeting, or is it better suited for on-demand training? If it is strictly for relating and building relationships, they would rather be in person. The event planner must just consider the cost the participants will consider before attending.

Second, as the world opened back up and people went back to work in person or remotely, they returned tired, somewhat traumatized, and unsure of how to navigate the uncertainty of the new normal. The Great Resignation began in the civilian culture as people left careers, chased bucket list opportunities, and switched to jobs that allowed them to be more present with their families and friends. With military families committed for a certain number of years or aiming for retirement, quitting was not an option for most.

In civilian jobs, part of the Great Resignation was to quietly quit, or do the bare minimum. Though they may not admit it out loud, quiet quitting was seen in the military force as well. Gen X, especially, had endured long years of difficult work and stressors on their family, and doing the bare minimum was tempting after striving for excellence for so many years. Considering that the tide was shifting toward protecting family time, leaving work on time rather than working later was a way to persevere without sacrificing more.

Third, since the pandemic, companies, especially in Europe, have been discussing the transition to four-day work weeks. In Belgium in 2022, a bill was passed "allowing employees to decide whether to work four or five days a week." This would not mean they have less responsibility but are willing to condense their work into four days instead of five. The United Kingdom did a six-month trial of four-day workweeks and "are now planning on making the shorter workweek permanent, after hailing the experiment as 'extremely successful,'"[316] with respondents reporting programs like this would "greatly enhance their health and happiness." Other countries across Europe are following suit. Work is becoming increasingly flexible, giving employees more freedom to make choices in how and where they do their work. This will be difficult for the military to compete with as they recruit Gen Z.

And what about the 50 to 80 percent chance of medical health breakdowns associated with stress? I think we can already see how it is showing up in mental health. If there are connections being drawn between medical concerns in the military community to the amount of stress they have endured, we are not talking about it. Research has long shown that as family morale declines, readiness will be an issue.[317] What if Gen Z's thirst for authenticity and holistic wellness is exactly what we need?

AN EXAMPLE WORTH REMEMBERING
Build morale through meeting basic needs.

Lexie Coppinger, a military spouse and family readiness support assistant, shares: "During the pandemic, there was a plethora of challenges that hit the military community. 1st SFAB had soldiers deploying in the midst of it all while kiddos and parents were faced with virtual schooling, spouses were having to quit their careers to further support their families through it all, and lessened social contact was having a very clear effect on folks. General [Scott] Jackson offered every ounce of support to keep soldiers and families involved and connected. A Facebook group called Spearhead Community was set up with weekly physical challenges and family competitions to encourage getting outside, staying connected, and a little friendly competition. General Jackson and Command Sergeant Major (CSM) Jerry Dodson even personally built a giant bookshelf to serve as a brigade free little library of sorts for soldiers and families. Their leadership was a shining example of great leaders recognizing the importance of simply caring for their soldiers and families."

Chapter 14

SHIFT OF AUTHORITY & INFLUENCE

FOR MUCH OF HUMAN HISTORY, those who held the most information held influence. Leaders like to quote Francis Bacon's "knowledge is power" as if to motivate themselves or others that influence comes with knowing more. Yet, as much as technology has shifted the way we connect and work, it has also shifted the way people learn. Anyone who has access to technology, regardless of age, now has the opportunity to become an expert in almost anything they desire. If influence is based on knowledge, then the playing field has been leveled, if not completely upended.

If we are going to understand how to lead, train, recruit, and retain the next generation, we must look closer at the ways access to information has shifted how people learn. Perhaps even more important, we must embrace new concepts that have shaped the way younger generations value authority—a topic of great concern for military leaders today. In a culture that is structurally designed to give authority to those who have climbed to ranks and positions of influence, even millennial leaders are asking how to lead when traditional definitions of authority are challenged.

EDUCATION AS WE KNEW IT

Higher education has long separated social and economic classes, professors from teachers, experts from laypersons, and those who know from those who do not. Education itself remains a luxury in many parts of the world. For generations who grew up in schools prior to the information age, learning was impossible without the knowledge of a teacher and a library filled with books written by experts. In order to learn, students were dependent on the information the teacher or professor knew. This usually required writing it down, memorizing it, measuring competency by regurgitating the memorized information onto a test, or writing a paper to show you could collect expert perspectives to support your understanding. While learning today still requires some level of memorizing and proving comprehension, technology has slowly, and more recently rapidly, changed how we learn and process information as well as where we prefer to get our information.

The calculator was one of the most controversial inventions in the world of education. Even though the slide rule was already being used, calculators entering the classrooms in the '70s evoked fear that students' "computational abilities would be ruined, that students would become too reliant upon machines, that they wouldn't learn how to estimate, that they wouldn't learn from their errors."[318] Not only did they create an unfair advantage for those who could afford them, but the main concern was that calculators removed the need for students to memorize certain steps of especially challenging math problems.

This type of learning, known as "instrumental understanding," is the mastery of knowing how without necessarily knowing why. As graphic computers were made more readily available, educators advocated that the calculators encouraged students' "relational understanding," a "kind of connected, conceptual understanding" where they "don't just know how to invert and multiply, they know why such a procedure results in the quotient of two fractions."[319]

It wasn't until 1994 that the Scholastic Aptitude Test (SAT) allowed calculators during the exams. That year, 87 percent of students brought a calculator with them, and even more did the next year.[320] Decades later, research has shown that "graphing calculators have a positive effect on students' relational understanding and a slight positive effect on their instrumental understanding."[321]

Gen X was in the middle of this calculator debate, with adults saying screen time would rot their brains and calculators would stunt their education. Today, technology is advancing faster than even educators can keep up with. As this is being written, there is a concern in academia around the invention and use of artificial intelligence. AI now has the ability to not only write research papers, film scripts, and books, but to do so in your own voice and style. Time will show how AI impacts the way students learn and enter the workplace.

Technology has a way of constantly innovating how we do things at home and work; however, paying attention to how it affects education for the youngest generation provides keen insights into how they will best learn in adulthood and how we must learn to lead them. The military is especially dependent on training and continued learning to develop leaders, teach doctrine, and ensure competency. Regardless of your role, interacting with the youngest generation will involve some level of teaching, whether it is providing actual training or simply mentoring. The challenge will be to capture the motivation of a generation that moves faster through information than previous generations.

More than five hundred years ago, the invention of the Gutenberg press unchained the Bible from the pulpit and printed it for the masses to read and interpret on their own. Similarly, with the internet, knowledge is no longer bound to libraries or even to the interpretation of the individual who spent their life acquiring it.

Every day, the internet is unchaining information from everywhere it has been held before. The ability to earn influence, authority, power, and respect has changed along with it. For leaders, the path to gaining respect and authority with millennials and especially Gen Z is no longer about who holds the most knowledge, information, or even experience.

THE WAY WE LEARN HAS CHANGED

Along with the rest of the world, Gen Z adapted their learning style quickly throughout the pandemic. Already digitally savvy and connected, they used laptops and platforms like Google Classroom and Schoology in many public and private schools. Teachers instructed children as young as elementary age to access class notes, homework, chats between teachers and students, and planning calendars, all online. Online learning platforms also allowed parents to get regular class and homework notifications if the teacher enabled it.

Before 2020, synchronous and asynchronous education was already being tested and used in continuing education, homeschooling circles, and higher education as an alternative to in-person facilitation. Textbooks, expensive to print and update, were moving toward digital versions, with many administrators and parents arguing they were less effective.

In her 2017 book *iGen*, Jean Twenge pointed out that Gen Z needed textbooks that included "interactive activities such as video sharing and questionnaires, but they also need books that are shorter in length and more conversational in their writing style."[322] As digital textbooks with interactive features continued to launch, most academic libraries retained a few printed textbooks for students who needed them. But during the pandemic's initial lockdown, printed materials were inaccessible and digital textbooks became critical for future learning.

Although there were initial concerns from older generations about Gen Z's ability to retain and comprehend information accessed digitally, the pandemic proved that students, especially in higher learning, adapted quickly and successfully to "screen-reading their texts."[323] Additionally, digital textbooks allowed publishers to provide more affordable options for students, especially given the pandemic's impact on family income. Reducing, if not eliminating, printed textbooks altogether could be imagined as a way forward.

While older generations in the education system were forced to adapt away from their preferred way of learning through printed materials, a large number of parents of Gen Z were unaware of the shifts it caused in schools and in their children's lives. With fewer or no textbooks, students were able to carry almost everything they needed for a day's work on their person. Textbooks and curricula were available online from a desktop, and in some cases from students' smartphones as well through various learning apps and platforms (if schools allowed it). For an increasing number of schools, this lowered, if not eliminated, the need for hall lockers. In some cases where school work could be completed fully online, paper and supplies were less needed as well. This further reduced the need for extra storage throughout the day and shifted the way students study and learn.

Our family moved during the pandemic and our Gen Z kids bounced between in-person and virtual school options throughout 2021. Both had been telling me how different it was for them compared to when I grew up. Like most parents, I was trying to parent them through the context of my own experience of being an adolescent.

Postpandemic, many schools are dealing with a shortage of teachers. I decided to step into their world and substitute in the local high school. While each school has different policies around phone usage, rules, and cultures, it was a shock to my Gen X memories of high school to see what my kids had been trying to tell me. Not only were there very few textbooks, but most of their curriculum, interaction, and textbooks were online. When it came to completing assignments, some students completed their work on their mobile devices.

Most classes of various levels tackled their work far faster than if they would have had to flip through the pages of a textbook to find answers. They simply accessed the information needed, scanned for answers, and knocked out the work quickly. Although less true for advanced classes, students in some core classes were allowed to use a search engine for answers, which enabled them to spend a large portion of class time freely searching or playing games online.

This, of course, threw me since as a student I had loved highlighting my textbooks and as a parent I expected the majority of class time to be filled with learning. For a brief moment, I had a substitute teacher identity crisis. Were they cheating? Should I say something? What were kids learning if some subjects didn't require memorizing or internalizing information? Where calculators shortcutted steps of the math equation, this seemed to skip large steps of learning. Or did it? The more I thought about it, what was the difference between finding answers in a textbook versus searching the internet?

Wondering if this had more to do with the fact that I was a substitute rather than the teacher charged with teaching the curriculum, I asked other teachers about their perspectives on education postpandemic. Many of them shared that the education system is in a significant period of transition and teachers are encouraged to be more facilitators of learning than gatekeepers of information.

Millennials had been the first ones to ask schools to adapt to using more technology and the internet. This initial shift created "new ways of thinking about teaching and learning a) from linear to hypermedia learning, b) from instruction to construction and discovery, c) from teacher-centered to learner-centered education, d) from absorbing material to learning how to navigate and how to learn, e) from school to lifelong learning, f) from one-size-fits-all to customized learning, g) from learning as torture to learning as fun, and h) from the teacher as transmitter to the teacher as facilitator."[324] Millennials also advocated for schools to bend to the social and emotional needs of the students. Teachers, or even textbooks, were no longer the sole source of information but considered co-collaborators.

Although moving our nation's education system entirely online was chaotic and traumatic at first for educators and Gen Z, it was incredibly impressive and proved that virtual learning is possible, when necessary. For likely every generation, but especially Gen Z, virtual education and interaction could now be adopted as another tool or resource, to be used in moderation.

Gen Z witnessed the social and psychological consequences of other extremes during their most formative years, developing strong opinions and values in the process. Strong polarizing debates around the virus, vaccinations, and their experience of school encouraged them to be more compassionate about holistic care, individual rights, and access to healthcare for all. Two wars and political division within the nation have made them less trusting of the government (64 percent do not) and more likely to value personal autonomy and global perspectives of humanitarian efforts. The impact of social media on privacy and in shaping the culture has motivated them to expand their portfolio of platforms and instead use them as tools. Similarly, the pandemic moved their entire world online, encouraging their value of online learning as a positive solution but only when balanced with in-person opportunities.

In many ways, having more options available to educators and Gen Z has made learning a more personalized experience. The online world is a completely personalized experience where you can swipe between apps, jump between articles using hyperlinks, and choose your own adventure on streaming platforms like Netflix and Amazon Prime. There are now countless ways to learn virtually, whether it is through a facilitator of a class, on-demand courses for every level, and platforms and forums like YouTube where you can troubleshoot problems with your random appliance or learn computer coding. You can learn about almost anything, casually, from anywhere, and at your own pace and convenience.

Artificial intelligence is also changing the way on-demand learning happens. It can personalize an individual's experience by learning their strengths and weaknesses. With pandemic-related reductions in teaching staff, "AI teaching assistants and automated grading systems are especially welcomed in most schools"[325] as it reduces the workload so teachers can connect with students in other ways. The question for leaders and instructors is, can the content be learned on-demand without facilitation? If yes, then Gen Z will question why it isn't when they are forced to sit through a briefing.

Yet, for Gen Z the value of human interaction is equally important when used appropriately. New trends in education include the role of student agency where students are encouraged to take ownership of their own education by asking questions they are most interested in regarding the subject, searching for the answers, and then teaching their peers.[326] By taking more ownership, students "have a voice in that environment through the collaborative creation of classroom norms." They are then able to "help improve the classroom culture" where students can "feel safe and know that when they express their thoughts, ideas, and concerns, that input is seen as meaningful and valued"[327] Gen Z wants to feel "empowered, so allowing them to make their own decisions, think critically, and offer their opinions will likely engage them further."[328]

Considering a majority of adolescents play at least one hour of video games per day, educators are also moving to gaming techniques that make the information immersive. About 51 percent of Gen Z surveyed stated they learned best by doing and 12 percent through listening.[329] By creating esports for academic learning, students are able to work on teams or in collaborative groups that compete for incentives. Schools are finding that students are not only more engaged but that it "benefits students' overall academic performance and social-emotional learning."[330]

Again, the question we can ask is, does the content warrant conversation or collaboration? While millennials appreciate more collaboration and dialogue, Gen Z would rather have efficiency over conversation. Again, Gen Z isn't necessarily trying to shortcut the process but they value time as a resource. If it requires collaboration, then it is worth it. However, if a task needs to be completed independently, they want to get at it, possibly without needing a full understanding of the why behind it.

For example, training that requires full and complete competence around topics like safety or operational security (OPSEC) or when leaders are looking for open dialogue, team building, or collaboration are all good examples of in-person opportunities that any generation would benefit from attending. Young family members, for example, are more likely to attend a predeployment training when it maximizes their time with military lawyers present for legal paperwork, education about OPSEC, and information about support resources. If the meeting were only about the support resources available to them, they would be less likely to attend since it could be researched online or distributed more efficiently in an email. Embedding that education into an event that addresses other important issues best taught in person increases the likelihood that they will catch the information.

The military career has long been marketed to the younger generation as a profession that develops leaders with skills that are unique and marketable. Gen Z largely considers financial security and professional development when making decisions and, after watching millennials graduate college with student loan debt and difficulty finding jobs, Gen Z wants to know that what they are investing in will pay off.[331]

However, they are more interested in "short-term outcomes, not necessarily on life-long careers in one place."[332] They want to know if this career is the most efficient means to reach their goals. Education, healthcare, and family benefits continue to be successful recruitment incentives. However, if the way in which the force operates does not reflect a modern, multigenerational approach to learning, advancement, and efficient use of technology, Gen Z may be harder to retain.

REFORMATION OF AUTHORITY

One of the biggest questions I hear from military leaders is, "Does the younger generation respect authority?" With a hierarchical culture, respect for authority is crucial to order and successful operations. The question isn't so much about the younger generation obeying orders. I believe it has more to do with the cultural norms and requests that leaders want to implement but that are often questioned by the younger generation. Questions like, "Why can't we put our hands in our pockets?" "Why should my spouse feel obligated to get involved?"

Back when millennials were just entering young adulthood, older generations interpreted their perceived sensitivity[333] and demand for flexibility as a lack of willpower and disrespect for structure and authority. As millennials entered the force, their learning style, which included a more relational understanding of the big picture, was a shift for older generations. Leaders were not accustomed to explaining why they asked a service member to execute a task. Also challenging was that millennials expected leaders and facilitators to be willing to "improve their knowledge, intellectual abilities, and their overall growth in learning alongside the student-soldiers."[334] It was a drastic shift from the "teacher knows best" model.

Millennials have since entered leadership roles and faced various life stressors of their own. They have not entered the workforce as entitled as the older generation expected. In fact, they consume far less alcohol and tobacco and have brought creative and collaborative solutions to outdated systems in the military. They are proving themselves to be not only reliable leaders but perhaps closer to the kind of leaders Gen Z needs.

Similarly, there have been concerns about Gen Z's views of authority given that social media has created direct access to the highest ranks, something that was once achieved only after hard work, loyalty, and time served. Fear of cancel culture has paralyzed even millennial leaders who now feel that power and influence have been flipped upside down with the younger generations having influence over their careers rather than the other way around.

In the workshops where I have invited Gen Z to share their communication style, they describe wanting to be seen as very different from millennials who, to them, seemed unnecessarily loud and overly sensitive. The difference, though, is that millennials used their voices long before social media became a tool that Gen Z uses to voice their own opinions. Gen Z has effectively learned to leverage their influence online without ever raising their voices or even leaving the couch.

Some could say that millennials paved the path for Gen Z's quieter approach to influence with their advocacy for the education system to adapt to their needs as students. In 2019 when Lukianoff and Haidt wrote *The Coddling of the American Mind* they stated, "Students and professors say they are walking on eggshells and are afraid to speak honestly"[335] around colleges in America. As a result, administrations and institutions are more passive and responsive to the needs of students today. Gen Z started entering college in 2015 and likely adopted a shift in power that they weren't fully aware they had been given.

An additional contributing factor is that the failings and vulnerabilities of institutions, brands, and leaders are more visible than ever. According to a case study by the Flamingo Group of Gen Z entering the workplace, "They've spent their entire lives online, and as a result, have been more intimately exposed to the failings of the powers-that-be to tackle world issues than any generation preceding them. From political scandals to #MeToo to climate change, all have contributed to a breakdown in trust toward traditional political and civic institutions that has shaken many people's sense of structure and place. Authority isn't in crisis, but it's being questioned, reshaped and reformed.[336]

News outlets were once considered vetted and trustworthy sources of information for Gen X, boomers, and older generations. Beginning with Walter Cronkite's rarely shared opinion shifting American politics toward Vietnam, opinion-based media has taken over and trusted sources for information are significantly harder to find. Now, everyone can be a source of some kind of content, but just because it is published online does not make it credible.

After reading an article that implied parents were concerned about Gen Z "believing everything they read," I asked a group of Gen Z high schoolers how they discerned whether the information they found online was true. Most of them replied that they do not trust the information they see online until they can verify it with other sources. They are guarded with blindly trusting experts or authority figures simply because of their status or connection to an institution and would rather test their character, authenticity, or reviews from their peers.

Gen Z's access to people and information has allowed them to enter debates, question authority from a distance, collaborate online with their peers, and share direct feedback with brands and sources. "Their simultaneous exposure to hierarchical frailty and ability to question it makes them a fascinating lens through which to understand how the world is changing. With traditional authority in question, Gen Z are effectively redefining it—both the way it's expressed and the way it's experienced."[337]

CURRENCY OF TRUST & AUTHENTICITY

Especially since 2010, the visibility of toxic military leaders and moral failings of leadership have contributed to the disintegration of public trust in the military and government. For many civilian brands and institutions, the constant threat of being canceled publicly was initially seen as a reputation killer. However, many have since embraced the influence of younger generations by answering their demand for authenticity and transparency, returning with an even stronger relationship with the public based on trust.

Considering the DoD tends to not be well-known for revealing vulnerabilities and cracks in the system and also lags behind cultural changes in the civilian sector, authenticity and restoring trust in relationships is not necessarily a strength. In addition, many of today's more senior military leaders grew up in a generation that was less open with their vulnerabilities and believe that failures negatively impact your career and reputation. With leadership decisions more likely to be anonymously submitted to Army W.T.F.! Moments rather than evaluated properly through the chain of command, people are understandably hesitant to take leadership roles.

Gen Z is not asking for perfection, nor are they eagerly waiting to ruin the careers of those who make mistakes. What millennials and Gen Z have shared is that when, not if, leaders make mistakes, they are far more respected when they are able to own them and grow from them. It is an opportunity to model humility, authenticity, and growth as a fellow human. It also gives those who are watching permission to be flawed and human as well.

Authenticity, however, does not have to mean you are airing your dirty laundry or showing instability. Instead, younger generations who are in the difficult seasons of raising a family and balancing work or deploying for the first time want to know that you have been there and understand.

Both millennials and Gen Z value the authority of mentors and leaders. In fact, both desperately want mentoring to grow in their abilities and skill set. Yet it is important for mentors today to understand that knowledge and time in are no longer enough to win their respect or loyalty. What these generations want is the wisdom of those who have experience and knowledge in the context of a trustworthy relationship.

Wisdom is not only about having knowledge but the ability to collect and extract meaning and perspective from the knowledge a person has gained. In that sense, it is something almost anyone can acquire. It is the humility to use the knowledge one has for the benefit of another.

Daniel Pink, businessman and *New York Times* bestselling author of *To Sell Is Human*, spoke to this idea of how expertise is shifting in the information age. He suggests that experts are now those who can curate the information rather than being the sole source of it. He says, "Information curation is more important than ever because we have more information than ever. While customers, prospects, and people on the other side of the persuasion table have the ability to get all of that information, they don't necessarily know how to make sense of it. That is where your expertise comes in … you can help them sift through information, find what is meaningful and reliable. Curating information is where expertise comes from, not accessing it."[338]

This can be freeing for leaders in that it is overwhelming and exhausting to try to be ahead or on top of all of the information available. There is no end to the amount of information on any given topic or perspective. What Pink is saying, and what younger generations are expressing, is that experts who are highly respected in their fields are those who can make sense of the information rather than being the sole source.

Pink goes on to give the example of how many of us approach advocating for our own healthcare. Most patients have the ability to research symptoms or concerns. Medical books and journals are unchained from medical libraries, informing us with a lot more information than we actually need. Doctors today, and leaders on every level, are more likely to win at relationships if they are willing to enter collaborative relationships where possible and, instead, take on the role of interpreter and guide.

This does not mean that we set aside what we know. Competence, through professional excellence and practicing what we preach, is a part of building trust and respect. However, it does require the removal of our egos. There is no longer room for ego nor privacy for it to stay hidden. While not every decision can be a collaborative one, taking on a more consistent person-centered approach to leadership makes the deposits necessary for when trust and loyalty are required.

There is a moment during Madi Hammond's TEDx Talk, "Living as Gen Z: From Fear to Positive Change," where she subtly describes the moment she and her classmates shifted their relationship and respect for their teacher. She started by listing the overwhelming amount of information Gen Z has access to that impacts their mental health, including "how to kill things, how to fight things, how easy it is to break glass, how many different ways people can invade your school, how many different ways you can die, how many ways people can walk past you, and how many serial killers you have met in your lifetime." She goes on, "Fifty years ago, my sixth-grade band teacher might not have promised to a group of eleven-year-olds that he would die for each and every one of them if a shooter attacked." She then implied that she and her classmates went on to "solemnly asked him what he would do in virtually every situation [they] had heard on the internet or heard on the news, just to make sure he would protect [them]."[339]

What Hammond's example shows is that authority for Gen Z is ultimately given after trust is established. It was not until the class verified that the teacher cared and saw their human value that they gave their loyalty in return.

Major brands have quickly picked up on the ineffectiveness of their ads targeted toward Gen Z due to how inauthentic they felt to the generation. Now brands are more likely to build loyalty with the generation by building a relationship with them, listening to feedback, and giving them a vote in the production of products, branding, and advertisements. This incredible amount of influence also provides Gen Z with buy-in, or partial ownership, increasing brand loyalty and participation in the relationship.

This is not new. We have always known (or experienced ourselves) the power of leaders who care about those they are leading. However, it is tempting to forget or bypass the people around us with the pressures of responsibilities and a system that teaches service members to passively submit to superiors regardless of what kind of leaders they are.

Especially in the military culture, there will be times (based on your role) when you will be expected to make decisions based on the information you have available to you. Even when you think you have the right answer, setting ego aside and leveraging humble confidence[340] will help you empower your team to bring their strengths in a way that ultimately informs your decisions.

People want to be seen as human, with families, dreams, concerns, and insecurities. A majority of our force wants to be treated holistically rather than as a commodity. Although they may not verbally challenge a superior, they are willing to vote quietly by not engaging the community outside of work hours.

WIN TRUST BY REDISTRIBUTING POWER

Winning trust most definitely happens through authenticity and competence; however, there are times when leaders can build trust nonverbally, simply by adjusting how they lead in the presence of power dynamics happening in the group. It is normal and natural for humans to desire control and influence over their circumstances, especially in group settings. No one likes to feel out of control, especially if patterns are being confronted or changed. Those who do feel out of control will eventually search for influence or power to regain a feeling of security. In addition, groups cannot function without some pattern or distribution of power.

There are various types of power within a group and each holds power in different ways. The most well-known and talked about five types of power are:

1. *Reward power*: Who in the group has the ability to reward others? Is someone's senior rater in the room or someone who has the ability to evolve a position or approve leave?
2. *Coercive power*: The opposite of reward power, who has the ability to use punishment to gain compliance from others? While this may be the same person who holds reward power, it can also be two different people if there is a bully or someone looking to elbow to the top.
3. *Legitimate power*: Who holds an actual position that obligates others to accept their influence? If a general were to walk in the room, although a service member's senior rater is in the room, there is now a secondary power player present.
4. *Referent power*: Who is considered wise, credible, or trustworthy in the room? They may not have a role that assigns them power, but people listen and are open to their influence.
5. *Expert power*: Who holds credentials or expertise on a specific topic? If you are trying to organize the funding of an event or purchase, the JAG legal officer has expertise and therefore power in how the decision is made.

A sixth has been added since the rise of the information age:
6. *Informational power*: Who holds the most information about a topic of interest? This could be information gathered informally through experience or formally through learning or information gathering.

I was recently asked by a commander to come and assist in increasing communication and productivity for a smaller multigenerational team. The generations ranged from millennials to boomers. As they entered the room and took their seats, I noticed that the boomer generation sat near the front where I would be teaching, while Gen X chose seats in the middle and millennials in the back. Knowing that seats were not assigned and far too organized to be a coincidence, I decided to pay close attention to the relationships in the room. Even though I wanted to believe the boomers were just eager to learn, I had a hunch it had more to do with the power dynamics of the group.

Over the course of the day, I realized my hunch was correct. One of the issues the team was having was a power struggle between the boomers (old guard) and millennials (new guard). The boomers, feeling challenged to shift by the millennials in other contexts before this training, subconsciously chose their seats to align themselves in proximity to the one they believed would have the most power during the training (me).

There are some power players who will be obvious, such as the commander in this situation who held both legitimate and reward power, however, others can be a little harder to find if you are coming in new to the group like I was that day. I would add two additional power players who often are attempting to influence others out of insecurity within themselves or in a power vacuum in relation to the group, especially if they feel threatened:
7. *Monopolizer*: This person will tend to seek power through commenting or talking more than others, filling the silence, or taking over discussions.
8. *Resistant one*: Possibly out of an analytical strength or being slow to trust others, this person will verbally or nonverbally challenge those who do not have legitimate, reward, or coercive power over them, to gain power over others.

I honestly do not believe the boomers had any malicious intent toward their colleagues that day. In fact, 93 percent of our communication as humans is nonverbal[341] and much of it is communicated subconsciously. When you walk into a room where you do not feel comfortable, you are more likely to sit away from the source of tension. People lean forward when they are interested, and lean back when they are thinking, deliberating, or unsure. When the generational groups walked into the room that day, they subconsciously chose their seats based on peer alliance (referent power) or in the case of the boomers' seat selection, I believe reward power.

Studies have shown that people who tend to sit in the front of the room are either looking for security or "consist of the zealous types" and are those "who thrive under the eye of authority figures."[342] As the workshop progressed, I realized that by positioning themselves close to me at the front, they attempted to align with the facilitator who could reward them with influence. This gave them the security to challenge and take a more assertive role by taking control of discussions of the group. They also shut down collaboration in the room with what they knew about policy and high-level information they felt the younger generation didn't understand.

In this moment, it was my role to lead as a curator of the dynamics at play in front of me. Rather than point it out, it would be more effective and respectful to break up the power dynamic in another way. I decided to get them out of their seats with a task that put them with peers in separate corners of the room. All were made equal by this exercise because they were all tasked to do the same thing at once.

Once they sat back down, I presented a large group discussion question, but this time shifted two millennials from the back of the room to empty seats in the front row. Both generations were confused by the rearrangement but I asked them to continue the discussion. Simply by shifting their seating, the millennials were more courageous to talk (feeling secure in proximity) and the older generation had to acknowledge them in the discussion. This was also a way to subtly shift the authority back to me by making changes to their seating arrangement rather than the participants defining it for themselves.

Late in the day, the group started to discuss solutions for how they could organize their communication and productivity, with the goal of reducing the number of meetings they were having. One of the front-row millennials spoke up and presented an online product management tool that had already been vetted by the military as a solution. Painfully, I watched the group dismiss his idea and continue the conversation.

I let the group continue as the young man tried again, presenting that the platform was able to do all the things they were looking for if they were only willing to try something different, and again they dismissed the idea. It was clear to me that the group wasn't purposefully trying to ignore the young man, but from their generational perspective, he did not know what they knew. Therefore, in their eyes, he didn't have enough experience to have authority in the conversation. They simply were not listening. Finally, a boomer said out loud, "I wish we could find a tool that would do all of these things!"

You can imagine the defeat the young man felt at that moment. Even though I had expert power for the day, I would be leaving. I knew that the strongest power player in the room was the commander, even though he had been letting the old guard have the power for some time in the organization.

At that point, the bigger and more long-term goal was to redistribute the power to the commander. When we don't know what to do, most people will jump in a try to force change. In this case, that would have only made the old-guard dig in deeper. What the commander needed was to win the loyalty of the rest of the group (notice Gen X had been mostly silent and invisible) by modeling for boomers his awareness in leveraging the strengths of the group.

The commander and I made eye contact and I nodded at him to signal it was his opportunity to lead, not mine. The commander interrupted the discussion and asked the young man to share his idea again, saying "Hold on, let's listen to what he's trying to say." Because of his influence, the room not only stopped to listen but actually heard the young man who had informational power. They were able to see how a small learning curve with a new tool might be worth the effort. In fact, the commander asked the millennial to look into the tool and brief the team on it when it was time to learn it!

Of course, the dynamics of that team or any other group are not going to change overnight or with one intervention. However, the lesson was that assuming authority and taking charge in an aggressive way is not always the approach to winning influence. Influence is earned through relationship building. Notice that the boomer generation was not corrected in public (which would have been humiliating and unkind). Likewise, the millennial didn't force his way in, which would have isolated him further.

The commander realized the value of collaboration in a moment where collaboration could lead to team building. In doing so, he developed another layer of trust with both generations by seeing the younger one for the strengths they brought to the team and valuing the older by still leaning on their wisdom.

Postpandemic, schools were faced with the important decision of returning to what they had always known or taking what they had implemented and evolving education forward. Renee Owen, an assistant professor of education leadership at Southern Oregon University, was quoted in an article that made me think about the military's opportunity to move forward after seeing and valuing the various shifts it has endured. She said, "In schools, there is a constant striving for improvement, but improvement—getting better at what we already do within the systems we already have—will never fundamentally change who we are or how we think. We will continue to get the same results unless we are able to see education in a completely new way."[343] Perhaps we need to consider the same warning for the future of our culture.

AN EXAMPLE WORTH REMEMBERING
Teach potential through words and action.

Air Force Senior Airman Elise Day-Barnett, a member of Gen Z, shares: "I will never forget the incredible leadership of Master Sergeant Samantha Maghamez. I had not received the 'normal' basic boot camp military or technical training my peers did due to COVID-19 hitting the US. Once into my first assignment to Minot Air Force Base, North Dakota, I slipped on the ice during a night shift and separated my shoulder. As a nineteen-year-old female and only ninety-eight pounds, the weight of the gear I was wearing caused tears in my shoulder requiring reconstruction, physical therapy, and a change in my responsibilities. It was a terrible time because I felt like I was not contributing to the mission, and I was less than my peers. It was then that I met my new noncommissioned officer in charge (NCOIC/E-7), MSgt Maghamez. She was one of the first leaders I encountered who was involved with her airmen and really cared.

MSgt Maghamez went out of her way to make sure her troops were taken care of, on the correct schedule for their family life, and just overall happy. She gave me room to grow, recognized my job knowledge, and gave me a position to lead in that job. She actually cared about my personal life and supported my wife and me when we needed it most. I am forever thankful to her for showing me to trust myself and be confident. Even in the midst of loss in her own personal life, she insisted on us holding a baby shower for the three pregnant women working in our section, her troops. She planned all of it, brought decorations and presents, all while standing there with a supportive and loving smile for those under her. The things she taught me will stay with me in my career and single-handedly shifted how I view the Air Force and the military as a whole."

SERVICE MEMBER CULTURE

"Human beings need three basic things in order to be content:
they need to feel competent at what they do;
they need to feel authentic in their lives; and
they need to feel connected to others.
These values are considered 'intrinsic' to human
happiness and far outweigh 'extrinsic' values
such as beauty, money and status."

—Sebastian Junger, *Tribe: On Homecoming and Belonging*

Chapter 15

PURPOSE & PERFECTION

I HAVE LOVED EVERY OPPORTUNITY to listen to the stories of service members over the years. Whether it is in the counseling office, over a beer during a social event, at our kitchen table with friends, or in hallways between training events, I have, like most military spouses, strained to envision their descriptions of worlds I will likely never see and experiences I'll never envy. One of the most significant moments of my life was the opportunity to travel with Secretary of Defense Ashton Carter as he visited troops in deployed locations for the holidays. In one whirlwind nine-day tour, I got to experience the entire force working together to move the needle on global conflicts and make national security happen. Like looking down on a massive chess match from the sky, I watched the SECDEF meet with prime ministers and generals. I listened in as journalists asked military leaders questions to guide their daily articles on the large-scale movements of global military powers.

I will never forget seeing the magic of such an incredibly complicated and extensive institution working as it was designed. Families and civilians rarely, if ever, get that chance. But for just a moment, I got to catch a glimpse of what service members spoke of when they talked about, if they talked about, the incredible world of deployment. I will never forget the familiar spark in the eyes of service members of every branch getting to do what they had long trained to do. I had seen it in my husband's eyes as well; I just didn't understand it until then, and it hooked me too.

I flew on C-17s, C-130s, and the Air Force's E-4B, and watched the Air Force crew perfectly execute the mission of carrying us where we needed to go each day. From changing a giant tire on a C-17 to sitting in the cockpit and watching an in-air fueling, I was awestruck at the talent and capability of those around me. I was taken back to my childhood and saw my own father, an Air Force pilot, through new eyes. Traveling to Incirlik Air Base in Turkey, just days before families would evacuate it, I spoke with other spouses who could have been me in another life.

I marveled at our Army soldiers as we toured Erbil, Iraq, and Forward Operating Base Fenty, Afghanistan, where my husband had been during his first deployment. I walked into B-huts, the USO, and the mailroom, all decorated with minimal strands of tinsel and battery-operated Christmas lights. The glow they put off paled in comparison to the pride and sense of purpose of those soldiers.

I also touched the memorial at Fenty that had the names of soldiers our unit had lost just six years prior. I felt the guilt of standing so close to the mountains where they lost their lives when it should have been their families visiting there instead. I also felt misunderstood by the civilians around me who didn't know why it all mattered so much to me.

No doubt, the service members who took me under their wing were all on their best behavior for our VIP trip, which was not only inconvenient but likely derailing months of operational plans. While the SECDEF took his meetings, I walked on gravel paths framed with T-walls, listening to anyone willing to share their story and tell me what they wished people understood. Even though I had never met them, I felt immediately that these troops were my tribe. They likewise described my visit as a bit of home coming to visit them.

Their stories and my listening continued as I experienced our Navy and Marines on an MH-53 helicopter over the Persian Gulf and walked aboard the USS *Kearsarge* and our coalition partner's *Charles De Gaulle* aircraft carrier. I wrote about the life-changing experience this trip was for me in detail in my book *Sacred Spaces*, and I will never forget seeing with my own eyes a small portion of what service members mean when they say, "I want to deploy again." While generations before might have been driven by patriotism and duty, what I've seen with my own eyes and heard in stories both at home and abroad is the sense of belonging and togetherness that no other setting or career can compete with.

After living this lifestyle among these families, comradery is likely the best word we can find in the English language to describe what no book has enough space to capture. It is so precious. This "in service for another" has long been the answer for so many who have been asked, "Why did you sign up?" It is for people that warriors go to war, and for people, they return.

As we look back on the topics of this complex problem of morale within the culture, every single one of them crosses paths with and deeply impacts the service member. Our history as a force has included generations of honor, duty, courage, and patriotism, inspiring millions of young people who wanted to be a part of something bigger than themselves. In some respects, the military has been a welfare state, changing family trees with promise, provision, and the "set apart-ness" that comes with wearing the uniform. Some of the scrappiest individuals have emerged from difficult circumstances and found a sense of belonging and self-confidence in the military. As one veteran pilot told me, "The anticipation of being able to fly an aircraft and have that responsibility and gratification outweighed the risk of having to serve in conflict. Where else could a twenty-two or twenty-three-year-old be given a supersonic jet to dance in the sky and later a multimillion-dollar jet to fly around the world?"

The variable of money doesn't just provide income for the service member, it provides training and the opportunity to learn new things, advance skills, and touch technology most could only dream about. Funding also provides a home for your family, high-quality childcare for your children, and a community for your spouse to thrive in. Funding is also connected to the provision of resources you need to be successful at your job and ultimately bring everyone home.

The question of "How are our service members doing?" is harder to answer after two decades of war and the turmoil of returning home. Perhaps this question is the crux of it all. To attempt to answer this question informs morale, readiness, retention, and recruitment. It communicates something vulnerable about our force, one that "must show no cracks" in its image to the world. Yet, its publicity in the last decade seems to speak for itself.

How are service members doing? It depends on who you ask. Each cohort will likely have a different answer. Many women and minoritized communities are still advocating against sexual harassment and discrimination, for more safety and diversity, and to be respected. On a higher level, if you were to ask someone nearing retirement, you might get an understandably disgruntled, sarcastic "Another day in paradise." Ask a new Gen Z, and their reply might be, "I'm still weighing my options." Somewhere in the middle is the mid-level service member who remains eager about the opportunities that await them and the belief that their contribution will make a difference.

With more than 1.3 million active-duty members serving in all branches and more in the National Guard and Reserves, there are just too many stories to tell. Their experiences over two decades are so varied that it is likely easier to get it wrong than right. While the stories of military spouses at home are also unique, there is a lot more commonality in the waiting, the leaving, and the stress of responding to the needs of the military. For service members, though, some have deployed, and others haven't, some military occupational specialties (MOSs) see combat more frequently, and many others provide much-needed support behind the lines. For some, there is nothing better (or more exciting) than climbing in an armored tank. For others, it is boarding a ship or controlling the subtle movements of an aircraft. The experiences and challenges of enlisted versus officers are also unique, not to mention the level of job satisfaction.

Since there is no shortage of content on military strategy and policy for the US throughout GWOT, and even more about the mental health challenges, suicide rates, or the exact opposite, how strong and ready our force is, I'd like to cover something I believe is missing in the written space. When you ask service members why they offer a sarcastic "Living the dream" answer, sometimes, if they feel safe enough to be honest, you'll hear "Nightmares are dreams too" or "To say otherwise risks my next promotion." I find that telling.

It is the unspoken words that lie beneath the surface, that I most want leaders to understand. Perhaps what we would hear collectively around the water cooler wouldn't be all criticism and traumatic war stories. Instead, if it were an open dialogue free from posturing and annual evaluation paranoia, I like to imagine it would be what you'd hear when you run into someone you've served with at the local taco dive. After a few exchanges of "Where are you now?" and "How's the family?" you might settle into a safe space to share what's really on your mind.

Sure, there will be funny deployment stories of chasing the USO guy dressed in a turkey suit on Thanksgiving, but there will also be moments of silence where your eyes drop in reverence to what this lifestyle has cost. You'll try your best to smile and nod, but purposefully, you'll leak out just a bit of the flooding emotion that comes with feeling seen by someone who knows. "For those who understand, no explanation is necessary; for those who don't, no explanation will suffice."[344] And maybe, if you felt safe enough, you would begin with, "We've been through a lot, haven't we?"

As I've thought about and talked with service members and their families about the state of our military culture, I've wondered how much of it is just the natural progression of any career, especially when so much of it feels so heavy. Covering two decades of any person's career is bound to follow them through various stages of optimism, passion, perspective, fatigue, and even cynicism. I still ask myself that question, wondering if the state of our culture has more to do with the this cohort destined to find their seat at the local American Legion bar sharing war stories while the next generation signs up to take their place.

Honestly, it feels pretentious to suggest otherwise. As if this season of US conflict was any different or worse than anyone else who also earned their seat. But then I listen to the stories of veterans from conflicts past and they seem to collectively share that everyone has a story that is sacred to them. Their wisdom, which comes from time and distance from what originally felt so overwhelming, is that it's normal to give your all to a career and then eventually find the clarity that in the end it was always, only, going to be about family and friends.

So I have asked myself over and over, "What's the point of sharing this story if this is just the progression of any military career? Especially if it was a choice we made for ourselves to enter into it?" But then I listen to a millennial leader in the middle of their career asking if they want to stay in an institution that feels very different from what they thought they signed up for. Then I think about my son, who is considering whether he wants to enter the Air Force with bright eyes, a mind that has been protected from the evil of war, and his entire future ahead of him. I believe I can say stories like these are my why. There are only a few who will ever be able to say they were able to actually change the institution. There are far more of us who still have the time and influence to make the institution a better place in which to serve.

PISS & VINEGAR & FULL OF PURPOSE

When unimaginable things happen, our minds want to assign a narrative to them. We are not built to easily accept hard things at face value with no meaning attached. So we make up stories, true or false, to somehow wedge into our larger story as if that chapter was always meant to be there. Our family joined the military because of a calling we felt, less to our country and more to this tribe. I believe that was a true narrative. Spending years apart to help the people of Afghanistan? That one is a bit harder to decipher.

There are some events we will be part of that feel far too inhuman to fit into our very human narrative. Abuse and trauma victims have this experience, but it doesn't always have to be trauma-related. I have seen deep disappointment at the hands of toxic leaders, the pain of infidelity, and even the unfairness of a medical diagnosis wage war within a mind and heart grappling for a way for it to make sense. Some, when all attempts fail, simply resort to blaming themselves.

Warfighters accept that they are likely to encounter the worst of what humanity can do to each other. They don't often expect the toll it can take when it comes about by their own hand. There are some things that no amount of meaning, no matter how hard we force it, will ever help it make more sense. Our mind, instead, attempts to reject or detach from what it cannot file away properly. We numb, distract ourselves with work and busyness, completely dissociate ourselves, or compartmentalize those memories as other. My husband calls deployment "bizarro world," where no part of it makes sense within the context of the normal world at home. It was his way of detaching from, but not forgetting, this place that was full of experiences that didn't fit within his narrative of husband and father but were too sacred to let fade.

Everyone needs a sense of purpose. It drives us into action, begins the healing process, casts vision on an unknown future, and harnesses courage in the face of risk. We simply cannot help but to look for it. At the beginning of 9/11, anger and a need for justice, and frankly revenge, fueled individuals to join and do their part. It was similar to World War II in the sense that war happened on a large scale on US soil, except this time it was to unsuspecting civilians rather than a military target. People wanted to do something about it. It also birthed a bravado or aggressive enthusiasm in especially the youngest generation that has long been a part of the warfighter's journey.

There are many theories of where the phrase "piss and vinegar" started. Some say it started in the early 1900s; others claim it was connected to General George Patton, who once said gas for his tanks was more important than food for his troops "since the men could eat their belts."[345] Either way, it has been a phrase that many service members have used to describe the scrappy emotion behind their drive to join the military or the motivation to deploy.

Grant McGarry, a 1st Ranger Battalion veteran, said of his decision to join, "Some were saying don't do it and others were not surprised. Their comments were just noise to me. I didn't care what anyone had to say. I had a goal and it was to become an Army Ranger. After finishing up my degree a few months later I shipped off to basic training and never looked back. I was full of piss and vinegar and chomping at the bit to get over to Iraq. In just two months after arriving at the 1st Ranger Battalion, I finally got my wish and deployed to Iraq."[346]

Almost everyone begins their career, but especially deployment, with some sense of purpose they assign to it, full of piss and vinegar, to test their limits or abilities or to make a difference in a positive way. Tatchie Manso said, "It was at that moment I realized that there was something I needed to be part of that was greater than myself."[347] Bradley Johnson wrote, "I arrived in South Carolina and began my adventure from citizen to soldier. Fourteen weeks later, I returned home ready to take on the world (or at least a part of it in the Middle East)."[348]

Such is the start of any young person entering adulthood. The beginner's tendency to enter into difficult challenges with overconfidence has been researched and it turns out it isn't just young soldiers eager to go to war. They call it the "beginner's bubble" and it happens to everyone. Researchers Carmen Sanchez and David Dunning found that when they asked participants to identify symptoms in a patient infected by a zombie (they chose a task no one could claim experience in), participants at first undervalued their ability. However, once they had a little bit of experience, they quickly overestimated their ability, believing they were 73 percent accurate when they were actually only 60 percent.

Sanchez and Dunning identified that when a beginner is given an initial education, they almost immediately respond with overconfidence. They wrote, "Participants far too exuberantly formed quick, self-assured ideas about how to approach [the task] based on only the slimmest amount of data. Small bits of data, however, are often filled with noise and misleading signs. It usually takes a large amount of data to strip away the chaos of the world, to finally see the worthwhile signal. However, classic research has shown that people do not have a feel for this fact. They assume that every small sequence of data represents the world just as well as long sequences do."[349]

I think about our youngest service members when I read this data, who sometimes come out of basic "chomping at the bit" to deploy. Yet, I also think about our response as a culture at the beginning of GWOT. Before the smoke of the towers had even cleared, once we had a direction and an enemy, we were ready to sign up and, in Johnson's words, "take on the Middle East." I'm not necessarily referring to the political decisions to go into Afghanistan and Iraq, although some more qualified to comment on that than I deeply feel we may have rushed into military action without enough information.

What I'm suggesting is that we may be innately designed to quickly create meaning for situations that warrant our confident and swift action. It is only natural to create a personal "why" or purpose to justify what we anticipate will cost a lot from us. I also think about the world of gaming, which continues to inspire the youngest generation with a completely different reason to join the military.

In 2002, the Army released an online game called *America's Army*, which cost an estimated $16 million to develop. Cheaper than television ads, its goal was "increasing recruitment for the US military and introducing potential recruits to the idea that their government's violence is used solely to protect freedom and the 'American way of life.'" It boasted over forty-three million downloads by 2009.[350]

This was just the beginning of the impact of video games on recruiting and perceptions of war. *Call of Duty* came out in 2003 as a first-person shooter game simulating World War II and eventually evolved into a multiplayer experience of modern warfare. Some worry that millennials and Gen Z are entering the force with a misconception about war and combat, and yet, the modernization of technology in our tanks, airframes, and in the operation of drones is proving to successfully utilize the multitasking ability and skills they have developed from gaming.

Reframing the meaning of war through gaming isn't just happening in the youngest generation. During deployment, service members are on missions and engaging in combat and then might decompress in their off time through combat-themed video games, concerning experts who wonder about the long-term effects of living in a constant mindset of combat. Service members' rampant use of video games during reintegration also has created rifts between couples. The bigger issue is that the new military-entertainment complex has normalized the state of war, creating an unrealistic expectation of combat, as well as desensitizing one's mortality, especially for younger, more impressionable audiences.

I recently listened to a podcast of a Green Beret Medal of Honor (MoH) recipient sharing his harrowing experience of surviving a battle during his second deployment to Iraq. His first deployment he described as reaching movie status. As the interviewer, a Green Beret himself, guided the conversation into his now-historic second deployment, I found myself holding my breath as if the guest could sense I knew we were entering into a more sacred moment of his story.

As the story progressed to the more tense moments of the battle, the guest described having to engage the enemy up close, brushing with death, while simultaneously attempting to help others around him get out of the situation alive. At the end of the interview, the interviewer validated the guest's valor in his courage to "create space and distance" for the safety of his brothers using only his sidearm, thus earning the MoH. He then said, "That was 'I'm putting it all on the line for my brothers, I know they are getting smoked up behind me, I'm going to create some space.' Everything after that is a video game."

Regardless of what our initial overconfident motivators may be, we can rarely sustain them over a long period of time. There is only so long a person can be fueled by piss and vinegar, chasing combat or heroism, or looking for the greater good when the mission becomes lengthy and costly (both financially and in the form of human lives lost). For those who have had the experience of brushing with death in combat, there is a sobering effect on the overconfident. In many cases, it creates a contradiction between your original motivation and the meaning attached to your mortality.

Johnson offers an example, "My time in Iraq produced an interesting combination of 'I'm a bad motherfucker' and 'Please let me get home to see my son.' It's an odd combination of 'Fuck you. Bring it on motherfucker' and 'I just want to go home.' On the one side … you are trained, ready, and willing. Above all else, you are fucking pissed off. At points, you go from feeling like you are wanted and needed by the locals to feeling like they don't want you there at all. So, you think, 'Fuck it.' You feel like equal parts Arnold in the end of *Predator* and Mel Gibson in the beginning of *Braveheart*. You'd rather not fight, but as long as you have to, you are going to rain down hell upon the enemy. I realize this all sounds cruel and illogical, but so is war. You truly stop caring. You don't want to shoot, but you will. And you won't think about it, until you get home that is."[351]

My husband describes the service members who entered into the service after 9/11 as "war babies," in that they (and he) began their military experience with the imprinting of war during their most formative years. Just as we have described the strong imprinting that occurs on whole people groups when they are at their most vulnerable, the beginning of an intense first decade of deterring global terrorism in two separate countries shaped an entire multigenerational cohort of warfighters. It shaped what they believe in their views of life, war, and possibly God, and also their view of the military community.

Elliot Ackerman, a Marine and author of *The Fifth Act: America's End in Afghanistan,* describes, "People have sometimes asked me, 'Elliot, how do you think the war's changed you?' and I've never known how to answer that question. Because the war in so many ways made me. I don't know how to unbraid it out of the knots that are me. But the friendships I have there, the memories I have from that time, of course, I think about and it's the time when I was growing up. I mean, I grew up there in the war."[352]

By 2006, the war in Iraq was evolving from "overthrowing Saddam into a vague cross between nation building and refereeing a civil war."[353] As American troops wrestled with the changing mission, as many as 35 percent were coming home with mental health problems and seeking care. The Army released a study that same year reporting that "27 percent of noncommissioned officers—a critically important group—on their third or fourth tour exhibited symptoms commonly referred to as post-traumatic stress disorders."[354] We also know that being able to create meaning is a powerful tool in trauma recovery. If a person doubts if their sacrifice was worth it or becomes demoralized, they are more likely to worsen their chances of developing post-traumatic stress and further entrenching its symptoms.

There is confusion when the meaning we assigned to something in the beginning does not fit the meaning that is revealed in the end, if meaning is ever provided at all. I believe this is also why there is a crash of sorts when enough time passes or we get more information. When reality sets in that you may not be able to do all you hoped during one deployment or when it is revealed that there are no weapons of mass destruction, what is there to do but question the meaning you gave to it in the first place?

As the force spread out between two conflicts throughout the first decade, service members described growing frustration that force strength was being diluted. One service member shared the following powerful metaphor: "Imagine you are tasked to build and finish a house. At first, everyone is focused on getting the house started and making progress. You have a lot of lumber coming your way, concrete is being poured, and everyone is all in. As far as you can tell, it seems pretty clear that the priority on everyone's mind is that this house is the most important thing for us to accomplish.

"After about four or five years, you begin to look around and realize that the people who were working with you seem to now be working on other houses. Now, there are only a few of you left to try to finish the house you started. In addition, you're not getting really great lumber anymore, either. The lumber you are getting is kind of warped, but the message you are still being told is 'Finish the house! Finish the house!' So, you get back to work, convinced that this effort must still be a priority.

"Then every day after that, you show up to work and find there is a little less lumber, a little less workers, and a little less support for the house. Suddenly, someone comes up to you and says, 'Hey, that house has to be finished in a couple of years.' Frustrated, you look at the house and realize that although you started off pretty strong, the plumbing is not in the right place. You call back and tell them, 'The supplies we're getting do not match the plans for this house. In fact, the plans for this house keep changing. We can't put a second floor in when our plans for the first floor didn't include the right roofing system to support a second floor.' After a while, it seems like people are just making up this house as they go.

"Finally, somebody comes along and says, 'Hey, the building inspector is showing up next week!' Panicked, you respond, 'What?! We've been calling you and telling you that the timeline of this house is not feasible, the supplies we are getting are not working, and the family that is supposed to live in this house are stopping by and telling us they don't really like the design of the house. In fact, we have people that are coming over at night and stealing our tools.' At some point, you just want to throw your hands up and say, 'Fine, whatever. The building inspector can show up, I just wanna get out of here. Thanks, and goodbye.'

"But as you leave, you're asking, 'What did I just build? What did we just do?' It is heartbreaking to go from everyone gathered on the capitol steps singing 'God Bless America' to things being so divided come 2016-2018 and into the pandemic. Everyone is attacking everyone. Everyone is supposed to build this house and yet they are hitting each other with their own lumber. It is very easy to say we deterred another attack on American soil, but when I read recently that 70 percent of the terrorism on American soil in 2021 was caused by extremist right and left-wing groups, the enemy just keeps changing."[355]

This topic of disillusionment has come up before. In the aftermath of World War I and the Spanish Flu, presidential candidate Warren Harding's "back to normalcy" policy was offered in a tone-deaf speech that left those who had gone to war lost. He declared, "There isn't anything the matter with world civilization, except that humanity is viewing it through a vision impaired in a cataclysmal war. ... Poise has been disturbed, and nerves have been racked, and fever has rendered men irrational. America's present need is not heroics but healing; not nostrums but normalcy. Not revolution but restoration; not agitation but adjustment; not surgery but serenity; not the dramatic but the dispassionate; not experiment but equipoise; not submergence in internationality but sustainment in triumphant nationality."[356]

Normalcy after World War I looked like the Roaring '20s and retreating from war so that America could focus on itself. That left much of the lost generation feeling they had "witnessed what they considered pointless death on such a massive scale during the war."[357] Harding's vision for America ignored the opportunity and responsibility of the nation to affirm meaning for those who had defended the nation on their behalf. Vietnam veterans experienced the same and it is likely many in the GWOT cohort are experiencing similar messages.

When an individual is unable to honestly live out their internal state of suffering externally, incongruence will eat them alive until they can find a place for their soul to find congruence again. For the lost generation who entered a society in a time of peace, this looked like recklessness. For those who faced additional years of GWOT, there would be no rest for a while.

According to the DoD, the US has lost 2,353 US military lives, and 20,149 more were wounded in action since Operation Enduring Freedom (OEF) began in Afghanistan in 2001.[358] There have been 4,431 US military casualties in the Operation Iraqi Freedom (OIF) conflict that began in 2003.[359] For many who served in the first decade after 9/11, OIF was a difficult and confusing time. Since its end, those veterans are more open than ever about the moral injury of civilian losses that are still not accurately counted or reported on but hover between 200,000 to 300,000.[360] Many Iraq veterans are part of the almost thirty-two thousand wounded in action.

Trying to capture the sheer amount of mental trauma service members endured due to OEF and OIF is not only overwhelming but difficult to do in a way that truly honors those affected. Since many do not access available resources, it's hard to know how many acknowledge symptoms of acute stress and/or PTSD. Of those who have accessed services through Veterans Affairs, a reported 29 percent of OEF and OIF veterans have experienced PTSD at some point in their lives.[361] Another study found that 60 percent do not seek help.[362]

Not every service member experienced combat or came home with mental health concerns. However, it is difficult to know how many do and have not spoken up due to the stigma associated with admitting a need for help. The military culture, especially during the beginning years of GWOT, was much different from today. Between 2001 and 2014, a systematic review of studies on why service members were concerned about getting help for mental health challenges found that 44.2 percent said, "My unit leadership might treat me differently" and 42.9 percent said, "I would be seen as weak."[363] Other reasons for avoiding care include a fear of having clearance revoked and worry a diagnosis might prevent chances of promotion. Many have shared with me that they have genuinely struggled to find the time or thought they could manage it on their own.

As the number of veteran and active service suicides started to climb, military leaders began to speak out about their own mental health journeys, normalizing the importance of getting help in order to reduce the stigma. Looking back, the authenticity of military leaders has hands-down been one of the most powerful and effective tools for shifting the stigma around mental health within our culture. It was not common back then for the boomer or even Gen X generations to expose vulnerabilities. Yet, the public authenticity of senior military leaders remains an exception rather than a rule despite newer generations casting the expectation. Why?

While the impact of OEF and OIF on the mental health of an entire multigenerational cohort is and should be one of our top concerns, there is another silent killer. In addition to their hesitancy to get help, few are willing to talk about it openly. In fact, to talk about it is to risk almost everything. It takes up residence in our homes and holds service members hostage. In many ways, it holds their families hostage, too, demanding the ransom of work ethic in return for rank. It is the dangling carrot of the promotion process, and it feeds on fear, perfection, and power.

DANGLING CARROT

All humans are created to grow and evolve. From the moment we are born, our health and wellness are measured by passing through developmental gates and milestones. We are evaluated from day one. Are we eating? Do we desire to play? Are we wanting to sit up and learn to walk? Not reaching these milestones sets off alarms that something is wrong. When a child does not want to learn or is completely unmotivated to move forward, we call it a failure to thrive.

This innate desire to grow and evolve is also what keeps us learning, provides a feeling of satisfaction when we accomplish our goals, and protects us from boredom. Businesses that tap into employees' desire to grow thrive as well. Those that incorporate some type of growth model or incentives in their personnel system are more likely to retain talent and reduce turnover. The employees, likewise, are looking to work for a business that values personal and professional growth and can offer a future in the company.

The military is well respected for its personnel system that inspires employees with leadership development and a path for future job security and incentives. These are, in fact, the recruiting messages that resonate with young men and women considering service. The seemingly clear path for promotion appears to provide certainty compared to the dog-eat-dog civilian business world.

When service members enlist or are commissioned, they start at the lowest level of their enlisted or officer cohort and climb the rank system over the course of their careers. Promotions are not necessarily predictable as the time in rank varies for each rank. For example, E-1 to E-3 ranks are currently offered a promotion if they successfully recruit another individual to join, whereas other ranks can range between one and five years of service before the promotion window is available. Each rank also has only so many slots available across the force, making it increasingly more competitive over time. So while there may be thousands of opportunities and assignments available for an E-2 or even a captain, there are fewer colonels and even fewer command sergeant majors and generals. Ultimately, the highest levels of leadership in the force culminate with the chairman of the Joint Chiefs and the senior enlisted advisor to the chairman, of which there is only one seat each.

Also affecting the promotion process is whether the force is in a state of growth or reduction. If the force is growing (as in the years after 9/11), the branch may open up more slots for specific ranks. Service members could find themselves promoted earlier than usual. However, if the force has downsized (such as during sequestration) there are fewer positions to fill at each rank, making it much more competitive to be promoted ahead of peers. What is happening within recruitment, retention, and even the economy can also impact the influx of personnel as well as gaps that need to be filled.

Of course, not everyone is aiming for the top seats of military leadership. Some are looking for continued job security for their families and some are simply waiting out their commitment. Others hope to stay promotable until they can reach the golden carrot of retirement at twenty years where they can collect a pension and retirement benefits. The amount of an individual's monthly retirement pension is calculated based on their highest three years of pay. To put it simply, the higher the rank when you retire, the higher your monthly pension payment. Until the recent Blended Retirement System, a pension was only available to those who retired at twenty years. This meant that if you left at seventeen years, you walked away with no pension to show for your years of service.

When it all comes down to it, provision and purpose are what most consider when deciding whether or not to make the military a career. Most feel that after the initial commitment, there is still time to begin an alternative civilian career. However, there is a window of time after about the five-year mark, where service members share they must decide if it is worth the risk to aim for twenty, knowing that if they get passed over for promotion they exit without the full package.

There is a shift I have heard several service members describe that happens about this time, after the first or second promotion. If they have committed to moving forward, it is a realization that there are only a couple of paths that can safely assure, but not guarantee, job security and reaching retirement. While each branch is different, for officers, taking command in some form is a must. Each assignment after that is an effort to broaden your scope and abilities, and make sure you are able to succeed in a position of leadership. But, there are only so many command positions available.

To distinguish service members from their peers and choose who gets promoted and who doesn't, annual performance evaluations—Officer Evaluation Reports (OERs) or Noncommissioned Officer Evaluation Reports (NCOERs)—help promotion boards assess top leaders. While service members wait for their next promotion window, they are required to complete the evaluations with the help of two raters who evaluate their performance and future potential. Called the rater and senior rater, they are the ratee's direct chain of command.

The "up or out" policy started in the Navy in 1916. It was later adopted by the Army after World War II and states that an individual can only be passed over for promotion twice before they are forced to exit the military. Looking at the opportunities and slots available in each rank, one only needs to do the math to see that almost everyone in the force will at some point leave, be passed over, or be forced out. Yet for those who depend on the military for job security or who are betting on promotion for a pension, holding off being passed over for as long as possible is the only way to stay in the game.

There have been a lot of opinions about this policy, with many arguing that it is outdated, costly to the military, and sabotages morale. If a business owner in the civilian world owns and runs a factory and has employees working on the production line, it would likely not be mandatory for a production line worker to promote to manager and aim for CEO or risk being fired. No, instead, the business owner knows that the talent of the production line employee is in some cases an expensive investment to replace. While incentives and opportunities for leadership are a great way to increase loyalty and buy-in for employees, the business owner knows that with every promotion, onboarding and training of a new employee will need to follow. In addition, some employees may not have the talent or desire for leadership roles.

Similarly, there are service members who may not want to promote, whether it is because their skills and abilities are better suited at a specific rank or a promotion moves them farther away from what brought them into the military in the first place. For example, pilots spend a good portion of their careers earning the opportunity to sit in a cockpit and fly. As with most other positions in the military, the higher up in rank you go the more leadership responsibilities you are given, taking you out of the cockpit and into administrative roles. I have heard of commissioned officers who gave up their commission in order to become warrant officers so they could keep flying.

Over the last few years, I've had a surge of clients preparing for retirement. All of them presented with some level of depression, apathy, or form of languishing over the deep loss they felt due to sitting behind a desk rather than doing what they loved. Many of their careers began during a time of war when their skills were given a deep sense of purpose. While some would say they were grieving their glory days or having a midlife crisis, these service members were not questioning what they wanted to do, or even their purpose. Instead, it was complicated grief from being redirected from their first love. For many of them, the deeper struggle was the cost of chasing the dangling carrot of promotion and security. Some even wonder if they chose this path willingly or if they were sold the idea that the noncompetitive route would have been less respected, putting their future in jeopardy.

The other concern is that during a high-growth time, movement within the force slows down performance and promotes individuals before they are ready, or asks them to leave positions in which they excel only to start over. For those who are older, trained, and qualified, a time of growth may force them out "in order to ensure the Army remains youthful and vigorous."[364] Either way, many military and policy leaders agree it is an expensive process that also reduces the overall quality of performance.

HIDDEN CULTURE OF FEAR & PERFECTION

The illness behind this system is that it creates a culture of fear and perfection. Even if service members do not want to promote or if they disagree with their branch personnel manager, they risk their careers if they speak up. The military expects loyalty or nothing. As one service member wrote in a public article to senior raters, "Officers will be completely honest with you about their career plans if 1) their career goals are genuinely to be exactly like you, 2) they are within a retirement window, or 3) they are ready to accept the (real or perceived) consequences of signaling their intent to depart active-service." He goes on to say, "Many officers harbor the (anecdotally valid) fear that they will be written off as a lost cause and taken out of the good jobs and away from soldiers if their uncertainty about remaining in the Army long-term becomes known. Leaders should not ostracize their subordinates or treat them as second-class citizens just because they have not yet committed to staying the full twenty. When you treat subordinates who plan to transition out of the Army poorly, you send the message that their worth is tied only to what labor they provide to you and their desire to leave the organization signals the end of that usefulness."[365]

So while some are eager to continue the path to stars or "kingship" (what some call general-makers), others hide the desire for anything other than what superiors tell them they need to remain competitive. They hide the desire to put anything else above the mission and quietly allow the system to steer them for them as long as it has a need for them. While most officers will have no problem making it to twenty years if they desire, competition for higher-ranking key leadership positions like command and general officer gets progressively harder.

Although the OER and NCOER have some key differences, the most common areas of concern are the rating and the narrative portion, or bullet-point comments section. Historically, the service member is rated with a block-check system that ranks them above, centered, or below their peers, permanently marking their profile. The narrative portion, or bullet-point comments section, allows the rater to include specific wording that seals the deal. Anything less than being rated above your peers creates fear that you may not be considered early and worst-case scenario, you could get passed over, even though the ratee might have five evaluations in their file by the time the board reviews for promotion.

This is where the culture of perfectionism becomes further entrenched in the institution. In order to promote ahead of peers, you need as many perfect evaluations as possible. When the system was first established, there was a problem with it being flooded with inflated and embellished evaluations. Some of this was due to raters feeling a sense of "combat loyalty"[366] to those they served alongside. Whether it was due to favoritism or connection, raters were less likely to rate their ratees lower after having just returned from a deployment. Others simply rated all of their ratees as high or excellent. Even though there have been more than fifteen revisions of the process, inflation continues to be a problem and service members continue to find ways around the system to benefit themselves or others they want to see promoted.

To address inflation, senior rater profiles and forced distribution of ratings were established to hold raters accountable and limit how many top blocks they could give from their rater pool at once. Still, leaders have found ways to manipulate their pool of raters in order to give more. As an example, one brigade commander pulled all sixty-three post-career course captains across the brigade into his rating pool so he had more top blocks to give out. Other ratees, however, may be at a disadvantage if their senior rater does not have the profile to give out a top block or does not know how to write an impressive top block narrative. The solution for many service members has been to write and embellish their own evaluations, filling out almost every section other than the checkboxes with the hopes of bettering their chances for a top block.

Inflated evaluations create another conundrum worth mentioning. What happens over time as service members draft their own inflated evaluations and consistently receive inflated comments from senior raters? Without an authentic assessment, service members develop a self-perception that may not be accurate, damned with faint praise. If everyone believes they are above average and senior raters rarely tell ratees they are not, it can be incredibly shocking and discouraging to get that first weak OER or be passed over.

An interesting contradiction to note is that it is actually extremely difficult to get fired in the military (unless the military is sizing down). One of the great frustrations is that, unless there is a criminal violation, even service members who commit moral failings get shuffled around until they are passed over twice. While the improbability of being fired does encourage some to test the waters of giving a healthy amount of energy to the job and possibly others to give the bare minimum, it does not reduce the expectation of perfection for those who are trying to make it long-term and get into coveted assignments. This creates frustration between those who strive, those who confidently coast, and those who have poor or even immoral performances but don't get passed over due to a spot that needs to be filled. Especially during a time when recruitment numbers are not being met, the force needs to retain those they have.

Military leaders share that, under the pressure of standards they must meet, a majority of their energy goes to managing problematic personnel rather than providing high-quality development and counselings for ratees. Timothy Kane, a Hoover Institution fellow at Stanford University, wrote: "Commanders don't have hiring and firing power they need and can't eject an abusive coworker."[367] Considering the number of toxic leaders and sexual predators revealed in the system, it makes for an excellent environment for low performers or destructive personalities to survive until they are passed over twice. Evaluations are therefore viewed as "time-consuming, excessively subjective, a cause of stress, demotivating and, it's been argued, detrimental to mental health."[368,369] Therefore, prewritten evaluations ease senior leader workloads and the pattern cannot easily be corrected without implicating themselves in their own lack of leadership.

You would think that seeing this system would convince those striving for perfection to relax and coast themselves. However, there is always another dangling carrot that promises a better assignment, new opportunities to show on your next OER, and another box to check to stay promotable. The "up or out" policy, then, ramps service members into a state of anxiety from evaluation to evaluation. The fear of not doing enough to earn a top block propels many into a state of workaholism or turns them into what many call "idea fairies," looking for that next great idea that their peers have not thought of or "chasing the next bullet point" that can be written on their evaluation. While some bullet points are noteworthy achievements that aim to build stronger teams, correct issues that plague the community, or even solve complex problems within their circle of influence, the greatest frustration is that once the evaluation is complete, the promotion offered, and the assignment given, finishing that long-term "great idea" is not enough for the next evaluation and can be easily abandoned. Leaving it for the next in line is not in the best interest for that person's evaluation. So it is in the next person's best interest to come up with their own innovative and unique bullet point. However, what some have described is that raters and senior raters can capitalize on the successes of their ratees' ideas if they can show on their own evaluation that they supported or aided in the project, solution, or achievement—if it falls within the window of their evaluation.

This is also affecting the health of our service members. For most of their years in service, service members run their bodies into the ground striving to max out physical training scores in order to stay competitive. Then, as their bodies start to break down, they hide the pain with prescriptions, delay surgeries, and avoid going on profile that would give them rest and medical care. All to avoid the perception of being broken and not competitive.

The path to twenty gets even more intense the longer you are in, the more perfectly you perform, the more loyal you are to the institution, and the more you are willing to give. At some point, it becomes clear (or personnel managers imply) that certain roles are being given to you to groom you for top positions. Getting these assignments implies that the institution sees star potential in you. All it costs is a little more.

To be fair, mixed into this anxiety and striving is a love for the job and the people. Most service members honestly want to lead well, make a difference, leave their mark, and even leave a legacy. The striving is not for selfish gain alone. This culture attracts those who have a heart for service, sacrifice, and work ethic that benefits a team. So the real question is, what impact does a system like this have on service members over time?

Is it any wonder why our culture continues to deal with the same issues decade after decade? Solutions are often rushed for the sake of bullet points or even avoided completely if seen as a vulnerability or inability to complete within the time constraints of an assigned role. Meanwhile, each dangling carrot offers potential rest from the current role, more influence, or more money. For many military couples, this eventually creates conflict when there is rarely rest on the other side of the next assignment. Instead, it brings even more pressure with the higher rank, and in most cases, more frequent relocations. Many military spouses struggle at this level to continue their own careers due to relocations and increased responsibilities in the community. Entertaining and social customs that the military does not reimburse for can create financial hardship but look good to the promotion board.

By the time military couples get to this place in their career and finally ask for help, they are exhausted and overwhelmed by the consequences of years of neglect. Their minds and bodies have endured years of stress and trauma.

Performance evaluations were designed to appropriately showcase performance and highlight where an individual has taken initiative, however over the years, the system has become a codependent, inauthentic, and inflated transaction between senior raters and subordinates. Worse, it has perpetuated a pace of work no person can sustain.

During the difficult years of the drawdown and pressure and tempo of ARFORGEN, the level of expectation, perfection, and dishonesty led Leonard Wong, a distinguished fellow, and his colleague Steve Gerras, a retired colonel and professor (both at the US Army War College) to write their 2015 brutally honest and viral monograph, *Lying to Ourselves: Dishonesty in the Army Profession*. In their fifty-two-page report, they shared that "the military as an institution has created an environment where it is literally impossible to execute to standard all that is required" considering "the deluge of mandatory requirements involved in the Army Force Generation (ARFORGEN) process." They continued, "an expectation of constant flawlessness in all aspects of performance is fulfilled only by deception from the ranks below, and denial or delusion from the ranks above."[370]

Then, they spoke of the "ethical fading" that has been permissible within the culture, saying it "allows us to convince ourselves that considerations of right or wrong are not applicable to decisions that in any other circumstances would be ethical dilemmas. ... The first time that officers sign an OER support form authenticating a counseling session that never happened or check a box saying, 'I have read the above requirements' when they really only glanced at the 1,800-word IA acceptable use policy, they might feel a tinge of ethical concern. After repeated exposure to the burgeoning demands and the associated need to put their honor on the line, however, officers become ethically numb. Eventually, their signature and word become tools to maneuver through the Army bureaucracy rather than symbols of integrity and honesty."[371]

The response was mixed. Those in leadership treated it with denial at first, then studied it in the schools, and then moved forward with little change. In a 2022 follow-up by Wong and Gerras called "Still Lying to Ourselves: A Retrospective Look at Dishonesty in the Army Profession," they admitted, "We purposefully avoided advocating self-advancement as a primary motivation for lying. Our logic was that more leaders would acknowledge the culture of dishonesty if we sidestepped the notion that many officers lie for self-serving reasons. In retrospect, we were too quick to provide an easy escape from the introspection we desired from each Army leader. Instead of encouraging culture change by urging individuals to examine their own motives, decisions, and actions, we overemphasized organizational and policy solutions. We should have pointed out that while Army policies and regulations create an onerous environment, the decision to lie is facilitated by an individual's aspirations to succeed in the Army. Competition between peers will always create underlying pressure to tell the system what it wants to hear."[372]

That same year, the Navy suspended the "up and out" policy for enlisted soldiers "in order to boost retention" through 2025.[373] The 2019 NDAA had similarly authorized officers to "opt out of their consideration for promotion to complete special assignments, advanced educational opportunities, or pursue required developmental milestones at their current grade level."[374] Further, Army representatives stated that "failure to be selected for promotion during the opt-in process will not negatively impact a soldier's career. ... All officers receive at least two considerations in and above the primary promotion zone before initiating a potential involuntary separation action."[375] During all the talk of "elbowing to the top" and perfection leading to burnout, Army Chief of Staff General James McConville declared in 2019 that it was time to make people his number one priority.[376]

PEOPLE FIRST

When McConville began his tenure as Army chief of staff in 2019, his People First initiative identified five priorities he planned to address: recruitment, a new talent management system and assignments process, a more ready and responsive army, modernization, and a commitment to improve the quality of life for soldiers, veterans, and their families. McConville shared that when he was with the 25th Infantry Division, he watched a captain miss the birth of his son when there could have been a way to get him there. This motivated him to pledge, "I will never let that happen on my watch, in any unit that I'm ever in."[377] He introduced a tool to aid leaders with guidelines on how to manage a healthier work-life environment.

The response from Army families to McConville's initiative seemed to be a hesitant "too good to be true" mixed with "it's about time." Families burned out from tough relocations, finagling their own paths to promotions, and intense operations tempo expressed that they finally felt heard. Military spouses expressed gratitude online that their family's immediate needs might finally be considered a priority over the needs of the military.

The intention behind People First resonated with the population. Executed through The Army People Strategy, various task forces were assigned to tackle some of the Army's biggest issues, such as talent management, army culture, and quality of life topics like housing, childcare, military spouse employment, diversity, and sexual harassment.[378] Its beginning succeeded in several areas. Housing issues and relocation challenges were being heard in public forums and the people responsible were held accountable. Families who had medical needs were to be taken more seriously. The talent AIM system worked almost like a dating match service where officers could upload their resumes and units could find their match in a free market. There was still competition with peers for assignments, but the idea was that a 1:1 match made an assignment more likely. This was a huge deal for families. Human Resources Command (HRC) would no longer be placing people based solely on the needs of the Army.

However, the talent system got off to a rocky start. Service members were encouraged to talk with families before making their selections. What families heard was that they would be part of the process. They expected the free market to include family feedback and the automated system to include extra information that would wave off assignments that negatively impacted spouse careers or children. But some units and branches didn't use the talent management system, even though they were expected to. The slow start also caused delays in processing orders, giving families short notice of upcoming moves. An article, "Army Human Resources Command's Online Talent Management System is a Bust for Military Families," was written by a military spouse who voiced frustration over the short notice and the impact the chosen assignment would have on her career. Feeling that it had not put people first, she wrote: "Military families like mine deserve stability and transparency in the selection and move process. If the Army really wants to improve efficiency and effectiveness in officer management, they would do well to remember the needs of the spouse and the family. We have given up, sacrificed, compromised and compensated for far too many things over the course of our own service as spouses and families. There is a minimal amount of dignity and respect required that would provide us ample notice for moves and that would at least pretend to take our needs into account. Talent retention is not simply an HRC game. Spouses have considerable influence."[379]

Still, people (at least those in the Army branch) really wanted to trust in McConville's new plan to take care of people. Their strong frustration had more to do with the state of need they were in when AIM launched than with the system itself. Making significant changes in an institution takes time and pilot launches are sure to have bugs. Many people were, and still are, pleased with the AIM system. A high percentage were able to get one of their top three choices of assignments. However, with highly sought-after assignments being limited and so many aiming for them, not everyone could get what they wanted or what was best for their families. Someone would have to take less popular assignments and locations. Yet, the thought of the military putting the needs of families first was too good not to believe in, so families did.

Soon after the launch of The Army People Strategy (or what most would call People First), the COVID-19 pandemic spread across the country and through the military community. With much of the force considered mission essential, people soon realized that the Army, through no fault of its own, was not going to be able to live up to the promises of People First any time soon, at least as it was first presented.

The years 2020 and 2021 were filled with the turmoil of the pandemic, battles over vaccine mandates for service members, debates around how and when to open up safely, and families reaching a place of extreme emotional, physical, and mental burnout. In 2020, the Vanessa Guillén murder at Fort Hood took center stage on America's news and social media channels, drawing public attention to the military's investigative and justice system. A congressional committee was brought in to investigate Fort Hood, highlighting sexual harassment and assault as well as the morale of soldiers and families.

The committee found soldier barracks with cracked foundations and moldy walls, and spouses who "were afraid for their husbands and wives for their overwork, their exhaustion, their misery, and depression, afraid they would come home to find their loved one hanging in the shower or dead on the floor."[380] Related to sexual harassment, a committee member stated, "We met with junior enlisted women who described a culture of sexual harassment, a culture of leaders watching as women and men were harassed before their eyes but kept silent, squad leaders and platoon leaders who seemed either unwilling or unsure how to help them. So their harassment became just another hazard of being a soldier, and no one was held accountable, and not one leader stepped forward."[381]

Our family was stationed at Fort Hood during the investigation. At one point, I was brought in by a local brigade for a military couple's date night. It actually would be one of the first events to open up in-person since the lockdown. I sat at one of the back tables with the CSM, who was attending alone, and struck up a conversation. Somehow the topic of People First came up and I asked how he thought it was going. He laughed under his breath and said, "People First is a joke. What they really meant was everyone except for leaders." He proceeded to tell me that he and other leaders had been worked into the ground making sure enlisted families felt first, meanwhile he was losing his. He and his wife were initially excited about People First, which made it all the more devastating when it was clear they did not count. After giving so many years and chasing so many carrots, he had decided he was done. This was his last assignment, and, to save his marriage, he would not be chasing any more.

This was one of those moments when I wondered how much of this was because of what was happening at Fort Hood, if there was a misunderstanding about McConville's initiative, or if it was Army-wide. As I talked with both enlisted and officer families, it was clear that many had not heard of People First, and many of those who had didn't trust it. It is possible that those who had heard about it developed an expectation of what it would mean to be put first rather than understanding what the DoD could fulfill. To be fair, the plan was vast and would need a considerable amount of time to make progress. While some may have been aware of the plan in its entirety, many people made assumptions based on what "quality of life" meant to them given their circumstances, including hopes for fewer deployments, cracking down on toxic leadership, considering family before the mission, or emphasizing quality of life for leaders. With families already in a deficit, it was as if they were in the middle of a twenty-year drought and they heard the news of People First as the solution. In reality, the plan's initiative to improve families' quality of life was to support achieving and sustaining a quick, responsive force.

Our body's ability to respond to trauma and stress through the autonomic nervous system is divided into two other systems, the sympathetic nervous system (SNS), and the parasympathetic nervous system (PNS). When faced with stress, the SNS is responsible for the well-known "fight, flight, or freeze" response. The PNS system regulates the body when not in danger and is referred to as the "rest and digest" branch.

Fight, fight, or freeze is often covered for service members in the context of deployment, however, the lifestyle itself does not offer many opportunities for the PNS to regulate the body back to a calm state. Neuroscientists studying hyperarousal have also added a fourth system stress response. Some call it fib, others fawn. The fib mechanism involves lying to protect oneself from the feeling of disappointing someone, deflecting a negative consequence, buying time, and self-preservation from a perceived failure.[382]

What happens if a person stays in a state of hyperarousal for a long period of time, perhaps a decade or two? When the body is unable to fully ground itself, hyperarousal becomes a constant state, making it difficult to function. The constant adrenaline required to perform under stress creates adrenal fatigue, exhaustion, irritability, and a number of other systemic issues, including insomnia. For service members, the availability of caffeine and energy drinks enables the constant need to stay alert and functional, while the acceptance of alcohol provides a depressant. Of course neither actually gives the body a chance to recover. Research also shows that chronic stress can lead to cardiovascular issues and overall inflammation that impacts almost every system in the body and contributes to cognitive aging and impairment.[383]

I find that last point interesting to note, considering the escalation of impulsive behavior, toxic leadership, and moral failings of military leaders over the last ten to fifteen years who would have served during a significant portion of GWOT before landing in coveted command roles. I don't believe the behavior that landed them in military media and shared across social media was always due to bad character. It is much more possible that the chronic stress of a two-decade career of perfection and unresolved global conflicts converged with believing their own OER press, contributing to poor impulse control and emotional regulation.[384]

Chronic anxiety and hyperarousal are also symptoms of a possible trauma bond. Just as we covered the relationship between spouses and the military, there is a unique bond between the service member and the military. A trauma bond does not have to only exist in physically or sexually abusive relationships, in fact, experts are showing some of the same states of arousal and anxiety in work relationships with psychological distress where there is financial abuse (threat of job loss or income), isolation from family or friends, threats of replacement, loss of autonomy in decision-making, as well as emotional gaslighting, name-calling, or humiliation. "Trauma-bonds almost always produce codependent dynamics where the objects of admiration of trauma-bonded codependents become redeeming saviors that can do no wrong—until they do. When enacted in a community, trauma bonds tend to create cult-like collective dynamics."[385]

Additional signs of trauma bonding, according to experts, include:
- Being unable to state your feelings, opinions, or desires without fear of upsetting the other person
- Altering your behavior in a way that violates your moral code in order to keep the other person happy
- Setting a boundary that is ignored or dismissed
- Feeling like you would be lost without the other person
- Being love bombed at the beginning of a relationship (over-the-top, extreme displays of attention, affection and romantic gestures)[386]

Even as I write this, my own dual relationship with the military tempts me to swing to the opposite, more positive extreme of seeing the good this relationship has provided: financial security, adventure, leadership development, and even a career that has enabled me to be in a position to write this book. I can hear service couples I know sharing the argument that the focus should be on the love of the people this career has offered, pride that comes with serving, sense of purpose, and character that is built from pushing your limits. And all of those things are true. So are the hard truths we are discussing here.

When I work with someone in coaching or counseling, it is not uncommon for them to share their frustration and then quickly switch over to positive language as if they are talking themselves out of what they just expressed. What they want is to have permission to be in a neutral place where both are true. Personal and work relationships can be both wonderful and frustrating, worth staying amid healthy boundaries, completely fulfilling yet weighing down. Living in congruence with ourselves is especially difficult to live out and extremely unpopular in a system like the military that offers belonging in return for compliance. So "incongruence becomes the currency of our belonging. The better we are at burying our pain behind positivity or productivity, the more it seems we're allowed to belong."[387]

I do not believe this bond, if it does exist, is created intentionally or with malice. Instead, it reveals more about what we bring to the relationship and what we allow. When I work with service members who carry the burden of balancing work stress with who they want to be at home, we find more ways for them to set boundaries at work than they thought were possible. Some fear that if they set healthy boundaries with others, it will create conflict. I can't promise it won't, but more often it manifests discomfort of a pattern changing. Nothing will make us fall back into an old dysfunctional pattern more than the fear of losing the relationship altogether, scarcity of the unknown, fear of missing an opportunity, or anxiety about regretting not taking the carrot.

Therapist and author K.J. Ramsey writes that perceived scarcity makes us feel like we "have to keep climbing over others to get to the top. We attempt to self-regulate by striving and subjugating anyone and anything that gets in our way on the path to the top." If we don't address the role scarcity plays in our lives, we will continue to strive for that carrot, believing it will offer freedom, only to find scarcity there waiting for us again. When our source of belonging and economic provision is at risk, we are more likely to allow ethical fading, ignore red flags, or abandon our own body's signals that something needs to change.

If the institution is to survive in the current all-volunteer model, leaders will need to maintain a healthy relationship with the institution. The best gift millennials and Gen Z bring is an unwillingness to sacrifice marriages, children, physical and mental health, and potential careers to serve their country. There is enough sacrifice on the battlefield, it need not happen at home as well, nor are they willing to put all their eggs in the military-career basket.

AN EXAMPLE WORTH REMEMBERING
What do you value most?

Steven Nisbet, president and cofounder of Shields and Stripes, formerly part of a Tier 1 unit in JSOC called 24th Special Tactics Squadron, shares: "October 8, 2019, in Boise, Idaho, as part of a special operations unit, my team and I were conducting high-angle rope rescue and rock climbing training. During the second day of training, one of the team members set up a rappel line from a seventy-foot face while I finished my climb. After the first operator got down safely, the second descended. He got halfway down when the anchor failed, and he fell around thirty feet. One of the other team members was at the top, tied into the rope, and was pulled off the cliff, falling seventy feet. After he impacted the ground, the team treated him for nearly thirty minutes until local EMS arrived and he was declared dead at the scene.

"During a six-month investigation, my senior enlisted advisor (SEA) was encouraged and even commanded by his superior officers to fire me and my team. He resisted during one particular meeting, describing how if he were to fire me, not only would I be a risk of suicide but the entire squadron would not trust the leaders of the community. He was dismissed from the meeting along with the squadron commander. All three of us were recalled the next day where we learned our fate. All of us would be fired by our group commander. My SEA and commander would be fired solely because they refused to fire me for something that was deemed an accident.

"You could not ask for a better leader than to sacrifice his career for the good of his subordinates. He chose to lose his identity for me."

Chapter 16

FINAL STRAW OR NEW BEGINNING?

OUR ENTIRE JOURNEY SETS THE STAGE for the state of military culture as it approached the end of the conflict in Afghanistan. It is hard to truly grasp, or even crudely sum up, the experience of millions of service members throughout OEF and OIF. Between 1.9 and 3 million service members have served since 9/11, and "over half of them have deployed more than once."[388] Some deployments were horrific, others overwhelmingly boring. Some deployed on ships, others in the remote mountains of Afghanistan. Some airmen experienced the conflict from the sky, while some soldiers looked the Taliban in the eye.

The institution is designed to support itself with almost every need fulfilled by enlisted or commissioned service members—doctors, lawyers, dentists, cooks, mechanics, chaplains. Not everyone has seen combat, nor has everyone come home with injuries or mental health concerns (although a reported 1.8 million have a recognized disability from serving in GWOT).[389] Each branch also varies in its approach to engaging warfare and the enemy and also varies in the policies and practices within its own subculture. In addition, the entire force has seen at least five different generations serve over the course of GWOT.

It is not my goal to address policy or exit strategy. But I can share what it was like as a military spouse married to a husband deeply affected by two deployments to Afghanistan, and as a friend to those who are living with injuries related to GWOT and those who lost their spouses in the war. I offer my perspective as a clinician who received many calls, messages, and requests for help from families trying to process and, finally, as one who has a passion for studying this culture. To me, the Afghanistan withdrawal is the culmination, peak, arc of the story. It's a historical marker that defines generations and will be in the history books as the close of another war. Eleven Marines, one Army paratrooper, and one Navy corpsman died during the withdrawal; all but one were Gen Z. Their loss added fuel to the question in every newsfeed: "Was it worth it?"

The question was in the hearts of military families, too. The civilian community was able to ask it out loud, and doing so seemed to validate the incongruence military families found hard to accept. It had to be worth it. If it wasn't, then what did we all just go through?

WINNING HEARTS & MINDS

In 2009, General Stanley McChrystal was assigned as the top US commander for Afghanistan with a new strategy to "win the hearts and minds" of the Afghan people. When the US invaded Iraq in 2003, many resources and attention were diverted from Afghanistan into Iraq. It wouldn't be until 2009 that President Obama would shift the resources and focus back to Afghanistan and task McChrystal to design a new counterinsurgency (COIN) strategy that would stop the spread of the insurgency and aid the US-backed Afghan government.

McChrystal's COIN strategy was based on the concept that "civilians are less likely to support violent opposition groups if the government provides public services and security."[390] The US was not seeking to occupy Afghanistan but to use a political strategy and economic strategy to build relationships with key leaders, help develop Afghan security, and invest money and education into the infrastructure. The goal was to help the people police themselves and reduce public support for the insurgency, thereby reducing terrorism that had been rampant and well-established for decades. McChrystal said, "If the people are against us, we cannot be successful. If the people view us as occupiers and the enemy, we can't be successful and our casualties will go up dramatically."[391]

At the time, McChrystal was asking for an additional twenty-one thousand troops. Congress and the administration deliberated on what kind of solution or strategy could lead to success for Afghanistan. Not continuing with the strategy McChrystal proposed would have warranted a typical counterinsurgency plan and could risk more lives. With the lessons from Iraq still fresh, Obama announced thirty thousand additional troops on December 1, 2009. Called "the surge," these troops would be sent to Afghanistan on top of the sixty-eight thousand already stationed there. By May 2011, Osama bin Laden was killed.

The phrase "winning hearts and minds" played an important role in the mindset of US troops and their response as we evacuated Afghanistan in 2021. Ironically dating to Vietnam and still carrying the stigma associated with the lost war, the phrase implies an emotional and idealistic objective. It has been repeated so many times it is a normal part of our cultural language, often used cynically.

For special operations forces (SOF), winning hearts and minds was a familiar approach to forming partnerships on a local level to meet an objective. Frustrated with the multiple deployments and lack of progress in the country, a group of SOF commanders developed the Village Stability Operations (VSO) and Afghan Local Police (ALP) programs in 2009, a bottom-up strategy used during Vietnam. The programs empowered local Afghans with governance by training and "arming pro-government militias to secure rural areas."[392] To be successful, special forces would embed into local villages and help locals resist Taliban control. Knowing that this was an approach that would take time and patience, SOF commanders developed relationships with key village leaders, promising to protect them from Taliban retribution and reassuring them that they were with them for the long haul.[393]

By 2011, the VSO/ALP programs successfully evolved and expanded, doubling the amount of special forces components in Afghanistan, and showed significant progress in "security gains, measured in both increased territorial control and a reduction of Taliban attacks on coalition and Afghan forces." By 2012, the program spread across Afghanistan to include thirty thousand ALP protecting more than 115 districts.[394]

Eventually, a combination of factors started to break down the progress, including political tensions, President Obama's withdrawal of troops in 2011 and 2012, and the inability to properly measure the program's effectiveness, as well as difficulty managing the rapid growth. For a majority of special forces, though, these years involved developing powerful relationships with Afghans, including translators, interpreters, and even Afghan commandos.

Service members on the conventional side had a different experience with winning hearts and minds. McChrystal's strategy included changing the rules of engagement for troops by telling them they were to not engage with the enemy unless they were absolutely sure they were a target. This was in an effort to mitigate the risk to more Afghan civilian lives, but it was frustrating for troops who were used to responding quickly to obvious threats.

Specialist Isiah James said, "My mission was very kinetic. I'm an infantryman by training, have been my whole nine years in the military. No, we were never taught to win hearts and minds. You can't really turn a grunt off, so to speak. We were taught to put, bluntly, two in the heart and one in the mind with our shooting drills. They teach us how to kill, kill, kill. But then, at the flip of a switch, you're told to go out there and to be nice to people who you have seen as a target your whole entire deployment. It was a hard transition to not look at everything as a threat. Every motorbike was an IED. Every pile of trash on the road was an IED. Every person giving you the evil eye, so to speak, at the market was waiting to target you. So, it was a very hard juxtaposition to be an infantryman on patrol every day, but then to know that your mission was to basically win hearts and minds and try to placate the situation."[395]

If there were civilians nearby, troops were told to reevaluate or not engage at all even if that meant losing potentially dangerous targets. McChrystal shared, "I think when we err on the side of maturity and caution, there is a cost. And I know that we're asking an extraordinary amount from them to operate with such restraint and self-discipline, but I think it's how we win the war."[396]

Winning hearts and minds continued through 2016, causing a rise in US deaths until Secretary of Defense James Mattis, "Patron Saint of Chaos," relaxed the rules. However, by that point, seven years was a long time for troops to restrict the use of force "so much that the risk of death [was] transferred from the civilian population to the troops, who hold their fire at their own peril."[397]

OPERATION ALLIES REFUGE

In April 2021, a firm date for the US military exit out of Afghanistan was announced for September. There were mixed feelings, but most military and civilian leaders agreed there was no easy solution. The general public was heavily divided, as was the military culture (both veteran and active duty). In an earlier survey, 59 percent of US adults said that the war in Afghanistan was not worth fighting, while 36 percent said it was worth it.[398]

US troops began their withdrawal in May and evacuated Bagram Airfield in July. General Milley soon after reported that "half of all districts in Afghanistan are under Taliban control."[399] As the Taliban began to reclaim territory, the White House strongly encouraged the evacuation of all US citizens, citizens of US allies, and Afghan allies through Operation Allies Refuge, denying any parallels to Vietnam's Saigon evacuations. The backlog for visas from the State Department was already on a heavy year-long delay, but Biden promised to accelerate special visas for Afghan nationals who had supported the US. Most service members who had served in Afghanistan were not surprised by the Taliban attempting to take back territory, but no one was prepared for them to take back the country in eleven days.

By August, the Taliban entered Kabul, the last stronghold US troops were using for evacuation. The US news was already circulating that Afghanistan was the "next Vietnam" as helicopters evacuated the Kabul embassy. When President Ashraf Ghani fled the country on August 15, the Taliban took over Kabul, triggering even more panic from the Afghan people. In what *The Washington Post* called "Two Weeks of Chaos,"[400] troops were isolated at the Kabul airport, and Afghan families from all over the country attempted to get to the Kabul International Airport and on the other side of its gates to escape the Taliban regime.

Meanwhile, outside Kabul, the Taliban had set up various checkpoints, hunting Afghans and their families who had assisted the US and were trying to evacuate. With reports of the Taliban beating and executing allies and their families, many of them went into hiding, seeking help from service members in the US they had served with before. With pressure on the administration to send in more troops to aid in evacuation efforts, Biden approved a total of six thousand while staying firm on August 31 as the final day.

Service members and veterans feeling helpless at home started to launch into their own heroic efforts of getting evacuees into the Kabul airfield and raising money for additional charter flights.

Marine Elliot Ackerman shared, "It was a collapse of hierarchy because as the war was ending in those days, I found myself on text chains and phone calls with retired four-star generals and admirals, some of whom had commanded the entire war, because no one could get anyone out because of the craziness. And because, for a brief window, the team that I was working with was having some success, we found ourselves serving in this collapsed hierarchy all working together. And that was surreal for me at times."[401]

Another example was the unofficial group, Task Force Pineapple, a volunteer group of former US special forces personnel, aid workers, and intelligence officers, who led a ninety-six-hour rescue mission. It began when an Afghan commando, who had helped Navy SEALs and other US special forces and who was being hunted by the Taliban reached out to Retired Army Lieutenant Colonel Scott Mann. The group was said to have successfully helped between seven hundred and one thousand Afghan people safely into the Kabul airport.

Mann shared, "This Herculean effort couldn't have been done without the unofficial heroes inside the airfield who defied their orders to not help beyond the airport perimeter, by wading into sewage canals and pulling in these targeted people who were flashing pineapples on their phones."

Another participant, Jason Redman, a combat-wounded former Navy SEAL and author, stated, "Our own government didn't do this. We did what we should do, as Americans."

Military spouses reached out during this time, sharing their concerns as they watched their service members on the phone for days on end fielding calls from evacuees and coordinating efforts. They were not only concerned about the lack of sleep but felt completely disoriented given that their spouses, some who had been retired for years or were finally in a better mental and emotional place, were now being pulled back into a war they had left long ago. Wanting to be protective, spouses felt the pull between wanting to support their veterans but also being angry that Afghanistan was invading their home again.

Back at the airport, thousands of Afghans flooded the airfield perimeter trying to evacuate with American troops. Stories of children being crushed, parents throwing babies over concrete barrier T-walls to troops on the other side, and paramilitary forces shooting into the crowd created a humanitarian crisis that service members have since described as morally injurious. There was limited food and water as families searched their phones for any kind of document or proof that would help them get through the gate to the other side of the airfield.

Families, now desperate, tried to climb the fences and swarm the tarmac as flights were trying to leave. The US Air Force had been cycling flights out of Kabul since July but were now packing as many people as possible onto flights to around ten safe-haven locations, including Ramstein, Germany, for processing through Operation Allies Welcome. Haunting images of Afghans hanging on to the outside of planes as they pulled away scarred in our minds. One service member shared how that imagery affected him after serving in Afghanistan: "We were legitimately prevented at times from engaging the enemy so that we would not have collateral damage. That's how much the Afghans meant to us, not just by rules of engagement. We understood we were winning hearts and minds. Collateral damage is damage nonetheless and has to be minimized at all costs according to the rules and laws of war. And yet, Afghanistan doesn't make sense anymore with people hanging off of planes as we are trying to leave."

On August 26, an ISIS-K member carried out a suicide attack at the Abbey gate of the Kabul airport, killing thirteen US service members and at least 170 Afghans, increasing the panic and chaos in the final days. In total, Operation Allies Refuge successfully evacuated 124,000 Afghan citizens, making it the "largest, most difficult and most dangerous humanitarian operations in US military history."[402] To date, 4,500 medals have been awarded, including ninety-six Distinguished Flying Crosses, twelve Bronze Star Medals, and one Gallant Unit Citation awarded to airmen who supported Operation Allies Refuge.[403] Among the stories were pilots who faced dangerous and difficult conditions as areas around the airfield deteriorated, civilian air traffic controllers fled, and landing lights and other navigational guides were not working. Stories from every branch of service have since been told including service members performing resuscitations and delivering three babies on C-17s. There were also around-the-clock repairs to airframes, constant in-air refuelings, and the coordination of feeding and caring for thousands of refugees in Operation Allies Welcome.[404]

Reading the incredible efforts and skill of our community, one can't help but be moved by their willingness to do the right thing for the sake of people, even under extreme conditions and even to the point of going outside of orders. Yet, this is who this tribe is through and through. From a young age, they are taught the values of selfless service, responsibility, the duty to stay until the job is done, and always executing with excellence. And they did.

AFTERMATH

If you were to ask anyone in the military community what they were feeling during the Afghanistan exit and then in the months after, you would have gotten a variety of answers, including anger, nausea, grief, or disbelief. It felt like a collective punch in the gut. Everyone experienced some sort of somatic response through words only the body could speak. In the aftermath, much of the culture sat in its own state of shell shock while others who were in the midst of refugee efforts were still holding off their own reactions.

Different from 9/11, people were now accustomed to watching events unfold live-streamed on almost every news channel and platform. Watching the replays of Afghan citizens begging for evacuation as they clung to C-17 tires and doorframes only seemed to sear the images deeper into our minds and made people feel even more confused and helpless.

What we felt in our own home, and what I heard from others was a sickening feeling of loss and confusion. Regardless of politics, the war needed to end. Yet, perhaps we expected to end with a level of dignity that matched everything we had given there. To watch it end in chaos as the Taliban systematically swept through and "erased" what had been achieved in just eleven days was extremely dysregulating. It pulled up stored memories and harbored feelings that surprised many service members and their families. Deep anger and sadness resurfaced over friends killed in Afghanistan as if people were needing to say goodbye to them all over again. There was also the grief of losing this war. What did that mean to walk away and say we tried but lost? It seemed as if none of us had even allowed ourselves to prepare for that scenario. The community encouraged each other to reach out to friends who had served, tell stories, and revisit efforts that they knew were purposeful, and at the same time try not to drink away what your body needed you to process.

When I think back on those months, it seems as if everyone's body was demanding a turn to be heard. Those who found ways to bring purpose out of their reaction found some traction to move forward. Checking on friends they had been to war with and family members getting involved with charities to aid refugees and women left behind in Afghanistan helped the healing process. The psychology and mental health field is still fairly new compared to the medical community. For decades, talk therapy and medicine management have been the most effective approaches we have had to treat trauma and mental illnesses. While there is certainly still effectiveness in these modalities, other approaches have gained more attention, like the very effective evidence-based Eye Movement Desensitization and Reprocessing (EMDR). Lately, however, trauma experts like Dr. Bessel van der Kolk are bringing new awareness to our body's ability to store trauma. In his book, *The Body Keeps the Score*, he "shows that the terror and isolation at the core of trauma literally reshape both brain and body,"[405] requiring individuals to pay attention to how stress and trauma are physically stored and how to release it.

After years of talk therapy where clients attempted to over-rationalize or "talk their bodies" into things they were not ready to do, somatic work is helping merge the two together. Somatic interventions such as yoga, iRest, and even acupuncture are giving people permission to listen to their bodies when under stress and respond to the anxiety, grief, and trauma that has been stored there for years. Somatic interventions are even helping individuals lead their parasympathetic nervous systems back into a grounded, calm state. It might sound a little strange to "listen to the body" but somatic work is the willingness to pay attention to the cues your body gives when it tenses up or feels unsafe, or even positive sensory messages that signal you are safe to feel attachment or joy.

In August 2021, the military community (especially those not in Afghanistan at the time) experienced a constantly changing somatic feeling: feeling safety with friends and family, to feeling possibly unsafe with the strong opinions of external family and media, to feeling the generation gap with younger service members who had never deployed to Afghanistan. For a moment, there was a clear "if you know, you know" feeling in the air.

The reaction of the media and civilian community made the aftermath another potential soul-wound for some in the community. As American citizens and news outlets put out consistently painful messages like "Was it worth it?" "This generation's Vietnam," and "We failed," service members valiantly offered understanding, intellectual perspectives, and commentary to help the country understand the complexity of a war that was so long Americans had stopped paying attention. I remember cringing at news anchors blatantly asking service members if they felt their efforts were a failure or not worth it. As difficult as that must have been, I recognized how empowering it was to find and articulate their own answers. As one female service member described, "I don't think anyone's feeling anything simple right now. Some of us expected it to end this way, but none of us wanted it to."

Talking through it is often one of the best ways to figure out how you feel, and everyone was doing that on some level. One veteran was ready to see the relationship end tweeting, "My take on Afghanistan has nothing to do with the Democratic Party. I'm concerned about the people who are left behind, but I'm sick and tired of American soldiers getting killed, families torn apart by PTSD, and veteran suicides, all for a country not wanting to defend itself. Enough."[406]

I wrote an op-ed piece I later tried to retract called "It's Time for Veterans and Military Families to Let Go of Afghanistan—But Not Forget."[407] It was titled by the editors but written before our troops had the chance to evacuate, before the bomb went off at the gate, and before so many Afghan citizens who had been promised protection had waded through sewers with the help of veteran volunteers. Admittedly, it was written from my perspective as a family member and clinician who had seen our efforts in Afghanistan steal so much. I was ready to see our family move forward, to see what our life would be like without war in the background. Premature? Yes, but also truthful.

A lot of military spouses reached out to me, angry. There was a lot to be angry about, but they were more angry at the way the chaotic withdrawal resurrected so much trauma, anger, and helplessness in their service members. Spouses, especially those from the special operations community who knew people personally who were in danger, were unsure how to help their service members process an ending they weren't proud of. Some of these couples had spent years finding healthy rhythms, sorting through trauma, and gaining a proper perspective of their part in the war. To them, it felt like all of that work was unraveling as some of them were trying to find their own way back to Afghanistan.

For a community that is taught to respect authority and utilize the chain of command when addressing concerns or complaints, the unbridled lack of hierarchy and the chaos of the situation opened the floodgates for many to express their dissatisfaction publicly, whether it was about the strategy or about top military leaders and the administration. In a survey following the Afghanistan exit, a majority of veterans ages eighteen to fifty-five years old described feeling betrayed and humiliated, more than half said the war in Afghanistan was not worth fighting, and seven in ten veterans believed that "America did not leave Afghanistan with honor."[408] Right after the suicide bomb killed thirteen service members, active duty Marine Lieutenant Colonel Stuart Scheller posted a social media video of himself in uniform, calling military leadership to take accountability for the failures in Afghanistan. He was immediately relieved of command, and within four days, his video accumulated 800,000 views on Facebook.

The reaction to Scheller's criticizing the chain of command was split at first. There were some who were unwilling to go near the topic, triple-checking their devices to make sure they hadn't accidentally shared the video for fear of jeopardizing their own careers, while others praised him for his honesty. After making a few more videos that caused some to be concerned about his well-being, Scheller resigned from his commission after seventeen years.[409] He was later formally charged with six violations of the Uniform Code of Military Justice.[410]

If you were to examine the culture's reaction to the Afghanistan exit out of the context of previous chapters, it may seem perhaps a little dramatic. However, what we were actually seeing was the undercurrent of mass confusion, repressed trauma, and decades of tension and exhaustion surfacing through the few who were unable to contain it or decided they were going to do something about it. The vast majority of service members (and their families) were going to do the right thing in whatever circumstances they found themselves. Veterans who were already out were willing to engage more openly in the debate. For those who were still in, there were plenty of opinions under the surface, but a few revelations were making the processing of this event even harder.

First, although service members knew that winning hearts and minds did not guarantee victory or even stabilization for the people of Afghanistan, it was a campaign nonetheless that shaped their minds toward the people for more than a decade. I believe some of the humiliation service members felt was due to decisions around the withdrawal that directly contrasted with what many of them had been ordered to do in their interactions with the Afghan people, even at the risk of their own lives. The contradiction of making promises to people only to abandon them was a statement many did not want to be associated with.

It is also possible that many saw the Afghanistan exit as a mirror of their own relationships with the institution. After devoting many years, so much effort and talent, and giving the institution the benefit of the doubt on fixing key issues that affected them and their families, the military did not seem to want to truly win the hearts and minds of its own people. Those who have crossed over into retirement have long tried to mentor younger generations that the military is not built to love you back, however, this event made it that much clearer to generations who were considering staying in for additional years of service.

Second, finding closure with Afghanistan was inviting service members and their families to find closure with all they had been through during the last twenty years. Not only was there significant physical and emotional trauma, but service members endured high amounts of stress and anxiety rotating in and out of deployments and trainings, advancing through the pressures of their careers, and trying to take care of their families during an incredibly stressful time. In the first decade, there was still the motivation and funding to feel like they were part of something meaningful and purposeful. The second decade offered no rest from operations, the anxiety of possible layoffs during budget cuts, working with limited resources, and trying to keep their families sustained and connected during increasing demands and frustrations at home.

There is a comradery that is deeply established in serving together but I have never met a service member who did not want to succeed first and foremost with their family. While it is often easier to give our best energy to where we feel most successful, which is usually work, service members who are married or who have children feel deeply affected by their desire to be seen as successful at home as well. As I heard one soldier say, "I can be great at my job and great at home, but I can't be great at both of them at the same time." The military has long known that the health and connectedness of families is a strength but also a vulnerability for service members.[411] It is why families have been called the strength of our force and have received so much attention, support, and funding. When families and marriages struggle, the force is vulnerable.

There is no question that two decades of war has negatively impacted the families at home, even though they found ways to thrive through the rich community that comes with shared difficulty. Similarly, service members admit the toll the lifestyle takes on their physical and mental health, even though we often don't hear about it until a much bigger problem emerges. Again, it is much easier to focus on where we feel most successful rather than admitting that one or more areas are disintegrating.

Starting around 2015, I was hearing more concerns than usual about military couples feeling disconnected and not successfully reintegrating. On the mildest level, couples reconnected but were beginning to accept that their lives were more different than alike. Other couples had dug deep trenches, where spouses at home struggled with giving up control and being the primary caregiver while service members escaped and coped through alcohol, video games, and pornography. More spouses than I would like to admit shared that they were afraid of their service members committing suicide. A few had even threatened to do so while intoxicated with a gun in hand. All of these families were considered high performers in our force.

As geo-baching became more popular, I noticed there was a generational component. Unless the family was geo-baching for a temporary reason or crisis, it was more common to see older spouses make this choice in response to feeling secondary to the military for so many years. It was also becoming an attractive choice for millennial and Gen Z spouses who wanted to build careers of their own and saw the service member's job as just that rather than as a lifestyle. Either way, the wellness and cohesion of the family have an impact on the service member's contentment in their own job and well-being.

Service member suicide had been an important public topic, especially with the "22 per day" veteran campaign. Around 2015, spouses began advocating for the DoD to start tracking spouse suicide as well. They did so starting in 2017. By 2018, suicides of active-duty military members were the highest on record since the DoD began tracking self-inflicted deaths in 2001. They continued to rise. In 2019, Air Force Chief of Staff General David Goldfein posted on social media, "We lose more airmen to suicide than any other single enemy."[412]

Thomas Howard Suitt, in a 2021 Brown University Costs of War report, estimated that 7,057 total US service members were killed in post-9/11 operations, and 30,177 had died by suicide. The findings show that although unique factors such as physical wounds, traumatic brain injuries, and multiple traumatic exposures have contributed, the "sheer length of the war has kept service members in the fight longer, providing more opportunities for traumatic exposure, and fueling a growing disapproval and ignorance among the public that has only enhanced veterans' difficulty finding belonging and self-worth as they reintegrate in society."[413] We also know that there is an association of suicide with intimate partner problems.

Every life matters, regardless of age or rank. The military had been talking about, teaching on, and briefing on suicide prevention since we had come into the Army, with the biggest concern being the majority (around 42 percent) enlisted men under the age of thirty years.[414] In 2019, along with the increase in reports of toxic leadership and moral failings, suicides of military leaders seemed to go up as well. Like the Costs of War findings, I was curious how much the increase in leader suicides had to do with the wellness and cohesion of families who had spent the majority of their military-connected years during GWOT.

The reasons an individual would turn to suicide are difficult to measure and predict. We know that attachment, belonging, and meaning lower loneliness and reduce hopelessness. However, unresolved family and partner stressors that feel insurmountable have led many to dark places. For those who are mid-career or facing retirement, surveys continue to show a growing military-civilian divide making the transition especially difficult. Although employers are willing to hire veterans, some service members struggle to find jobs (or second careers) that offer a similar opportunity to use their specialized skills sets.

I will never forget one officer I met just a few months prior to his death. He was not only a leader in the community, he was a pillar, well respected and highly decorated from his involvement throughout GWOT. We shared a panel discussion on military family wellness, and to make conversation, I asked about his family. I remember noting the look in his eyes as he described the struggle of trying to show his family that he was choosing them over the dangling carrot in front of him. The media rightly did not cover his death out of respect for the family or at the risk of lifting his life above any others who had served. In the community, however, it was a significant blow that shook many who served with him and looked up to him. Some of the shock and confusion around his death and the deaths of other leaders is not knowing fully the circumstances surrounding their decision. Without information, the grieving community may assume he was concerned about transitioning out of a career he thrived in, or perhaps that he felt the push and pull between his career, the need for self-care mentally and physically, and the family who had reached their own limit.

Service members who came in as part of the surge for Iraq in 2003 are hitting their twenty-year golden carrot finish line. We will not know the attrition or retirement numbers for 2023 for some time. However, there is talk in the community of a mass exodus of service members since the end of Afghanistan, leaving at the end of their commitment or choosing to retire earlier rather than giving more than twenty years. There is a Mother Teresa quote that says, "If you want to bring happiness to the world, go home and love your family." It seems a lot of military leaders and families are ready to do just that.

QUIET REBELLION

Since the pandemic and the end of the Afghanistan conflict, I have seen a considerable increase in Gen X service members turning in retirement packets, even though their branch was offering another level of leadership that would have given them higher pay and more influence. In many cases, officers turned down the opportunity for a star to leave the military and work on their marriages and families. I have also seen a considerable number of leaders turn down great opportunities because they had already begun to make poor decisions due to the accumulation of years of stress and neglect on their minds and bodies.

In a 2022 DoD Army survey examining the reasons service members chose to leave, 40 percent of separating enlisted soldiers cited morale as their top reason. Concern for family was the top reason Army-wide. Those leaving due to changes in Army policy spiked during the five months surrounding the "drawdown of US and allied forces in Afghanistan and the DoD-mandated COVID-19 vaccine requirement."[415]

For those still serving, it is much harder to measure their reasons for staying. Since 2017, the DoD has stated that retention rates remain stable, rising in 2020 due to the pandemic and an uncertain job market. As stated earlier, DoD active duty demographics from 2020 to 2021 show the portion of service members ages forty years and older increasing from 8 percent to over 12 percent, with retention of those close to retirement or avoiding the transition until the economy calms down.

Civilians came out of the pandemic with a new perspective on how they wanted to return to the workforce. As individuals realized how much they had devoted their best energy to their jobs, upon returning to the workforce many decided to instead give the bare minimum while collecting a paycheck. Perfectly paired with millennials' value system toward better work-life balance, some quit their careers completely (the great resignation) while others who did not want to walk away from their jobs simply decided to "quietly quit."

It is far more difficult to measure the number of service members who may have chosen to quietly quit, but those staying in are taking a different approach to serving. There are many aspects of this lifestyle that service members and their families have accepted they cannot control. Yet, every human who feels out of control will eventually find a way to assert their autonomy, whether it is a toddler throwing a tantrum or adults quietly quitting. People need a sense of certainty in the things they value most that add quality to their lives. After years of putting the military and mission first, many at the expense of health and family, and after a pandemic and chaotic end to war, families of every generation are making up for lost time by putting their family first whenever possible. I call this the "quiet rebellion."

This means that military families are less likely to want to participate in the military community outside of work hours, even if it is a morale-boosting event like a military ball or family organization day. Almost with a "give Caesar what is Caesar's" mindset, service members and their families are devoted to the basic expectation of their commitment to the force, but are unwilling to offer more than what is expected simply because it has always been done that way.

While baby boomers are still serving in the highest levels of leadership, Gen X is in upper leadership and has always wanted a better work-life balance, many of them not knowing how to ask for it or set proper boundaries to ensure it happens. Those who are still in are holding out for retirement and not eligible for the new Blended Retirement System are caught between leading with the imprinting of the military they once knew and the new cultural shifts that encourage more family time. Millennials, now in command roles, are helping make that shift happen and are far more comfortable ensuring it happens. Gen Z is not only supportive of this mindset, they will not stick around if it is not part of the package.

In an article to senior leaders, one millennial shared: "It's ok that we're not a family at work. We can be congenial and even highly effective coworkers without being a family (calling us a family is actually shown to be a contributor to burnout). This distinction is important; I will do this for five or ten or twenty years but I will be with my actual family for much longer. I fully grasp how my entire family is serving but if I'm a junior officer, then I'm probably less sure if I'm staying in the Army for a full career. With the Blended Retirement System, it's more reasonable than ever to leave at the four, eight, or even twelve-year mark and begin a new career or go back to school. ... If I'm not chasing field-grade command or stars, then it may make more sense to depart sooner rather than later, for the sake of myself, my family, and the aspirations and norms of my next career. It's a decision I don't take lightly as I don't forget the impact it has on my family."[416]

This perspective can be jarring for older generations, especially those who served prior to 9/11 and the first decade post-9/11. This exposes the significant, yet often unspoken, generational gap within the culture. As one older Gen X stated in response to the above quote, "This strikes me as both naive and narcissistic—and misunderstanding the dynamics that make the Army different from just a job. There have always been me-first people in the Army, but frankly they tended to be ostracized in the units I was in. It is hard to build the kind of trust it takes to be successful in combat units with this perspective, and it creates a weird bright line between who is family. My Army peers (the ones I truly care about) are as much my family as my blood kin. I will never have friends as deep and connected as those with whom I served. But it takes the right environment to build that."

How does a generation that considers the opportunity to serve a "calling" and those they have served with "family" recruit and lead a generation that does not value or want the same experience? The culture and lifestyle are deeply personal and worthy of protecting. It is understandable that the solution for them would be to teach, mentor, and try to inspire the younger generation to want the same experience. Yet, is it possible to recreate that without the "right environment" of war?

In addition to the quiet rebellion of choosing family, service members are less willing to take leadership roles due to the risk of having their mistakes aired on social media or being canceled or disqualified by the younger generation. Although it is a bit controversial, there is also fear of being misunderstood, getting behind on the changes that are happening culturally in the areas of diversity and inclusion, or being wrongfully accused of something that could irreparably taint or sabotage their career.

The investigation of Major General Donahoe's comments on Twitter was a good example. He was investigated for bringing "negative publicity to the Army, for engaging in social media arguments over the military's COVID-19 vaccine mandate for troops, and criticizing [Tucker] Carlson's critique of gender-based military reforms." From his perspective (and that of many others who followed him) he was defending the troops he was leading.[417] Even though he retired honorably without reprimand, *Task and Purpose* stated, "Donahoe's saga shows how the Army has struggled to demonstrate that it cares more about supporting female soldiers than getting criticized by cable news pundits and prominent lawmakers." One soldier was quoted as saying, "Intentionally or not, this whole thing showed women that we are not worth defending. If he can get slapped for this, why would anyone defend women in public?"[418]

AN EXAMPLE WORTH REMEMBERING
Why do we stay?

That is the question even the DoD is asking as they evaluate retention rates and the power of current leaders to encourage the next generation. It is also a question that almost every family has asked themselves. During days that are especially difficult, when the cost of war is too great, it is easy to ask ourselves if it is worth it. That was also the heaviness of the media's question when they asked if Afghanistan was worth it. Although the media was asking service members their opinions on whether our US involvement in the Middle East was valuable or effective, it invited a deeper reflection into the cost of the many years they had devoted their lives and families to career that did not presently feel victorious.

In 2010, 3-61 CAV and their families were asking this very question. Our service members had just come home from an incredibly difficult year in Afghanistan and we had gathered together at a military ball to celebrate. We had lost eleven soldiers during that deployment and many soldiers had come home injured. We were exhausted emotionally and physically, and morale was still teetering from the trauma our squadron had endured. We were eager to get together with this family that had shared our highest highs and very lowest lows. We were also eager to get lost in the music (and alcohol) that could both give us the opportunity to feel alive while numbing the grief and guilt over those who were not there to celebrate with us.

There were a lot of things that felt unfair, that made many of us question if this institution was worth what we had just experienced. If anyone had the right to give up on it altogether, it was our commander, (then Lieutenant Colonel) Brad Brown, who I mentioned at the start of this book. One of our troops was assigned to Command Outpost (COP) Keating, a remote US Army outpost in the valley of the Afghanistan mountains that put us at a dangerous disadvantage. It had already been set to close, making it a prime target for Taliban soldiers. On October 3, 2009, between 300 and 400 Taliban ambushed the post. After a long and difficult battle, fifty-two American troopers took back the base. Eight US service members died in the battle and another twenty-two were wounded. From that battle were awarded two Medals of Honor, two Distinguished Service Crosses, nine Silvers Stars, twenty-one Bronze Stars, thirty-seven Army Commendation Medals with "V" for valor devices, and twenty-seven Purple Hearts.

An investigation found two commanders to blame for "inadequate measures taken by the chain of command." Brown was one of those commanders, but ask any soldier who served with him and they will say the target was on the wrong back. Despite the reprimand, Brown humbly served out his deployment and command with 3-61 CAV. We were all asking, "Was it worth it? What was all of this for?" Knowing this, Brown gave a speech at the end of that deployment ball that many of us have revisited when we have felt the cost of this lifestyle. It is important enough to share in its entirety:

"One of the beauties of not having a separate farewell is that I get to write one less speech, and ya'll get a few hours of your life back. For those of you who have not had to endure it, speechwriting is one of the most trying ordeals in life. As a speaker you commit hours and sometimes days to prepare something that is generally of absolutely no interest to your audience. And in the attempt to make it more interesting, you just tend to make it longer—never a good thing with a captive audience that has been drinking since early afternoon.

"Last night somewhere around 0230 I came to the subconscious realization that the speech I had spent the past several days writing was absolute crap, and I was going to have to start over from scratch. So I've pretty much been in crisis typing mode ever since. I wrote this knowing that in the history of man there have been billions of speeches given, and maybe a dozen that are truly remembered. And I certainly realize that this won't be one of them. So why commit the time and stress for something that serves so little purpose? Because the Army is full of contradictions, and that is part of what makes it special.

"One of my favorite contradictions dating back to my days as a scout platoon leader was the order to 'maintain contact, but don't get decisively engaged.' When you are equipped with light skin HMMWVs and you are going up against BMPs, any contact is likely to be decisive if you are the HMMWV engaged. As a young ROTC cadet, I first heard a senior officer say that a leader must put mission first and people always. Back then I thought this meant that a leader had to make sure his troops were fed and equipped and got a little sleep during the mission. Life was easier when the only thing being shot at you were MILES lasers.

"As a battalion S3 in Iraq in 2004, I came to fully appreciate exactly how contradictory that saying could be—because I was writing orders that sent soldiers into harm's way, and I knew that eventually someone was going to be killed on one of those missions. It is one thing to understand and do your duty, it is another thing to explain to a soldier, or junior leader, or a family member why a task has to be accomplished—even though the mission is minor, and it might cost him his life.

"A few months ago the spouse of one of our fallen heroes asked me: how can you live with yourself sending soldiers on missions where they might get killed? That's a tough question, and I've spent a lot of time thinking about it. In the three years I've spent deployed, and most particularly in the past year in Afghanistan—there was not a single individual mission—no cordon and search, no key leader engagement, no ambush, no recovery—that was worth the life of a soldier. But if you follow that idea to its logical conclusion, then you would never do anything—because every mission has risk. Even doing nothing is risky, because if you do nothing, the enemy will come to you—on his terms. So leaders are placed in that contradictory situation where they must balance the needs of the mission with the potential risk in soldiers' lives. And there is nothing you will do in life that entails greater responsibility.

"It is easy to look back at an event and ask: why did that soldier have to die on some insignificant mission, in some minor town in a distant province—and reach the conclusion that it wasn't worth it. It isn't right, and it isn't just. And this highlights the contradiction of the Army—that as individual soldiers, we have subordinated our freedom, our time, and in some cases our lives to the greater good of an institution. And as an institution, the Army does not feel compassion, or friendship, or loss. The Army is committed to fighting our nation's wars, and that is a mission which is more important than the life, the career, or the family of any individual soldier.

"And this is the paradox: why would anyone choose to serve in an organization that by design must consider every person expendable in the name of mission accomplishment? I don't know who coined the phrase, but I've used it often over the years: you can love the Army, but it isn't always going to love you.

"The answer to this riddle is that the Army is really two distinct, but inseparable entities. The first is the institution—it is the history, doctrine, the rules and regulations, the tasks, conditions, and standards. It is represented by the uniforms that you wear, and the symbols on your lapels. It is ranks, branches, and colors. Soldiers come and go, but the institution endures, and will continue so long as it serves its purpose in defense of the country. It is impersonal, and to an extent indifferent to the individual. It is that way because it has to be—because no Army can function if it places the needs of the individual over the good of the whole.

"That institution is the bones of the Army. But its flesh and blood are the people that carry the symbols and wear the uniforms. The Army is a great organization because it is filled with folks who are willing to subordinate themselves to a greater purpose, and to those with whom they serve. Scientists will tell you that people are basically rational animals that act in their own self interest—and this is generally true. But the Army is full of irrational people—it takes a special breed to run into a hail of gunfire to render aid to a wounded comrade. And this auditorium is full of them—soldiers and spouses. They will volunteer to watch your children with no compensation but a thank you. They donate their time and energy without pay. They will offer their homes to a traveling family they haven't seen or spoken to in ten years. They will cut their neighbor's grass while they are deployed. And they will place themselves in harm's way on a routine patrol that is not worth the life of a single soldier—knowing that their sweat and blood contribute in some small way to the success of the greater institution.

"Many of you are at the crossroads where you are trying to decide what to do with your life—whether to reenlist or take orders to the career course—and take that next step of commitment to a career in the Army. There is no doubt that the Army is a hard life—and it is a hell of a lot harder when the nation is at war. You will look at the pros and cons, maybe talk it over with your spouse, and do a cost-benefit analysis. I hope most of you will stay even though it seems to make no sense.

"The fact is that the Army can be a contradiction because life is not a math problem. I love serving in the Army because it is full of people who care more about each other than they care about themselves. That's not to say the Army is full of saints. We drink too much, like to gamble, swear a bit, and sometimes do dumb things. You can find an easier life and a more financially rewarding job, but you will never be in a community of better people. If you have a problem, you will never have a peer group more willing to go out of their way to lend a hand. When I was at that crossroads, I looked at the people I was serving with and realized that I wanted to be like them. I have never regretted that choice.

"As it does with all of us, the time is approaching quickly when the institutional side of the Army won't love me anymore. But the friendships Sue and I have made will endure long after I take off the uniform for the last time. Not friendships of convenience or proximity, but the bonds shared with selfless people who care for each other. That is something I would not trade for any rank, or any amount of money. Enjoy your time and your service, and take care of your people—because they are what makes the Army the greatest institution in the world today.

"Thank you for your time, and thank you for your service.

"Led by love of country, Destroyers!"

—Lt Col Robert "Brad" Brown

As we enter a time of peace, the youngest generation will not have the same experiences, stories, or understanding of war as the older generation. The gap of experiential understanding will grow wider and the community will feel less bonded in common stories. However, Brown's words, even back in 2010, offer some key insights for leaders to help bridge the generational gap.

First, words matter. If you find yourself fighting to find the right words, even staying up into the early morning hours, chances are something important needs to be said. There are a lot of things Brown could have said, but the deeper question was, "What did our squadron actually need to hear?"

Second, he authentically verbalized what everyone was thinking and feeling. He had been a leader who was not willing to ask his soldiers to do something he wasn't willing to do, and because of that, he was keenly aware of what questions they were asking. In essence, he was willing to walk through the rubble of a bruised and bloodied squadron and acknowledge that we could not move forward until he validated where we were. By talking the audience through the contradictions of the institution, he modeled for us how to live with them and bring our own meaning, building even more loyalty and trust.

Third, he saw this last opportunity to speak to his soldiers and their families as a chance to vision cast rather than lament. There were a lot of things that didn't make sense about that deployment and some experiences we will never be able to package up and tie nicely with a bow. But we did have control over who we will choose to be for each other and how we are willing to pass that along if we are willing to not give up.

Finally, to the younger generation who might be reading these words, those who have been to war and have experience navigating the institution have an earned wisdom. While it is normal to want to dismiss the past or an outdated way of doing things, maturity is being willing to humbly admit when you could be wrong and be open to additional ways of thinking. There are some things we must be willing to try or continue, not because we understand them or because we relate to them, but because of the greater value of investing in people.

That assignment with Brown as our commander was our first. We felt like a couple of kids who were asking ourselves what we got into, and so were others. My husband and I have pulled out this speech many times over the years, reflecting on our time with leaders like Brown. Most of the people who served in that unit stayed, including Sergeant Major Stephen LaRocque, who was shot in the knee during that deployment and returned to finish out the deployment with his men. He is currently serving as the Command Sergeant Major of US Army Garrison Rheinland-Pfalz, the primary evacuation hub for Afghanistan refugees during Operation Allies Refuge. When he was interviewed about his role in Operation Allies Welcome, he said, "I've been to Afghanistan numerous times. I've had a lot of Afghan soldiers, Afghan families, and interpreters help out a lot of US citizens over there, as well. So it's an honor to help get them to safety and to bring a lot of people back home. It means everything."[419]

Two years later, CSM LaRocque stood beside his Gen Z son, Landon, as he was sworn into the Army, following his father's footsteps. I asked him why he has continued to serve more than twenty-seven years despite the hardships. His answer, although completely in line with being a Gen X, is likely what you would hear from others who care deeply about this life: "I absolutely love our country. I recognize that is it far from perfect, but I honestly believe the ideals upon which we were built, the ability for the average person to become whatever they want, is incredible and rarely exists anyplace else. To be able to serve my country gives me meaning." He also shared that he felt a sense of purpose in taking care of soldiers and families in his current role. "The Army has been good to my family and me. I don't necessarily feel like I owe the Army anything anymore, but I'm not ready to quit yet, either. Nicole recognizes how important this is to me and allows me to continue on. If she ever put her foot down, I would be done in a flash and she knows it."

It's worth it. We may not have had those words at the moment the media asked after the chaos of the Afghanistan exit, but we eventually found them. Our two decades of war began when a group of people chose to use planes as weapons to ignite fear in the hearts of a country and exert power over its leadership, but that is not how it ended. Twenty years later, our military community used planes as a vessel of hope and rescue to the most vulnerable, igniting empowerment and giving people a future, wishing they could have rescued more.

That is what I love most about this culture. We deeply care about people. So much so, we are collectively devoted to doing the right thing no matter the cost. There will always be outliers in the community who turn destructive as they are tempted with power or simply make poor decisions that harm others. A majority of this community will distance themselves from service members, regardless of rank, who do not align with the values of selfless service or any of the other values shared within their branch. We don't put up with bullies.

It is absolutely magical that the character displayed in our force during the Afghanistan withdrawal is not an anomaly. This is who we are, this is who our families are, too. There is a shared value system that unites generations of those who put on the uniform and take an oath. They are willing to find their way back to a country they long left to fulfill a promise and will find a way to get another plane in when there are more people to save. When the mission is over and they have a moment to breathe, they are also the ones who will grieve when they feel they could have done more.

There is a lot of pain in our story. In our commitment to live the call of selfless service and courage, we have lost some of the best of us. We've lost marriages, sacrificed time with our children, and, in some cases, sacrificed parts of our souls for something far bigger. And while we may not know the ultimate reason, we know it was worth every second for the friendships we have gained.

In the words of Brad Brown, "Not friendships of convenience or proximity, but the bonds shared with selfless people who care for each other. That is something I would not trade for any rank, or any amount of money."

WHERE DO WE GO FROM HERE?

"We cannot solve the problems we have created with the same thinking we used in creating them."
—Albert Einstein

Chapter 17

MORE THAN A NUMBER: RETENTION & RECRUITMENT

THE MILITARY CULTURE WILL THRIVE AGAIN when the people thrive. We can continue to blame recruitment challenges on the nation's disconnect or the military's attempt at progress, but until we are honest and willing to get our own house in order, the military will not have a competitive advantage over the civilian marketplace. Time and distance from the stress of war and pandemic and change will certainly help, but people won't forget. Their stories are etched into the fabric of who they are and inform their daily decisions, including whether they are encouraging the next generation to follow in their footsteps.

Leaders today have a significant challenge ahead of them. How do you simultaneously recruit the next generation while healing the current one? Thankfully, the force is large enough that most leaders are not individually tasked with both at the same time. This means that regardless of your role in the system (political, military leader, spouse, civilian, or otherwise), your part matters and contributes to shifting the military culture again—in a positive direction.

HONORING THE CULTURAL NARRATIVE & THOSE IN IT

Our lessons learned from Vietnam are not that far away. We cannot be a nation that sends its people to war on our behalf and then be unwilling to hear their stories or be unwilling to tend to the wounds they inherited. Being a person, a leader, who is willing to sit with and respect someone's story is singularly the most powerful way to be a healing agent. This doesn't necessarily mean you need to listen to the story of every person you encounter. It means that you begin to see each person you encounter as someone who has a story to tell.

Depending on who you are leading, you will come across both the individual narrative as well as the collective cultural narrative. Both are important and, as we have seen, have a way of informing and influencing each other. Our personal narratives drive our decision-making, propel us into the future, or trap us in the past. Our cultural narrative shapes our identity with strong connections to the past and creates lifelong connections with those around us in the present. Cultural narratives also provide the structure of patterns and traditions, healthy or unhealthy, that are inherited and passed down to future generations.

Stories remind us of our humanity, inspire authenticity and transparency, and have the power to connect us through a shared or common experience, even when our individual stories are different. During the pandemic, for example, people felt powerless, afraid, angry, and a number of other emotions. We did the best we could to process it in the moment, but once we had the opportunity to move back into a normal state in society, we were so desperate for certainty and connection that many of us never looked back to process what it meant to have that experience be part of our story. Our bodies remember, though. When we sit at a table with friends, the pandemic is a collective memory. We can share our individual stories and leave seen, just a little bit lighter and more grateful.

As a powerful teaching tool, storytelling helps us remember lessons already learned and victories already won. It is a connection to our past as much as it is a connection with others who share the past with us. In essence, it is about the power of remembering. Our minds have a fascinating way of forgetting the things we should remember and remembering the things we wish we could forget. By telling the stories of past victories, we remember our successes and where we got it right, which is especially helpful when we are in the midst of challenging moments or facing potential defeat. Stories of failure, difficulty, or defeat are just as important, if not more important, to help us remember where our priorities were off, where humility may have changed the trajectory, or how hindsight can help illuminate future decisions.

This ability or willingness to recall stories of our past, even those that are hard, is what historians beg us to work on. History does, indeed, repeat itself in families as much as it does in cultures. Yet as time keeps us moving farther away from the source, it fades from personal memory and is logged as a part of history that begins to feel less relevant. This is great news for those who are working through trauma, grief, or disappointment and need the comfort and distance of time. "Time heals all wounds" is true in that details will eventually fade, leaving only a scar as a reminder. In this way, not remembering, or forgetting, has a healing element to it in that you are still able to share the story of the scar, even feel the phantom pain of trauma, but time has allowed perspective and meaning to fill the gap between then and now.

This coping skill of forgetting, however, has a way of numbing some parts of the story that are dangerous or destructive to forget. There are some things about ourselves or others that are too painful or that we would rather forget. Remembering those details, though, is what keeps us from repeating the same mistakes. For example, looking back on World War II, we agree that the Holocaust, killing an estimated eleven million people (including six million European Jews), was not only morally wrong but devastating to an entire culture and to the nations that fought against the National Socialist antisemites and Adolf Hitler. Yet, time and distance have clouded the initial unethical micro-steps that deviated the party to such an extreme trajectory of evil and hate. These micro-steps are equally important for us to understand and remember, as they are what keep us from repeating history, even something evil or vile. While many social scientists and scholars have studied the "road to evil" of the National Socialist antisemites, Anthony Kauders, professor of modern history at Keele University, suggests that Cognitive Dissonance Theory could be an additional explanation.[420]

When an individual makes a decision that goes against their moral conscience, even a small one, it creates dissonance or incongruence that they will naturally seek to resolve "in order to retain a sense of oneself as good, decent, reliable, or sensible."[421] However, particularly when an act is especially irreparable, the dissonance is even stronger. "The more costly a decision in terms of time, money, effort, or inconvenience and the more irrevocable its consequences, the greater the dissonance and the greater the need to reduce it by overemphasizing the good things about the choice made."[422] Beliefs are a lot easier to change than behavior. So, we tend to seek out media, information, and even script cultural narratives that validate our actions, that create consonance within ourselves, to the point of defending our actions rather than challenging them. Kauders concludes that Germans embracing the Nazi morality "afforded a sense of self-consistency in times of radical upheaval: it was easier to change one's views than to question the state or to intervene on behalf of a minority."[423] The work that scholars and historians have done in studying human behavior during the Holocaust is revealing an enlightened understanding of how people's small decisions can contribute to how a culture evolves its collective narrative over time and embraces ever-expanding cognitive dissonance.

Although the focus of our story has been the cohort of service members and their families who have served since 9/11, the history of our force all the way back to the forming of our Constitution reveals the micro-steps that set us on a trajectory of where we are today. While some of these decisions have made a positive difference in the quality of life for the military culture (such as opening active duty to married men, benefits, healthcare, and more), we can see how other decisions, many also positive, opened Pandora's box for issues that have been quite challenging for both families and the military to navigate.

Switching to an all-volunteer force was one of the most pivotal moments in the history of our military's development. Although there were other significant moments, like the build-up of World War II, this moment shifted the military's attention and efforts to recruiting service members rather than relying on a draft. Soon after, the family's support of the service member was not just key to retention, but the military would need to recruit families to meet recruiting goals. Recruiting families, however, meant that the DoD needed to provide a more supportive lifestyle that would make it possible for service members to know their families would be taken care of while they were deployed or away on orders. In all fairness, the very thing that the institution was concerned about happened. Families were willing to commit to the lifestyle of the military in return for the military's commitment to provide for their basic needs, creating a dependency but also a partnership.

Throughout the Cold War, the demands on families grew, but families also found ways to thrive by building their own community. The social culture was good for the military and gave military spouses a sense of identity and purpose. The subtle micro-step of shifting the DoD's message of "taking care of families" to "taking care of families so that they can take care of themselves" shifted the responsibility of the institution for the culture's wellness without clearly communicating that shift to the people.

By the time 9/11 happened, the programs, benefits, and resources the military provided to military families became widespread enough that privatizing their oversight through outside contracted companies helped share the responsibility. The real dilemma, and the difficult conversation that needed to happen, was how the family's dependency was growing beyond what the institution could provide. Even though it was a positive solution, not fully communicating how it would change the culture's relationship with the institution may have created more relational confusion. From the family perspective, the institution was still their provider. If the privatized company failed them, the institution failed them. Although there were many, many micro-steps taken along the way, it is possible that faced with the choice to have a hard conversation with military families about what the institution could or could not provide, it was much easier to make changes without vision casting.

The institution wasn't alone in struggling to have hard conversations. As much as our story to this point has seemed like a Greek tragedy (and in many ways, it has been), what is truly inspiring is that service members and their families won't allow it to end that way. As one of the pilots said of their thirty-hour mission during the evacuation of the Kabul embassy, "At that point, your instincts just kick in to make sure the Americans on the ground are safe, and you push your exhaustion and your stress and your worries to the side to get the mission done."

Service members and their families have a talent for staying positive, finding the good, and leaning on what is most important: their family, those they have served with, and the values that have created a common ethos among them. This grit factor is what enables them to embrace the suck and overcome the challenges. Rather than complain, many military families adopt the philosophy that if it is not something you can directly influence, then there is usually a way to push through it, even if that means just waiting for the next assignment.

The institution would call this resilience, but I call it grit. However, the story of this tribe over two decades of war reveals the cost to a culture when they are constantly assuaging internal or familial dissonance by choosing an overly optimistic view or overemphasizing the good. We have seen how the promotion system, family culture, and institution encourage an environment that rewards positivity, productivity, and in some cases, silence with belonging.

On the opposite extreme from the National Socialist antisemites, Viktor Frankl was a Jewish-Austrian psychiatrist and survivor of the Holocaust. In his book *Man's Search For Meaning*, Frankl shares how in order to survive the torture of the concentration camps and the depressive state his fellow prisoners adopted in their circumstances, he gathered rocks and envisioned them as future students, teaching them how he had survived. Frankl's approach to dissonance is to lean into it and see it as a teacher.[424] If you experienced dysregulation, discomfort, or even dissonance with this story, then I deem my storytelling a success. Experiencing our own discomfort with this story as well as with the difficult parts of our own stories, helps us become comfortable with the questions that arise.

Most want to push dissonance away because it feels conflictual. However, in the face of toxic positivity (overly avoiding, rationalizing, or rejecting negative emotions),[425] Frankl suggests tragic optimism, "the ability to maintain hope and find meaning in life, despite its inescapable pain, loss, and suffering."[426] In other words, embracing difficulty and suffering as part of our story, learning from it, and asking important questions of ourselves and the situation, helps us bring meaning to the suffering rather than ignoring it, which brings more suffering.

Avoidance isn't the only reason we may have missed the broader story or not seen it unfold. Those who have been actively living under the stress of GWOT easily missed a majority of this "wicked" problem because they were already having difficulty enduring this experience. Rarely can a person zoom out to see the complexity of the situation when they are just trying to survive it. In those moments, positivity and gratitude are what help keep your head above the water rather than ruminating on what might be pulling you under, especially if doing so questions what brings a sense of belonging and provision.

Additionally, there are so many layers to this problem that they are too difficult to see or accept all at once. No one family or service member was simultaneously experiencing all the variables we have presented here. The division and clearly contrasting experiences of the officer and enlisted lifestyle, alone, is another example of how the story of our culture may have shifted under our noses. While the military lifestyle can bring similar stressors to both groups, there are specific challenges in each. Enlisted families are more likely to experience financial hardship, issues with housing and childcare, and food insecurity, making them much more susceptible to negative experiences as well as being taken advantage of by the community or privatized companies.

If senior leaders are not living in the same neighborhoods or dealing with similar stressors, they are less likely to experience the issues and may live and make micro-decisions based on their own experience or experience bias. When subgroups within a culture have very different, but still difficult, experiences of the lifestyle, it is harder to see how war, significant quality of life issues, and other stressors are impacting the broader culture. Your own experience is all you know, and stories of past generations, unfortunately, fade or are forgotten if they are not revisited or told.[427]

When we allow ourselves to broaden our view, we see a wicked problem that has emerged after multiple decades of decisions that were made in the moment or during seasons of feast or famine that have compounded over time. There were also things we could not have anticipated, such as the influence of technology and how global events have shaped the values of younger generations. In many cases, the consequences could not have been predicted, in others, competing priorities were positioned before people, resulting in a culture that is in great need of respite and repair. When you love your community and depend on it for belonging, meaning, and purpose, it is difficult to see the hard truths that could leave you disappointed or, worse, needing to advocate for yourself and not know how.

When I listen to the stories of service members, spouses, or couples, I invite them to temporarily "sit in the pocket" of what has been difficult. Many, including military spouses and service members of all branches and ranks, have difficulty sitting with these hard truths and often negative feelings. Doing so, though, honors them as valid experiences that we cannot and should not ignore. They are valuable parts of our story, and our experiences deserve our attention, even for a little bit. Sitting in the pocket does not mean we build a house there, but it is valuable in that it helps us have a more realistic and healthy perspective of our story in order to move forward.

With this book, I invite you to sit in the pocket of some of the most difficult parts of the military family story over the past two decades. My intention is to present a more accurate picture of the state of our culture as it faces a significant opportunity. Millennials, but especially Gen Z, need and respect open dialogue about incongruence, discrepancies, and dissonance. They have lived in and grown more comfortable with the dissonance of going to a school that should be safe but feels unsafe. They have endured the mass media messages of perfection while facing their own imperfections. They have something to teach us about accepting the vulnerability of our humanity if we are willing to listen.

REPAIRING & RESTORING RELATIONSHIPS

We have evolved before, we can evolve again. Since Wickham's white paper, the DoD has built out incredible programs, including master resilience training, counseling programs and training on suicide prevention, substance abuse programs, spouse employment initiatives, and quality childcare centers. People First is a continued response to requests for even more resources to improve the quality of life for service members and their families. The updated Army's Developmental Counseling Form in March 2023 included questions about the service member's work-life balance, revealing efforts to make improvements.

These improvements are making a difference, and despite the bumpy start of People First, it has slowly gained credibility as consistency of messaging and action have been proven over time.[428] However, Wickham made a key point in 1983 regarding relationship building that gives insight into who we can be as change agents today: "Family problems caused by stress, and those conditions which produce stress, are cumulative—they become more severe over time and are costly to correct. In the past, we have generally attacked the problems only after they have become severe and the impact obvious. For both humanitarian and readiness reasons, we need to shift the emphasis from a focus only on families already experiencing problems to programs designed to help families cope with stress by building better stability and adaptability."[429]

The DoD's development of programs around resiliency has taught generations of military families how to leverage new and better coping skills, and the DoD is currently trying to address some of the bigger problems that have attracted congressional attention. However, there is an important step that was missed between addressing current problems and preventing problems through teaching families how to cope.

Relationship building requires vulnerability and the active desire to repair the relationship when needed. There is no way around it. This is true when the relationship is thriving but exponentially true when trust has been broken. Resolving incongruence with ourselves requires sitting in our own skin and listening. Relationships are no different. It is tempting to think that the larger the people group and the more formal the enterprise, the less vulnerability is needed, but it simply is not true. Sure, there is work to be done, problems to fix, missions to execute, and difficult asks to be delivered, but as long as there are people coming to work every day to engage with other people to make the enterprise happen, relationship building will make or break the whole thing.

Honoring relationships, even in business, means valuing the relationship as worthy of repair. The important step that was and is missing is the authentic acknowledgment that the relationship is fracturing. Someone once asked me, "Who in the relationship gets to decide if trust has been broken?" It is a valid question, although the answer might seem obvious. When you don't mean to break trust or your intentions are positive, it is easy to become defensive or even call the other party overly sensitive if they feel betrayed. However, if we see the relationship as worthy of repair, then whoever feels wounded or betrayed has a right to express their concern. If we wish to keep the relationship, we must pursue the other party and work toward repair.

Too often, in the face of hurt and disappointing others, we skip the very difficult step of validating the pain and, if we aren't ignoring it, moving straight into, "I'm sorry, I'll fix it." If we are not careful, we will believe the myth that relationships and culture will heal or improve organically on their own rather than listening to the dissonance within us that makes us want to do the intentional work it requires to repair and restore the relationship.

The military institution has a vulnerability problem. Of course it does. It is not built to reveal cracks in the system or in its design, which is why Colonel Brown's message is so important to understand: "The Army does not feel compassion, or friendship, or loss. The Army is committed to fighting our nation's wars, and that is a mission which is more important than the life, the career, or the family of any individual soldier." Millennials and Gen Z will need to grow into that understanding as they build a relationship with the institution. However, the people within the institution and surrounding the institution do not have to share or adopt the same struggle with vulnerability. That is where our opportunity lies, and Gen Z will hold us accountable.

When former President Bush describes his visit to Ground Zero, he describes it in a very multisensory way. He remembers the smell, the ash, and the way the air stung his eyes. Ground-level leadership is being willing to not only listen to the story of the people but to truly open your eyes to their reality, even if that creates dissonance within yourself. Your ability to sit in the pocket of discomfort only stretches you to become a better, more informed leader. As the military makes improvements in the mold crisis, there have been images shared on social media of senior leaders walking in the barracks and going into the homes of families who have been begging for repairs. These ground-level leaders have been willing to hear the stories that are difficult to hear but are just as valid as those of military families thriving.

Leaders have definitely made good and productive steps to show this humility throughout the mold crisis. A good example was the open apology of military leaders to Congress and military families that they had let families down.[430] Colonel Michael Greenberg, commander at Fort Belvoir, stated, "We are taking the actions to earn back the trust of our housing residents and holding ourselves and privatized housing companies accountable to provide safe and secure housing for our service members."[431]

Yet, to Wickham's point, the relationship will not heal by simply showing we can focus on fixing the problems. Similar to the reaction of the culture to the initial launch of People First, the institution's and leaders' willingness to own the bigger problems on the surface feels to families as just the beginning. Repairing the relationship will look like taking this approach on a deeper level to the culture that is weary from war, but also the readiness for war. For those who have endured the majority of the GWOT years, they are looking for leaders who will say, "There is no mission without the trust and loyalty of people. Lose that and the mission will fail. The mission begins with people."

The Native American culture has long used storytelling and ceremony to bring the community and the warrior together, especially after long periods apart or after battles have taken innocence from the warfighter and replaced it with otherness. The ceremony welcomes the warrior back into the community by asking the warrior to share their story of war. By sharing it with the community and having them intentionally hear it, it reminds the listener that not only is war costly, but that we all share in that cost as well as the responsibility to help heal and restore the warrior as those who sent them.

In the service member culture, we could do more to help reconstitute weary and overutilized service members. While we offer behavioral health and chaplain support, it is underutilized and still has a stigma attached to it. Lieutenant Colonel Stoney Portis, commander of B Troop during Keating's 2010 battle in Afghanistan, studied his platoon's ability to reconstitute quickly after such an exhausting battle despite the squadron needing the platoon for additional combat operations. The brigade had tasked a "combat and operational stress control (COSC) team—including a psychiatrist, a psychologist, and internal chaplain support—to the troop to oversee the grieving and debriefing processes." He wrote, "We were fully prepared to resupply, re-man, and retrain the force, but the necessity to address psychological stress presented unique challenges to the success of our reconstitution efforts.

"The most profound decisions I made as a troop commander during reconstitution were those relating to the combat and operational stress of my soldiers. Successful combat and operational stress control (COSC) efforts increased [return to duty] rates among veteran troopers, which improved B Troop's personnel status in terms of unit strength and mission capable personnel. When I could not prioritize COSC efforts due to mission constraints, the conditions of my soldiers declined, which was adverse to unit strength.

"Interestingly, I sometimes observed that combat stress interrelates with intangible indicators of unit effectiveness, such as cohesion and morale. For example, as the conditions of my soldiers improved, 'intangibles' like individual and unit morale increased. Furthermore, the strong bond between soldiers within the same platoon seemed to serve as coping mechanisms for those who suffered from combat stress. Each of these dynamics supports the conclusion that the combat effectiveness of B Troop varied with the steps we took to incorporate COSC into the reconstitution plan."[432]

In other words, getting a group back to a healthy place where they are ready to engage the fight again comes down to not only resupplying, not only depending on the strong bonds of community, but providing the soul care that comes from caring leaders, debriefing, storytelling, and respite. The brigade, including the COSC team and leaders, received the platoon back into the community by being present witnesses to their story, giving space to allow and validate their reactions during debriefings, and ultimately acknowledging the warfighter's experience before asking them to return to battle. Some of the COSC team even joined the troop on combat missions to earn trust. This was not just about providing confidential behavioral health, it was the COSC team and leaders working together to take ownership of their reconstitution, earning the warfighters' trust. The trauma the warfighters experienced was not considered a problem, because it was what the warfighter endured at the request of leadership.

On a much larger scale, soul care for the military culture would require us to ask some hard questions. While privatized behavioral health, nonmedical counseling through DoD programs, and resilience training are beneficial and needed for addressing current problems or Wickham's recommendation for coping skill development, they do not build or repair the relationships or address deeper systemic issues that continue to cause significant stress on service members and families. Soul care, for an institution in the business of war that will inevitably wound the soul, should not be a side thought or even a hidden one. It should not be one too quickly outsourced to those who do not hold appropriate cultural competency, either. It should be integrated into the business of what the military does, based on what they ask of people.

Imagine the military culture as one warfighter and think back on the journey we have just gone through. Somehow, a twenty-year war in the Middle East ended without receiving the warfighter back into the military community at home (much less the nation), and no acknowledgment of what had been endured at home or abroad. Instead, he was quickly reconstituted for new emerging threats and asked to ramp up adrenaline and momentum to be competitive. He was told behavioral health is available as needed for any trauma or stress incurred on the side if he is able to slip away from work or is willing to fight the red tape and a six-month waiting list.

As a TRICARE provider and one who has served as a Military OneSource provider, I've found the programs beneficial and needed. However, we have missed crucial steps in relationship building and soul care for the culture if we are hoping that organic healing will happen through treating a wicked problem as though it were a simple or even a complicated one. Even if trust has not been broken, but the relationship is in need of repair or investment, there are some behaviors that repair and some behaviors that invest. Stephen Covey, author of *The Speed of Trust,* offers thirteen behaviors that are worth further exploration, training, and implementation for any leader. They are equally important for the organization, institution, or business to model. These behaviors must be balanced by each other as one pushed to an extreme can become destructive.

Thirteen Behaviors of High-Trust Leaders:[433]

1. Talk Straight
2. Demonstrate Respect
3. Create Transparency
4. Right Wrongs
5. Show Loyalty
6. Deliver Results
7. Get Better
8. Confront Reality
9. Clarify Expectation
10. Practice Accountability
11. Listen First
12. Keep Commitments
13. Extend Trust

It is tempting for businesses and institutions to expect these behaviors from their employees but fail to hold the business itself accountable to model them. Apologizing and doing the right thing to fix problems is part of righting wrongs in a relationship and is a behavior that leads to repair. The military and senior leaders have shown the ability to also talk straight on certain issues by acknowledging where they let people down and aiming to deliver results. However, Covey shares that all of these behaviors are important for leaders who earn high trust from those they lead. Cherry-picking the ones we are most comfortable with or only using them when there are problems to fix is not going to help restore the relationship so that it can weather challenging seasons, such as when the military must make a withdrawal.

There are hard questions to ask. The story of our culture brings up a few that deeply affect retention and relationships with the present culture as well as future generations:

▷ How does privatizing or outsourcing parts of the institution's relationship with the military culture prevent or deflect opportunities to build and repair relationships with its people? How can the institution and its leaders reclaim the relationship while still letting privatized companies oversee the delivery of services?

▷ In the past, the institution utilized peacetime as a time to restore, provide respite to its people, and modernize for future capability. How will the institution rethink restoration and respite as peacetime is replaced with deterrence and gray zone operations?

▷ How is the institution doing with Wickham's recommendation to balance modernization efforts, funding for needed training and capability, and taking care of the key partnerships found in families?

▷ In what ways can the institution take more public ownership for the requests that it made of its people (above problem-solving) and begin efforts to build relationships rather than manage them?

▷ What is or will be the cost of low trust within the culture of the organization over time?

▷ What does soul care and respite of a culture look like when the institution does not or cannot rest?

LEADING THE NEXT GENERATION

The reality is that most of us, inside or outside the military culture, will not inherit high-level influence to shape the broader trajectory of the force. In the moments when we feel small and largely affected by decisions that impact our lives and those of families around us, it is easier to point to the institution rather than accept our role in the system. Displacement is a coping skill often used when people feel out of control of their circumstances. It allows them to discharge negative feelings onto something or someone else that feels safer. When we have a bad day at work, we might accidentally displace our anger toward our spouse and kids, knowing they will forgive us. Similarly, we displace our frustrations toward the job or toward the military when we feel out of control. The military institution is technically a safer place to direct those feelings when we know the institution won't answer (unless we have displaced our frustrations publicly). However, the deeper frustration is the dissonance within us that says we are unhappy or feeling out of control. This is the moment where, as legendary football coach Vince Lombardi said, "Leaders are not born, they're made."

I will never forget sitting down with my husband and two of his superiors at a lunch table in the Pentagon and asking them how we could do our part to improve the morale of soldiers and families. We were still new to the military, eager to serve, and on the ground level seeing the need of the community ten years into the war. Their answer, which I would not recommend, was "Go home and bloom where you are planted."

"Bloom where you are planted" is an overused cliche that was once an encouraging way to say "make the most of your new assignment." Over time, it has become a phrase to push off the energy and motivation of the next generation who could complicate the already established plan.

As deflated as we were that day, there were two truths that inspired most of my career after that. One, there was partial truth in the misused phrase that we each have a circle of influence we have already been given. If we tend it well, including the people who are in that circle, that circle will grow and flourish. And two, each experience we have with other leaders, good or bad, helps us become the leader we want to be. Weak leaders as well as toxic and unhealthy leaders exist everywhere, mostly developed out of bad habits, unprocessed trauma and stress, the need for power, underdeveloped leadership skills, and a myriad of other reasons. The suffering we endure at their hand is an opportunity for growth in our own leadership style.

We also encounter inspiring leaders who value the people they serve and see them as worthy of investment and repair. These leaders not only make a considerable difference in everyone's present circumstances, their behavior and kindness toward others multiplies when the next generation is inspired to pay it forward as leaders themselves. I hope you have been inspired by the many stories of positive and impactful leadership throughout this larger story, even though the story itself was at times discouraging. These leaders are examples of bright beacons of light to the people they served during their darker days. I encourage you to reach out to such leaders and mentors in your life and thank them for the difference they made in your story. You never know the impact it will have in theirs.

Regardless of your role or responsibilities, you are likely surrounded by a multigenerational team. You have incredible influence within your circle, even if you have superiors of an older generation with different leadership styles than you prefer. As the institution changes and evolves, engages in war, peace, or somewhere in the middle, you can influence both the organization itself and the experience families have while serving. It is for people that warriors go to war, and for people they return. Service members will follow a good leader into battle, but they come home to those they love. In order to win at leading the military culture, win both arenas. A few ways to begin:

▷ *Know your own story.* What are the key historical markers that shaped you into the person you are today? What key historical military markers shaped your perspective of leadership and the military lifestyle? What experiences or cognitive biases do you believe and are therefore enslaved by, preventing you from seeing a deeper, possibly more complicated truth? We all have cognitive biases that prevent us from seeing what others see. Knowing and processing your story well will help you find them, root them out, and expose them when possible.

Especially if your story was impacted by the two decades of war after 9/11, I would encourage you to "sit in the pocket" and, as Frankl said, let it teach you about yourself and who you want to become because of those experiences. How have previous generations (like your parents and grandparents) and their motivations shaped who you are? Giving words to these questions helps improve your communication as you speak to other generations and can inspire your ability to cast vision, build and develop stronger working relationships, and even help repair and restore when necessary.

▷ *Listen to the stories of those who are different from you.* Each generation came into the military culture and possibly imprinted on their first experience of military life and deployment. Older generations (boomer and Gen X) remember a simpler time that was not that long ago. Social media and the internet changed almost everything about how people relate, work, and connect. Be patient and willing to learn from those who have wisdom to share about their imprinting of the military culture, what worked and what didn't, and how especially postpandemic, face-to-face interactions in the right context are more welcomed than ever. Likewise, other generations, like Gen Z, are bringing in stories of an entirely different experience of society, mental health, and those they trust. The nation they grew up in is not the same nation boomers or even Gen X lived in or wanted to serve.

▷ *Be curious. Ask more questions.* Edgar Schein, a social psychologist from MIT, and Peter Schein, a strategy consultant in Silicon Valley and the cofounder of the Organizational Culture and Leadership Institute, wrote *Humble Inquiry, The Gentle Art of Asking Instead of Telling*, a must-read for every adult. Their book inspires leaders to ask more questions than we are naturally predisposed to. Schein writes, "It is leaders who will need humble inquiry most because complex interdependent tasks will require building positive, trusting relationships with subordinates to facilitate good upward communication."[434]

While I decided to take the "bloom where you are planted" comment on the chin, a simple question of "Tell me more about what inspires you?" or "What morale issues are you seeing that concern you?" could have invoked a useful fact-gathering conversation for that leader and an opportunity to hear about the concerns of those "boots on the ground." Humble inquiry could have also helped mentor me to self-discover *how* to bloom where I was, solidifying a positive experience and building potential partnerships for the future.

▷ *Be open to change* and the coming shifts brought by the next generations. Leading the next generation is not about coddling or catering and doesn't have to require completely changing systems, policies, or procedures. Millennials and Gen Z hold influence as a majority of the force and those the force needs to recruit and retain. They are already shifting the way the culture works, socializes, and communicates and, although it has shifted more quickly than it would have without the digital age, it is not likely to slow down. Staying curious about the shifts in communication, requests for feedback, and where collaboration could be useful is a potential trust-builder and strengths approach to team building. Situations in which the military culture cannot (yet) evolve are opportunities to educate and mentor.

The military relies heavily on its current force to bring in the next generation of recruits. According to the Army Recruiting Command, "83 percent of the young men and women coming into the Army are coming from military families."[435] While the dust is still settling on the Afghanistan withdrawal, the military community is dealing with an adolescent (Gen Z) mental health crisis and attrition rates that show that millennials and Gen Z are willing to leave for quality of life reasons. Families are also now not locked into a twenty-year commitment to collect benefits for their service toward retirement. The new Blended Retirement System allows service members to invest in their retirement Thrift Savings Plan (TSP), and the government will match their contributions. "Military members can leave the service at any time and have an existing retirement fund that they can take with them anywhere. Even if they get out of the military before completing twenty years, they would keep the money they have in their TSP fund."[436]

The success of an all-volunteer force depends on retention, making retention more important than recruitment. In 1995, 40 percent of youth said they knew a family member who served. By 2017, only 15 percent knew someone serving[437] and as of 2022, only 1 percent of the American population was or is actively serving. The civilian-military divide continues to widen, making recruiting efforts that much harder when civilians do not have any connection or reference point for the military or the lifestyle.

When between 88 and 92 percent of the force is consistently made up of the youngest two generations (currently millennials and Gen Z), the focus must be on providing an internal culture that people want to work in and that their families want to be a part of. Shortly after the military moved away from drafts and became an all-volunteer force, recruitment messages centered on the military being a family-friendly career choice. Even though the military tried to pivot from that message in the '90s, the idea stuck. Wickham saw the importance of the family partnership and the investment that needed to be made in the internal culture: "The Army is an institution, not an occupation. Members take an oath of service to the nation and Army, rather than simply accept a job. As an institution, the Army has moral and ethical obligations to those who serve and their families; they, correspondingly, have responsibilities to the Army. This relationship creates a partnership based on the constants of human behavior and our American traditions that blend the responsibility of each individual for his/her own welfare and the obligations of the society to its members. Our unique mission and lifestyle affect this partnership in ways rarely found in our society. Since we are in the readiness business, we are concerned not only with the number of people in the force but also with their degree of commitment—their willingness to not only train, but also to deploy and, if necessary, to fight—their acceptance of the unlimited liability contract. The need for reciprocity of this commitment is the basis of the partnership between the Army and the Army family."[438]

In the 2021 MFAN survey, families reported that they would not recommend military life, saying that it is not family-friendly, due to: "the pay is low compared to the stress of the work; bad leadership; benefits like health care are not worth the struggles of military life; and the frequent moves and deployments."[439] Additionally, the report highlighted a disconnect between officer families who described their families' health as excellent while families at lower ranks or in marginalized racial and ethnic groups were more likely to not.[440]

We cannot pretend that the only issue in recruiting Gen Z is the claim that they are less patriotic or less prepared mentally or physically to join. It is also not the complete truth that families are less willing to participate in the military lifestyle because spouses are working full-time jobs. Much of our story has been shared, circulated, talked about in forums, and validated by military families online. The youngest generation has seen and heard about the issues of toxic leadership and sexual harassment, watched the Afghanistan exit live-streamed, and are daily weighing their options of what kind of organization they want to be a part of.

"There's a hole in the bucket, Dear Liza."[441]

Gen Z military kids are not only experiencing their own burnout from GWOT, but they have also watched their Gen X military parents experience burnout. Where the institution cannot offer respite, the people will take respite. As GWOT came to a close and millennials and older Gen Z filled the ranks, they brought their strong value of work-life balance. They are putting family first where possible, more so than any military generation before them. Although this is making older generations nervous as social customs and traditions become more difficult to arrange and sell to the younger generation, this shift is the culture's way of prioritizing their self-care. Especially if mental health or personal coaching becomes more difficult to obtain or becomes a luxury service, military families will take respite by taking leave when possible, saying "no" where there is an option, and not involving their children and spouse if it asks more of them than necessary. The military said, "We take care of families so that they can take care of themselves," and this is what it looks like when the culture does just that after being asked to sacrifice more than they intended.

As leaders, you can encourage service members and families to take respite when possible, and model yourself that taking time off is important. There are older millennial and Gen X service members who have three months of leave saved up and either are afraid to take time off, afraid to let team members down, or are working with leaders who decline their leave requests.

If you want to create opportunities to build relationships, do so during work hours or create events that are worth the cost of energy, time, and resources that your younger families will have to incur to attend. Ask yourself, "Would I come to this event if I had to force my toddler to behave when it is past their bedtime and when I would normally have some quiet time with my spouse?" If the answer is no, consider inviting younger generations to the planning process to share their ideas, strategies, and creative solutions.

With younger generations shifting their view of authority to their search for authentic leadership, transparency is crucial—and no longer optional. Being humble and vulnerable is not a weakness. Vulnerable leaders create healthy boundaries that build trust in competency and character, reveal humanity through shared personal life experiences, and impart wisdom through authentic mentoring.

These shifts do not have to rob the military culture of our heritage and the traditions that make us who we are. Again, the goal of any culture or business is not to shift completely around the youngest generation but to remain curious about the requests and values they bring with them. Older generations can share the stories of traditions that have been meaningful, and how they carry the weight of generations past. Encourage them to see the incredible lineage they have become a part of without necessarily expecting them to adopt the same values or motivations. Each generation brings their own why to carry on values and traditions that enable them to be part of something bigger than themselves.

MORE THAN PLASTIC ARMY MEN

I watched my husband once explain military operations to a room full of spouses using plastic military toys. It was brilliant. Spouses, who came with a perception somewhat limited to their service member's role, stood in a circle around a cake table as he leveraged all land, air, and sea military assets to close in on the objective (the cake). He positioned the toys, including two-inch green soldiers, across the table, explaining the difference of location between those considered tip of the spear in theater compared to supportive roles at forward operating bases. This answered the important question, "How close will my soldier be to the front line?" To make it a little more fun, he also taught the slang that each branch uses to affectionately play up their rivalry with other branches. "Crunchies," what tankers love to call foot soldiers, got the biggest laugh.

Those little green army "crunchies" sell in a value pack of thirty-six and have inspired boomer and Gen X children to "play army" since the '50s. They were in *Time* magazine's list of one hundred most influential toys[442] and continued to inspire millennials and Gen Z in Disney Pixar's *Toy Story* films throughout the '90s. Although various alternatives have been made to introduce female soldiers, branches in different colors, or more modern equipment, the traditional army green soldier is still the top seller.

As I watched the circle of adults that day allow themselves to play and giggle, I couldn't help but feel a bit of nostalgia settle in. Play is so important for adults, as well as kids, to temporarily remind the body of its natural stress-free state. For just a moment, this lifestyle felt a bit lighter.

As I reflected later, I thought about the sad fate of a toy that was made popular for how dispensable it was. They can be purchased in bulk and are less expensive than the variety pack of toy tanks, helicopters, and airplanes. Each soldier is molded into one of only a few positions, feet sometimes fixed into place, waiting for imagination to bring them purpose. Part of the joy of playing army with these little soldiers is that they can easily be flicked across the room, lost in the sandbox, or rolled over like bowling pins with tanks or bouncy balls. If and when you run out, you simply get more.

In the military community, it is frequently passed around that the sooner you realize the military won't love you anymore, the less likely you are to get "chewed up and spit out." It is a way of displacing the dissonance within yourself that we fear the joy of "playing army" has instead become a cog in the machine. It is very possible that this is partly what this story was really about, a full cohort of military service members and their families who gave their all over two decades of war until they could give no more.

If that is the case, then the institution, which has every right to be in the business of mainly and necessarily deterring war and ensuring our nation's security,[443] could surely let this cohort find respite in the sandbox and crevices of the car seats and seek the marketplace for a new bucket of recruits. This is the right of the institution, of any business really, to define its mission statement and run that business in a way that meets that objective.

I do not believe that the DoD treats or desires to treat service members as cheap or disposable. On the contrary, branches invest billions of dollars into training and broadening assignments to build up stronger leaders. One of the many reasons service members enlist and commission is for the leadership development this career offers. Tomorrow's future leaders depend on the experience and talent of the leaders today. The defense budget has even committed to paying the medical bills and pensions of retirees for life. I genuinely believe that the institution wants to develop and retain the leaders they have invested in, even for no other purpose than it is more cost-effective.

I do believe, though, that it is easier to let this cohort slip quietly into the night like so many generations before them without true reconciliation and acknowledgment. So I choose to believe that this story matters and that the people who are still in and committed matter. They are still a valued possession that is not easily replaced and are too expensive to lose. My fear, and my reason behind telling this vast and complicated story, is that we are on the verge of losing the loyalty, passion, and talent we have built over two decades because we were so focused on the cake and hoped the numbers would naturally be there when we turned around.

On every level, today's military leaders are faced with the nearly impossible task of recruiting the next generation while simultaneously restoring trust and loyalty to the current one. Leaders on every level will do well to keep front and center the value of people over the capital those people provide. It is far too easy to get lost in the responsibility of people management rather than the often messy dynamics of leading and caring for people.

In today's open-source digital space, the youngest generation can see the person behind the curtain before they ever walk into a recruiting office. They are listening first to their peers and then to the reviews current service members are leaving on their social media accounts, blogs, and live streams. Peers and authentic voices are louder forms of marketing than any slogan wrapped over a city bus or cinematic video that sells purpose or value. We have no choice but to do as we say and practice what we preach.

In 2023, the Army revealed its new rebranding message "Be All You Can Be" which resurrects the recruiting messages of the '80s. It is brilliantly done, as some of the best marketing is to capture the attention of the one who holds the purse strings, the one who has the power of "yes." For the military, that is most often parents. What could be more powerful to Gen X parents than resurrecting the nostalgia of playing army with little green soldiers or pretending to be GI Joe until the street lights came on? The song played over and over and is a reminder of how the heart once swelled at the thought of becoming something bigger, bigger than your backyard or imagination could have ever provided. And yet, this is the same generation left lingering with the question of whether they gave too much to an institution that prioritizes advancing the objective above all else.

There is, unfortunately, a good chance that history will repeat itself. There are already topics looping around that seem new but have been around before. Conversations around unfit housing, the strain on spouse employment, and retirements and bonuses come and go. Within the next ten years, a majority of the generation who served in GWOT will leave or retire, taking much of their experience of deployment and advocacy with them. Generation Alpha will have a new name and will bring in new perspectives that likely give millennials (their parents) a run for their money.

As I continue to serve the families who have given so much over the years, my hope is that we do not repeat what we just endured. When an institution runs on the energy, motivation, and selfless service of people, there must be respect for the limits of those people. And where we must ask more of them and their families than any one person should or have the capacity and strength to give, I hope we will have learned from this season that we cannot take advantage of humans without weighing the cost. They will give as much as they are given and will care as much as they are cared for.

In World War II, aviators started to use the acronym CAVU (Ceiling and Visibility Unlimited) to communicate when visibility was unrestricted both up and out. In 1948, just a few years after the war had ended and the tension of the new Cold War began, Charles Anspach, president of Central Michigan College of Education, inspired a young generation as they started the next chapter of their story. Perhaps we can hear the same encouragement as this chapter ends and a new one begins: "During the last war the term CAVU [Ceiling and Visibility Unlimited] was born. ... The first time one hears it, he finds nothing unusual in the word; but if memory is given its freedom, he recalls a multitude of happenings from the last war, with the anguish and heartbreaks accompanying these events; and then suddenly the significance of CAVU flashes across the sky of one's imagination. Out of the darkness, discouragement, and despair, there comes the promise of unlimited visibility and a ceiling without limitation. When one is in the thick of the storm, the flash which brings the word is most welcome. There is more than the dramatic in such an experience; there is a renewal of purpose and unexpressed gratitude for another chance."[444]

~

There is no mission without the trust and loyalty of people. Lose that and the mission will fail. The mission begins with people and you must be the one to lead them.

LEADERSHIP TIPS AND QUESTIONS

CHAPTER 1: GENERATIONAL PERSPECTIVE
Tips for Leaders
- *Pause before offering solutions.* Daniel Kahneman, author of *Thinking, Fast and Slow* states that when we are quick to offer feedback or solutions, we are likely pulling from information we can recall quickly rather than actually addressing the question or topic. By doing so, we risk coming across as overconfident and insensitive to those around us. Pause first, then tap into the deeper-think system of the brain so you can offer a more relevant and thoughtful response.
- *Leverage humility.* President Bush's 9/11 example is a reminder that transparency is not a vulnerability for leaders. People want an example they can live up to.
- *Leverage curiosity with those who are different from you.* Consider viewing people from different generations as a puzzle to figure out. Commit to asking more questions rather than leading with assumptions.
- *Ask team members what life markers have defined their identity.* What historical, technological, or cultural experiences contribute to their motivation, work ethic, or set of values?

Questions for Understanding & Perspective
1. What was your experience before or after the Great Culture Shift of 2011?
2. In what ways do you clash with younger or older generations around you?
3. Which of your values and beliefs are firm and unwavering?
4. What topics brought up by other generations signal their desire for change? Are there any that make you uncomfortable or defensive?
5. What generational shifts in the military culture make you most uncomfortable and which make your role in leadership more challenging?

CHAPTER 2: HONOR, DUTY & PATRIOTISM
Tips for Leaders
- *Listen for motivating keywords.* Every generation has keywords or themes that motivate or help define their values. Listen to your teammates, children, and parents for words or themes that come up again and again.
- *Be consistent with your most important messages.* During the world wars, presidents chose two to three key messages and delivered varying versions of them consistently, without the internet, and reached into people's homes to inspire honor and duty. Mission statements and values often get lost because leaders fail to communicate them consistently over time.
- *Consider your external stakeholders.* World War II was, in a sense, every American's war in that civilians participated in war efforts. This buy-in created loyalty and support. Think of ways to communicate to civilians ways they can

serve the country or military families by maintaining OPSEC, making room in their communities, and opening jobs for military spouses.

Questions for Understanding & Perspective
1. What similarities do you see between the lost generation and today's generations? What values and traits came out of that time that could inform what will be important to Gen Z?
2. What values are important to your generation compared to the values you hear about in those younger than you?
3. When can messaging and vision casting (even in leadership) cross the line into biased propaganda?
4. Today's service members often view their military commitment differently from previous generations. What motivations have you heard from today's service members about why they joined the force?

CHAPTER 3: FORGOTTEN & VILIFIED
Tips for Leaders

▷ *Revisit traditions.* Military events and social customs are an important part of military cultural tradition. As new generations join the culture, invite them into the planning process and be open to brainstorming ways their perspective may help reach younger members.

▷ *Bring meaning and purpose through connections from the past.* Invite veterans of past conflicts, such as Vietnam veterans, to share their stories, wisdom, and motivation. Facilitate safe, open discussion about differing motives for service and the importance of mentoring.

▷ *Offer additional enrichment training events.* Many enrichment opportunities for military families center on financial management, marriage enrichment, and communication. If your team has older generations, consider training on retirement, empty nesting, aging parents, and parenting adult kids.

▷ *Know how your work style and communication affect others.* How do others experience your values, motives, and communication? Consider a 360-degree assessment or anonymous survey to solicit feedback on ways you can lead better.

Questions for Understanding & Perspective
1. Is patriotism still an important value to you? Is it a driving motivator for those on your team? What other values do your teammates share that are equally if not more motivating to them?
2. Based on a multigenerational poll of service members, the top values that all generations look for most in a leader are: integrity, compassion, humility, and empathy. What actions model each trait, and how are you doing? Which one is hardest to embody and which is a top strength for you?
3. If Gen Z has similar values to the silent generation (e.g., frugality, being less vocal, uncertainty of beliefs), what kind of conflict or contrasting perspectives can you anticipate as they continue to join the ranks?

4. While the consistency of propaganda messaging from World War I and World War II influenced patriotism and community buy-in, it is possible that Vietnam's public messaging sparked questions, debate, and conversation. What consistent messages do you see today and what kind of influence do they have on the country's perception of government, the military, and global involvement?
5. What similarities, if any, do you see between post-9/11 veterans' reactions to the withdrawal from Afghanistan and Vietnam veterans'? How can you lead those still serving differently in light of those similarities?

CHAPTER 4: FOLLOW THE MONEY
Tips for Leaders
▷ *Educate yourself in order to educate others.* Many leaders feel they don't have the time to be an expert in all things, especially government funding. However, having a basic understanding allows you to educate others when funding changes impact their work and families. When needed, provide educational events or meetings to explain when there is (or is not) funding.
▷ *Budget for, and make room on the calendar for, recuperation and reconstitution.* It is tempting to view events that provide respite and boost morale as secondary to mission-essential line items, however they are extremely important to force readiness. Especially in the aftermath of the withdrawal from Afghanistan, creatively budget sustainment, family enrichment, and team-building activities.
▷ *Model healthy work-life balance when possible.* It is easy to assume people will take leave (time off) for their own respite when they need it, but many tend to follow their leader's example.

Questions for Understanding & Perspective
1. What do you feel America's role should be in global stability, especially in a world of finite resources?
2. Does operating in gray zone environments during wartime and/or peacetime reduce threats or create more conflict?
3. Do US military activities and presence justify the need for sustained and increased funding when there would naturally be a conversation around reduced spending in the absence of war?
4. If there is no end to global conflict and no peacetime for a country and its defense, how can leaders ensure reconstitution for those whose jobs and livelihoods exist to defend the nation? What role, then, does respite play in maintaining a ready force?
5. Looking back on leaders who made a difference in your career, in what ways did they ensure positive morale and motivation? Did they encourage time off or self-care?

CHAPTER 5: LAW OF DIMINISHING RETURNS
Tips for Leaders

- *See those behind you and those ahead of you in light of their stories.* Those who experienced our country and military after 9/11 may still carry a strong sense of patriotism and service to the country; however, younger generations may clash with this perspective after a different experience of war and national division. Allow both stories to be true and value the strengths of each generation as strengths to the team. The goal is not to convince others to share the same "why" but to be curious and work with better understanding.

- *Look for ways to make important first impressions during vulnerable moments and transitions.* Find out which service members are in their first assignment or are in a vulnerable season of life (e.g., growing family, first deployment) and create a plan for outreach. Your leadership style may make an impression they will never forget.

- *Find key motivators through storytelling.* Ask others around you to share early experiences that imprinted their view of what military life should look like, what is or is not important to them, and how funding impacts their experience.

- *Model contentment and gratitude in any season.* When there is an inflow of funding, people groups can easily slip into taking resources for granted. Likewise, budget cuts can increase anxiety. Regardless of the current season, look for ways to vision cast how money is shaping the current environment and educate on the funding process to create understanding.

Questions for Understanding & Perspective

1. How would you describe the difference between influence and leadership?
2. Is a leader's role to manage other influential groups within military culture or to embrace them?
3. Is your generation in the majority or minority of those represented?
4. In what ways can younger generations positively influence your leadership style?
5. In what ways did the first decade after 9/11 shape you and those you work with?
6. How does an inflow or lack of funding influence your emotional state as you try to lead others?

CHAPTER 6: THE GREAT CULTURE SHIFT
Tips for Leaders

- *Embrace your "why."* Younger generations are often more comfortable questioning the purpose behind a leader's decisions. Look for appropriate opportunities to engage in discussions that evolve higher-level thinking for all generations. If you cannot expand on the question, be authentic or mentor/teach if the answer requires trust and obedience without further explanation.

- *Extend grace to leaders, even without full understanding.* Regardless of our level of leadership, there will always be someone superior to us who sees the situation from another angle. Blaming others is often a way to deflect or diffuse

our frustration and is different from accountability. Model a willingness to learn from those in a position of more authority to help you gain perspective.
▷ *When your decisions cannot prioritize people, prioritize people in other ways.* When hard decisions are going to affect people in ways you know will be difficult, over-communicate rather than under-communicate. Invest in the relationship, leverage empathy, and as leadership expert John Maxwell says, bring them up to the thirty-five-thousand-foot level to understand the why.
▷ *Ensure your messages are received.* Messaging is not just part of the equation, it's everything. As communication platforms expand, leaders must constantly evaluate and consider new ways to communicate. Consider expanding to a multi-disciplinary team approach to cut through the noise. Do not rely on service members to communicate critical information to families.

Questions for Understanding & Perspective
1. What part of the congressional/defense budget funding process is new for you?
2. How has funding shaped your story and in what ways does it presently impact your role or ability to lead?
3. Leaders are often caught between making decisions that are in the best interest of the people and making other decisions that negatively impact those same people. How can leaders make decisions confidently while still building trustworthy relationships with those they serve?
4. As strategic warfare changes to a constant state of deterrence and response, what are your thoughts about the president's continued ability to leverage military force without congressional declarations of war? Given the risk of downsizing the force, should Congress step in to pull back the number of missions that are spreading the force thin?
5. How do we rehabilitate the military culture while also making up for two decades of readiness degradation?
6. What can you do within your circle of influence to reconstitute the culture of service members and families?

CHAPTER 7: BROKEN PROMISES
Tips for Leaders
▷ *Do not ignore a problem.* Thinking a problem will go away on its own is rarely a solution. When you do not know how to solve a problem or it is not your responsibility to solve it, be willing to authentically validate the concerns of others.
▷ *Recognize shifts in disguise.* Using the cyberbullying example, a better response from leaders when the community asked for help could have been listening first and then evaluating the online community for its usefulness in helping families connect, find resources, and get important information. Discerning shifts when they are happening can provide clues for innovation, opportunities to create order, and shepherding the community through the adjustment of change.

▷ *Look for the truth in other people's stories.* Patti Davis, daughter of former President Ronald Reagan, came to regret publishing a scathing autobiography that discussed her parents. She said, "I've learned something about truth: It's way more complicated than it seems when we're young. There isn't just one truth, our truth—the other people who inhabit our story have their truths as well."
▷ *Allow room for other people's personal perspectives and emotions.* If we are too quick to offer our perspective, we risk emotionally invalidating the other person, often disqualifying ourselves as leaders in their eyes. Listening and being able to hear the emotions of others is the gateway to earning the trust of those around us.

Questions for Understanding & Perspective
1. Is there any part of the military culture story that makes you uncomfortable or brings up a desire to invalidate the reactions of another generation?
2. What generation is most different from your experience?
3. Is empathy (valuing the emotions of others as a necessary part of the equation) a strength for you or is it a challenge to leverage? What can you do to improve your use of compassion instead?
4. Who do you believe held the most responsibility in the mold crisis, Congress for their oversight of the DoD, the DoD over contractors, or military leaders over the installation?

CHAPTER 8: CONSTANT CHANGE
Tips for Leaders
▷ *Create more margin to assess your people.* When your plate is full of tasks and deadlines, it is difficult to build time to assess how decisions impact the people connected to your organization. Not doing so, however, risks the success of the mission and the team.
▷ *Be intentional about your plan and its impact.* The more complex the issue, the more leaders must be prepared to listen rather than assume the blind trust of people. If organizational change must happen, intentionally communicate the plan for implementation in a way that builds trust and helps everyone adapt together. For large-scale changes, give people early notice and consistent reminders throughout, including how the change may impact their daily experience.
▷ *Learn to see the power and influence of cognitive bias.* The human brain tends to simplify and process information through a filter of personal experience and preference. This is especially true in situations of heightened emotions, motivations, and social pressures. Think through problems first on your own in order to avoid groupthink (desire for harmony or conformity) or anchoring (relying on initial information to make a decision), then brainstorm with others to check for your own bias that limits your view.

▷ *Balance where you apply your best energy and talents.* When your cognitive energy is low, it is easy to default to giving your talents where they feel most successful. For example, war planning may feel easier than managing people, families, and emotions. The same is true with work life and home life. Make sure you save some of your best energy for people and places that are important to you, even if you don't feel most confident or successful there.

Questions for Understanding & Perspective
1. What lens of experience shapes your focus as a leader or those who lead you?
2. How can leaders better process their experiences from war or other significant moments in order to expand the lens of their leadership beyond that experience?
3. How does the current force generation cycle affect service members and families under your leadership? What are some ways you can creatively offer education to help reduce confusion and stress?
4. If you were part of the force during ARFORGEN or SRM, how did the switch impact you, your fellow service members, and the families in your community?
5. What culture-wide problems do you see service members and families dealing with today? How can you encourage creative opportunities for them to bond and offer support to one another in the midst of them?
6. What steps can you take to surround yourself with others who are willing and able to challenge you to see different perspectives?

CHAPTER 9: TWO FOR ONE MODEL
Tips for Leaders
▷ *Lead with an open mind.* Generations who had a positive past experience of the culture supported by spouse volunteers will struggle with shifts that move away from that model. Be willing to ask the difficult question, "Is this shift long overdue or is it a battle worth fighting?"
▷ *Value the families who come to you.* While it is easier for leaders to focus on the mission and let families take care of themselves, leaders are still tasked to oversee the well-being of families as a support to the mission.
▷ *Break negative cycles.* You have incredible influence in the often confusing relationship between the DoD and families. With the lines between families and the DoD muddied, be graceful with spouses and family members who are asking for help or frustrated by the lack of government response to their request. Be willing to bring solutions and education rather than continue the cycle.

Questions for Understanding & Perspective
1. What responsibilities do you have toward military spouses and families?
2. Was the DoD's decision to support families in a way that encouraged dependency a sunk cost (investments of time or money that are now irrecoverable) or do you feel it was the right decision that needs continued investment?
3. Should the programming and cultural traditions built and supported by military spouses continue as in the past or evolve into something new?

4. If social programming supported by military spouses went away, what structure of support, if any, would need to be provided in its place?
5. How important is it for the military culture to have a close-knit community that supports itself through the stress of the lifestyle?
6. How do you typically respond when military family members bring up needs or express frustrations or complaints? How has learning more about the history of the DoD-spouse relationship helped you see their request differently?

CHAPTER 10: TRAUMA BOND
Tips for Leaders
- *Avoid the status quo trap.* Breaking free from the status quo can feel dangerous, especially when promotion in the institution rewards strong leadership that doesn't rock the boat. Examine alternatives and opportunities as well as risks of keeping things the same. Evaluate then make a choice to act.
- *Practice assertiveness as healthy communication.* Be clear, direct, and kind in your leadership style and invite others to do the same.
- *Support respectful expectations of military spouses.* Stay conscious of unhealthy expectations that can perpetuate the cycle of spouse unemployment. Are leaders around you leaning on volunteers for program success? Are events planned around work hours for those who may be trying to build a career?

Questions for Understanding & Perspective
1. What level of Maslow's ladder do you feel you are on or have reached? What insights can you draw from seeing service members and spouses attempt to achieve purpose and self-actualization?
2. Were you taught from a young age how to be assertive and direct when asking for help?
3. In what types of situations do you default toward unhealthy communication patterns (passive, passive-aggressive, aggressive)?
4. Which type of communication do you observe when service members interact with superiors versus subordinates?
5. What similarities do you see between the military culture and a welfare state?

CHAPTER 11: SOCIAL MEDIA & COMMUNICATION
Tips for Leaders
- *Keep learning.* When you hear about a new platform, try it out, join in, and learn how others are using it. You don't have to become a content creator to stay up to date on how the youngest generation is using technology.
- *Ask the younger generation for outreach ideas.* Ask how they would handle messaging, or information distribution using new technology. Have them brief their ideas and consider running a pilot.
- *Create monthly lunch-and-learns.* Regularly scheduled meetings are not the best time to learn new skills. Set aside monthly meetings that are lighthearted,

judgment-free, and fun where employees can learn more about the current platforms and technology.
- *Cut through the noise.* Especially for events, send direct communications that are to the point and include all necessary information. Avoid links that require multiple clicks to find out more. Follow up with reminders as appropriate.
- *Seek efficiency, but not at the expense of people.* Technology will always offer shiny new ways to work and be productive, however successful leaders make sure the most important aspects of a healthy internal culture are not forgotten. Make time for innovative discussions on efficiency and workplace morale. Put hierarchy aside to allow for honest dialogue.

Questions for Understanding & Perspective
1. With the military culture online and in the local community, which programs and traditions should continue versus those that can evolve? What does it look like to reimagine the force with respect to the way people work, recreate, and connect?
2. What was your generational experience of technology? How has that influenced your opinions and motivations toward the internet and social media today?
3. How did the rise of social media shape your own experience of the military culture? In what ways did you choose to connect online rather than in person?
4. What platforms are you using for outreach and communication? Are they reaching a full multigenerational audience?
5. How have technology-related shifts in the culture shaped your leadership style in positive and negative ways?

CHAPTER 12: THE ENEMY IS NO LONGER A COUNTRY AWAY
Tips for Leaders
- *Leverage the strengths of your people.* Invite younger team members to present on new ideas, platforms, and software that can be incorporated into training and education events.
- *Offer exciting incentives.* Every generation enjoys being rewarded for hard work. Offering incentives or rewards increases motivation and taps into competition for older generations and gaming strategies for the younger.
- *Choose connection over disconnection.* Knowing what is happening online does not mean you have to engage more than you want to. Stay aware of online trends and topics you may not otherwise hear about.

Questions for Understanding & Perspective
1. What values do you see Gen Z bringing into the workplace based on their most formative years?
2. How do cancel culture and the quiet influence of the younger generation impact the way you choose to lead with confidence?
3. How do you feel about the increased use of technology and how it impacts your personal life and role as a leader?

CHAPTER 13: CULTURAL BREAKDOWN
Tips for Leaders

▷ *Lead with eyes wide open.* If you are leading military families, they may be carrying more chronic, compounded stress than you realize. Think of ways to inspire resilience while also respecting families' requests for time off, respite, and room for self-care.

▷ *Welcome mental wellness into the conversation.* While the entire nation has been through significant events like political division, civil unrest, and the pandemic, Gen Z experienced most of it in their most formative years. Gen Z is looking for mental wellness to be positively supported in work environments, especially those that are high stress and place demands on their family.

▷ *Evaluate your expectations of work ethic.* Expectations of work ethic are largely shaped by your generational perspective. As younger generations challenge long work days and sacrificing health and family for the job, pay attention to your own reaction and ask if their request is healthy or in need of mentoring.

Questions for Understanding & Perspective

1. What impact has stress had on your life? Are there significant events that have had an accumulative effect on your health and perspective?
2. When looking at the military culture today, what issues have been labeled or viewed as isolated incidences that may actually be evidence of long-term chronic stress?
3. What presenting issues could be ripple effects from the pandemic's impact on the military culture?
4. What large-scale changes can be made in the organization to reduce stress levels in the force? Similarly, what changes can be made within your circle of influence that could reduce stress on leaders still recovering?

CHAPTER 14: SHIFT OF AUTHORITY & INFLUENCE
Tips for Leaders

▷ *Keep abreast of educational shifts.* Periodically research how education is shifting in the country. How are high schools and universities adjusting to the way students learn? The military culture can be isolating from the rest of the culture and easily lag behind civilian competitors.

▷ *Make in-person experiences meaningful.* If suddenly all of your tech were to break down, would you still have an engaging presentation or training? People appreciate when new technology is used, but hope it is worth showing up for.

▷ *Don't overthink the power dynamics in the room.* The goal is to be aware of power players in the room so that you are more aware of your own subtle shifts toward them.

▷ *Be humble and authentic.* You're not as bad as your worst press and not as good as your best press. Humility and authenticity is the best form of leadership.

Remember that you are not expected to be perfect and imperfections make you human. Regularly own areas that you could improve, and team up with others who can complement your weaknesses.
- *Be confident in your strengths and weaknesses.* In what ways do those you lead see you as an authority based on a relationship of trust and respect rather than your title or rank?

Questions for Understanding & Perspective
1. The future of education is changing rapidly, especially with artificial intelligence. What are the dangers associated with technology assisting humans with thinking, problem solving, and decision-making?
2. Having a desire for power, authority, and control is normal when we feel insecure or out of control. Where might your need for control or power be fueled by insecurity or fear?
3. As leaders, insecurity in the strengths of those younger than us can make us feel threatened. Are there areas of your life where you push away, block, or fail to lift up others who may have strengths in information, healthy risk taking, or even relationship building? If so, how could that be limiting your team or your leadership?

CHAPTER 15: PURPOSE & PERFECTION
Tips for Leaders
- *Consult with your eighty-year-old self, often.* Whenever you need a good perspective shift, imagine yourself at a much older age. What wisdom is that version of yourself sharing? What is most important, worth striving for? What would that elder tell you they wish they would have known?
- *Process each carrot with your loved ones.* Especially if you are married, the cost of the job, including striving for promotions, affects your family as well. Couples who make decisions together do far better than those who decide on their own or bottle up feelings resulting in resentment.
- *Listen to your body.* The military is making a long-term investment in you. Take the time to be well throughout the journey. You can't give your all to the institution only to be unwell during the years that truly matter with your loved ones.
- *Set healthy boundaries.* Boundaries can be scary, but are also respected by others. We are often afraid to say "no" to working late or a task we don't have the resources to do because we don't want to be seen as weak or not a team player. Boundaries, however, reveal your humanness, encourage peers to be human, and prevent perpetuating an unrealistic work environment.

Questions for Understanding & Perspective
1. What motivates you in your career, role, or leadership position? How has your "why" changed over the years?

2. In what areas of your life have you allowed ethical fading to occur? What prevents you from leading with more authenticity?
3. Fear is a strong motivator. In what ways does fear play a part in your work and relationships?
4. What prevents the system from being challenged and evolving into a more healthy system of motivation and promotion? What will happen if it does not change?
5. What do you foresee shifting in the institution as younger generations come in and value their wellness more than the mission of the institution?

CHAPTER 16: FINAL STRAW OR NEW BEGINNING?
Tips for Leaders

▷ *Be a leader people stay for.* The decisions of the institution and other leaders do not always have to be a reflection of you. When the institution cannot or does not put the needs of one or many first, find ways to counter that message in the lives of those you serve with.

▷ *Physician, heal thyself.* This phrase, originating from an ancient Greek proverb, is a call to look at one's own hypocrisy before serving others. We cannot lead others if we are not willing to do the hard work of healing that which prevents us from leading well.

▷ *Regularly check in with others around you.* Members of each generation have their own levels of stress, regardless of their time in deployment or combat. In a culture where vulnerability is hidden as weakness, lead with a balance of strength and compassion. Make a point to reach out to see how others are doing.

▷ *Plan now to end well.* Regardless of how long you might serve, regularly revisit your "why" to vision cast how you want to leave or retire. Live and lead in light of that vision now.

▷ *Know what to do with your first draft.* We can value our first reaction as somewhat truthful. Learn to identify it, process it appropriately with the right people, then write the better draft or message that elevates the vision and morale of others.

Questions for Understanding & Perspective
1. If you served during the years of GWOT, how did the withdrawal from Afghanistan impact you, your perspective of the military, and your part in it?
2. What examples have you seen of people choosing integrity and the greater good that have inspired you as a leader?
3. In what ways has your body tried to call your attention to the stress and/or trauma it has stored?
4. In light of what you see from the service members and families around you, what changes or shifts in the culture do you foresee happening next?
5. What is one practical action step, one behavior, you can implement on a daily basis that can influence those around you in a more meaningful and positive way?

CHAPTER 17: MORE THAN A NUMBER: RETENTION & RECRUITMENT
Tips for Leaders

- *Lead with the whole story in mind.* It is easier and more convenient to forget or numb the parts of the cultural story that hit a little close to home or for which we don't have solutions. Choose to see the different generations around you in light of their experience rather than your own.
- *Lead more often with tragic optimism.* When others bring you problems and concerns, instead of offering a simple solution or an overly optimistic response, address your cognitive bias, be willing to sit in the pocket of difficult questions, and lean into curiosity and empathy before launching into solutions.
- *See the relationships around you as worthy of repair.* The institution may not be able to step into the vulnerability and dissonance of repair with its people, but you as a leader can. Learn to grow more comfortable with conflict by seeing the opportunity to restore the relationship as more important than the discomfort of internal dissonance.
- *Become a ground-level leader.* As the culture continues to shift and change, be willing to walk in the rubble, listen, and adapt your own perspective where needed.

Questions for Understanding & Perspective

1. Which generation did you enjoy learning the most about? In what ways can you lean in with curiosity to get to know them better?
2. What shifted your personal perspective or understanding over the course of this story? How will you live and/or lead differently in light of it?
3. What negative narratives do you believe about yourself? Are they actually true? How would your life be different if you believed a more positive, truthful narrative?
4. How would you identify if you were feeling incongruence between your internal and external world? When do you find yourself treating incongruence with an overly optimistic perspective rather than leaning into the dissonance?
5. Zoom out and consider the various experiences of the generations in the force today. What is your plan to reconstitute the people in preparation for what the institution is currently asking of them?

ENDNOTES

1   FORSCOM Public Affairs, "Army Officials to Discuss How ReARMM Will Synchronize Readiness," Oct. 1, 2021, https://www.army.mil/article/250753/army_officials_to_discuss_how_rearmm_will_synchronize_readiness

2   Patricia Kime, "Military Teens Are Struggling With Mental Well-Being, Food Insecurity, Survey Finds," *Military.com*, Oct. 7, 2021, https://military.com/daily-news/2021/10/07/military-teens-are-struggling-mental-well-being-food-insecurity-survey-finds.html

3   Terry Spencer, *The Associated Press*, "Military Families Angry About Damage, Thefts During Moves," *Military Times*, Oct. 7, 2018, accessed Aug. 18, 2022, https://militarytimes.com/news/your-military/2018/10/07/military-families-angry-about-damage-thefts-during-moves

4   John Ismay, "Military Families Say Base Housing Is Plagued by Mold and Neglect," *The New York Times*, Dec. 13, 2019, https://nytimes.com/2019/12/13/us/military-base-housing-mold.html

5   Emma Platoff and Shawn Mulcahy, "Fourteen US Army Leaders Fired or Suspended at Fort Hood," *The Texas Tribune*, Dec. 8, 2020, https://texastribune.org/2020/12/08/fort-hood-vanessa-guillen-army-investigation

6   Gina Harkins, "Inspectors Said Her Toxic Leadership Was 'Worst Seen in 20 Years.' She Just Became a 1-Star," *Military.com*, Aug. 19, 2020, https://military.com/daily-news/2020/08/19/inspectors-said-her-toxic-leadership-was-worst-seen-20-years-she-just-became-1-star.html

7   Daniel Johnson, "The Military Has a Suicide Crisis. Its Leaders Bear Most of the Blame," *Task & Purpose*, Oct. 15, 2021, accessed Sept. 19, 2022, https://taskandpurpose.com/news/military-suicide-crisis-blame

8   Meghann Myers, "Military Suicides Up 16 Percent in 2020, but Officials Don't Blame Pandemic," *Military Times*, Sept. 30, 2021, accessed Aug. 19, 2022, https://militarytimes.com/news/pentagon-congress/2021/09/30/military-suicides-up-15-percent-in-2020-but-officials-dont-blame-pandemic

9   Nikole Killion, "Senate Report Finds 'Mistreatment' of Military Families by Housing Companies," *CBS News*, Apr. 26, 2022, accessed Sept. 19, 2022, https://cbsnews.com/news/military-housing-senate-report-balfour-beatty

10   Rebecca Kheel and Thomas Novelly, "One Year Later, Troops and Veterans Involved in Afghanistan Exit Grapple with Mental Scars," *Military.com*, Aug. 26, 2022, accessed Sept. 19, 2022, https://military.com/daily-news/2022/08/26/one-year-later-troops-and-veterans-involved-afghanistan-exit-grapple-mental-scars.html

11   Eleanor Watson, "Navy Investigation Finds Hawaii Water Crisis Exacerbated by "Unacceptable Failure of On-scene Leadership," *CBS News*, Jul. 2, 2022, accessed Sept. 19, 2022, https://cbsnews.com/news/navy-investigation-hawaii-water-crisis-red-hill

12   *The Associated Press*, "Pentagon Links Leadership Failures to Violence, Harassment at Military Bases," *PBS New Hour*, Mar. 31, 2022, accessed Sept. 19, 2022, https://pbs.org/newshour/politics/pentagon-links-leadership-failures-to-violence-harassment-at-military-bases

13   Courtney Kube and Molly Boigon, "Every Branch of the Military Is Struggling to Make Its 2022 Recruiting Goals, Officials Say," *NBC News*, Jun. 27, 2022, accessed Sept. 24, 2022, https://nbcnews.com/news/military/every-branch-us-military-struggling-meet-2022-recruiting-goals-officia-rcna35078

14   Lara Seligman, Paul McLeary, and Lee Hudson, "Lawmakers Press Pentagon for Answers as Military Recruiting Crisis Deepens," *Politico*, Jul. 27, 2022, accessed Sept. 19, 2022, https://politico.com/news/2022/07/27/lawmakers-pentagon-military-recruiting-00048286

15   Stony Brook University, "What's a Wicked Problem?" accessed Jul. 28, 2023, https://stonybrook.edu/commcms/wicked-problem/about/What-is-a-wicked-problem

16   Ibid.

17 General Stanley A. McChrystal, *Commander's Initial Assessment*, National Security Archive, Aug. 30, 2009, accessed Jul. 28, 2023, https://nsarchive.gwu.edu/document/24560-headquarters-international-security-assistance-force-kabul-afghanistan-gen-stanley

18 Elisabeth Bumiller, "We Have Met the Enemy and He Is PowerPoint," *The New York Times*, Apr. 27, 2010, accessed Jul. 28, 2023, https://nytimes.com/2010/04/27/world/27powerpoint.html

19 Roxana Tiron, "US Military Services Face Biggest Recruiting Hurdles in 50 Years," *Bloomberg Government*, Sept. 21, 2022, accessed Jul. 28, 2023, https://about.bgov.com/news/us-military-services-face-biggest-recruiting-hurdles-in-50-years

20 Rikke Friis Dam and Teo Yu Siang, "What Is Design Thinking and Why Is It So Popular?" Interaction Design Foundation, accessed Jul. 28, 2023, https://interaction-design.org/literature/article/what-is-design-thinking-and-why-is-it-so-popular

21 Lydia Saad, "Historically Low Faith in US Institutions Continues," Gallup, Aug. 17, 2023, https://news.gallup.com/poll/508169/historically-low-faith-institutions-continues.aspx

22 K.J. Ramsey, "Naming Spiritual Abuse," *Church Hurts*, accessed Feb. 12, 2023, https://podcasts.apple.com/us/podcast/naming-spiritual-abuse-kj-ramsey/id1494531948?i=1000485299764

23 Ibid.

24 Dave Philipps and Tim Arango, "Who Signs Up to Fight? Makeup of US Recruits Shows Glaring Disparity," *The New York Times*, Jan. 10, 2020, accessed Jul. 10, 2023, https://nytimes.com/2020/01/10/us/military-enlistment.html

25 Kube and Boigon, "… Military Is Struggling to Make Its 2022 Recruiting Goals," (n13).

26 Tiron, "US Military Services Face Biggest Recruiting Hurdles in 50 Years," (n19).

27 William Strauss and Neil Howe, *The Fourth Turning: An American Prophecy*, (Crown, 1997).

28 Sheila Callaham, "Pew Research Center: New Stance on Generational Labels, with a Caveat," *Forbes*, May 28, 2023, https://forbes.com/sites/sheilacallaham/2023/05/28/pew-research-center-new-stance-on-generational-labels-with-a-caveat

29 Kim Parker, "How Pew Research Center Will Report on Generations Moving Forward," Pew Research Center, May 22, 2023, accessed Jul. 29, 2023, https://pewresearch.org/short-reads/2023/05/22/how-pew-research-center-will-report-on-generations-moving-forward

30 Jeffrey A. Hall, "How Many Hours Does It Take to Make a Friend?" *Journal of Social and Personal Relationships*, 36(4), (2019):1278–1296, https://doi.org/10.1177/0265407518761225

31 Edward A. Gutiérrez, "America's Forgotten Wars: Debunking the Myth of an American Lost Generation," National WWI Museum and Memorial, accessed on *YouTube*, https://youtube.com/clip/UgkxIbLk81UX2QT5zrj0MMxMoiEKTCSq8nmM

32 There are strong and valid opinions from those who served in Iraq regarding decisions made after 9/11. As leaders, it is important for us to learn from key moments of strength as well as highly debated decisions judged by time.

33 MasterClass, "President George W. Bush Teaches Authentic Leadership," accessed Sept. 30, 2023, https://masterclass.com/classes/president-george-w-bush-teaches-authentic-leadership

34 Erika I. Ritchie, "Patriotism Fueled Military Enlistment After 9/11, Remains Strong 20 Years Later," *Orange County Register*, Sept 7, 2021, accessed Sept. 24, 2022, https://ocregister.com/2021/09/07/enlistment-fueled-by-patriotism-surged-after-sept-11-attacks-remains-strong-as-opportunities-factor-in

35 President Woodrow Wilson's Declaration of War Message to Congress, April 2, 1917, Records of the US Senate, Record Group 46, National Archives, accessed Oct. 11, 2023, https://archives.gov/milestone-documents/address-to-congress-declaration-of-war-against-germany

36 Jim Garamone, "World War I: Building the American Military," *DoD News*, Mar 29, 2017, accessed Oct. 12, 2022, https://www.defense.gov/News/News-Stories/Article/Article/1134509/world-war-i-building-the-american-military

37 Woodrow Wilson, "July 4, 1914: Fourth of July Address," Miller Center, accessed Oct. 15, 2022, https://millercenter.org/the-presidency/presidential-speeches/july-4-1914-fourth-july-address

38 Molly Billings, "The 1918 Influenza Pandemic," Stanford University, Jun. 1997, modified Feb. 2005, accessed Oct. 15, 2022, https://virus.stanford.edu/uda

39 Aaron Kassraie, "Spanish Flu: How America Fought a Pandemic a Century Ago," Sept. 21, 2021, accessed Oct. 31, 2022, https://aarp.org/politics-society/history/info-2020/spanish-flu-pandemic.html

40 Byron Farwell, *Over There: The United States in the Great War 1917-1918* (New York: W.W. Norton & Co., 1999).

41 Editors of *Encyclopaedia Britannic*, "Bonus Army," *Encyclopedia Britannica*, Dec. 4, 2019, https://britannica.com/event/Bonus-Army

42 "The Cost of US Wars Then and Now," Norwich University Online, https://online.norwich.edu/academic-programs/resources/cost-us-wars-then-and-now

43 "Defense Spending as a Percent of Gross Domestic Product (GDP)," US Department of Defense, accessed Oct. 17, 2022, https://www.defense.gov/Multimedia/Photos/igphoto/2002099941

44 Testimony of Paul V. McNutt, Federal Security Administrator, in US Congress, House Committee on Military Affairs, Hearings, *Allowances and Allotments for Dependents of Military Personnel*, (59th Congress, 1st Session, 1942), 2-3.

45 "Research Starters: US Military by the Numbers," The National WWII Museum, New Orleans, accessed Oct. 17, 2022, https://nationalww2museum.org/students-teachers/student-resources/research-starters/research-starters-us-military-numbers

46 US Army Chief of Staff, "The Army Family, 2," Bureau of Labor Statistics estimate, in William M. Tuttle Jr., *Daddy's Gone to War*, 31.

47 Brian Becker, "Military Spending, Economic Crisis and Imperialism," *Liberation*, May 5, 2005, accessed Oct. 17, 2022, https://liberationnews.org/05-05-01-military-spending-economic-cris-html/

48 "The Cost of US Wars," Norwich University Online, (n42).

49 President Dwight D. Eisenhower's Farewell Address, January 17, 1961, National Archives, accessed Nov. 2, 2022, https://archives.gov/milestone-documents/president-dwight-d-eisenhowers-farewell-address

50 Erin Blakemore, "How the GI Bill's Promise Was Denied to a Million Black WWII Veterans," *History*, Jun. 21, 2019, updated 2023, https://history.com/news/gi-bill-black-wwii-veterans-benefits

51 Natalie Walker, "Truman and Women's Rights: The Women's Armed Services Integration Act," Truman Library Institute, Mar. 1, 2021 https://trumanlibraryinstitute.org/truman-and-womens-rights

52 Donna Alvah, *Unofficial Ambassadors*, (NYU Press, 2007), 62; Betty Sowers Alt and Bonnie Domrose Stone, *Campfollowing*, (Praeger, 1991), 110.

53 "Oleta Crain," Colorado Women's Hall of Fame, accessed Jun. 30, 2023, https://cogreatwomen.org/project/oleta-crain

54 Virginia Culver, "Retired Army Major Fought, Lived Through Bias," *The Denver Post*, Nov. 21, 2077, accessed Jun. 30, 2023, https://denverpost.com/2007/11/21/retired-army-major-fought-lived-through-bias

55 "People: The Younger Generation," *Time*, Nov. 05, 1951, accessed Oct. 17, 2022, https://content.time.com/time/subscriber/article/0,33009,856950-3,00.html

56 Jessica Pearce Rotondi, "Missing in Action: How Military Families in Tortuous Limbo Galvanized a Movement," *History*, May 20, 2020, accessed Oct. 19, 2022, https://history.com/news/missing-in-action-mia-vietnam-war

57 Ibid.

58 Paul M. Cole, *POW/MIA Issues*, (RAND, 1994), National Defense Research Institute, accessed Oct. 19, 2022, https://rand.org/content/dam/rand/pubs/monograph_reports/2006/MR351.1.pdf

59 Joshua D. Angrist, "The Draft Lottery and Voluntary Enlistment in the Vietnam Era," *Journal of the American Statistical Association*, 86(415), (Sept. 1991): 584–595.

60 Morley Winograd and Michael Hais, "President Biden's Generation: Silent No More," *Brookings*, Jul. 8, 2021, accessed Dec. 30, 2022, https://brookings.edu/blog/fixgov/2021/07/08/president-bidens-generation-silent-no-more

61   "Essay: The Silent Generation Revisited," *Time*, Jun. 29, 1970, accessed Nov. 2, 2022, https://content.time.com/time/subscriber/article/0,33009,878847-5,00.html

62   Louise Hidalgo, "Dr Spock's Baby and Child Care at 65," *BBC*, Aug. 23, 2011, accessed Jul. 13, 2023, https://bbc.com/news/world-us-canada-14534094

63   James Wright, "A Generation Goes to War," *HistoryNet*, Oct. 17, 2017, accessed Sept. 14, 2023, https://historynet.com/generation-goes-war

64   "Final Words: Cronkite's Vietnam Commentary," *All Things Considered*, Jul. 18, 2009, accessed Jan. 22, 2023, https://npr.org/2009/07/18/106775685/final-words-cronkites-vietnam-commentary

65   Devin Fisher, "Vietnam Vet Shares Coping Skills With Combat Warriors," *Fort Carson Mountaineer*, May 15, 2009, accessed from: https://www.army.mil/article/21185/vietnam_vet_shares_coping_skills_with_combat_warriors
Average annual days of combat cumulatively experienced by veterans: WWII 40; Vietnam 240; Global War on Terror 1200.

66   Wright, "A Generation Goes to War," (n63).

67   James Wright, *Enduring Vietnam: An American Generation and Its War* (Thomas Dunne Books 2017), 319.

68   Steve Holland and Nandita Bose, "Biden Defends Afghanistan Decision, Blames Afghan Army's Unwillingness to Fight," *Reuters*, Aug. 16, 2021, https://reuters.com/world/us/biden-says-us-mission-afghanistan-was-never-supposed-be-nation-building-2021-08-16

69   Carmen Ang, "Ranking US Generations on Their Power and Influence Over Society," *Visual Capitalist*, May 6, 2021, accessed Aug. 7, 2023, https://visualcapitalist.com/ranking-u-s-generations-on-their-power-and-influence-over-society

70   Wayne S. Sellman, "Military Service in the 1980s," Defense Technical Information Center, accessed Dec. 30, 2022, https://apps.dtic.mil/sti/pdfs/ADP000822.pdf

71   "Hours of Video Uploaded to *YouTube* Every Minute as of February 2022," *Statista*, accessed Jul. 13, 2023, https://statista.com/statistics/259477/hours-of-video-uploaded-to-youtube-every-minute

72   LZ Salley, 101st Airborne, "Vietnam War Statistics," Vietnam Veterans of America, Chapter 310, accessed Jan. 22, 2023, https://vva310.org/vietnam-war-statistics

73   The Associated Press, "Lawmakers Unveil a $1.7 Trillion US Spending Bill as Shutdown Deadline Looms," *NPR*, updated Dec. 20, 2022, accessed Jun. 3, 2023, https://npr.org/2022/12/20/1144365502/congress-spending-omnibus-shutdown-bill

74   "National Defense Authorization Act for Fiscal Year 2023," Jul. 1, 2022, accessed Nov. 9, 2022, https://govinfo.gov/content/pkg/CRPT-117hrpt397/pdf/CRPT-117hrpt397.pdf

75   Bryant Harris, "Congress Authorizes 8% Defense Budget Increase," *Defense News*, Dec. 15, 2022, https://defensenews.com/congress/budget/2022/12/16/congress-authorizes-8-defense-budget-increase

76   "Interpretation & Debate: Declare War Clause," National Constitution Center, accessed Dec. 26, 2022, https://constitutioncenter.org/the-constitution/articles/article-i/clauses/753

77   David K. Henry and Richard P. Oliver, "The Defense Buildup, 1977-85: Effects on Production and Employment," *Monthly Labor Review*, Aug. 1987, Bureau of Labor Statistics, https://bls.gov/opub/mlr/1987/08/art1full.pdf

78   Center for Preventive Action, "Methodology," *Global Conflict Tracker*, accessed Dec. 26, 2022, https://www.cfr.org/global-conflict-tracker/methodology

79   Mackenzie Eaglen, "US Military Force Sizing for Both War and Peace," America Enterprise Institute, Mar. 15, 2015, accessed Dec. 6, 2022, https://aei.org/research-products/report/us-military-force-sizing-war-peace

80   Steven Feldstein, "Do Terrorist Trends in Africa Justify the US Military's Expansion?" Carnegie Endowment for International Peace, Feb. 9, 2018, accessed Jul. 14, 2023, https://carnegieendowment.org/2018/02/09/do-terrorist-trends-in-africa-justify-u.s.-military-s-expansion-pub-75476

81   Ibid.

82 Proverbs 29:18, *Amplified Bible*, Zondervan.
83 Andrew Lisa, "50 Insights Into the US Military-Industrial Complex," *Stacker*, Dec. 12, 2020, accessed Dec. 5, 2022, https://stacker.com/military/50-insights-us-military-industrial-complex
84 Chuck Moran, *War Industry Muster*, accessed Dec. 5, 2022, https://warindustrymuster.com
85 *Field Manual (FM 3-0) Operations*, Headquarters, Department of the Army, 2022.
86 "Contracts for Dec. 7, 2022," US Department of Defense, accessed Mar. 12, 2023 https://www.defense.gov/News/Contracts/Contract/Article/3239197

Google Support Services LLC, Oracle America Inc., Amazon Web Services Inc., and Microsoft Corp. were awarded "a hybrid (firm-fixed-price and time-and-materials, indefinite-delivery/indefinite-quantity) contract with a ceiling of $9 billion, to provide the DoD with enterprise-wide, globally available cloud services across all security domains and classification levels, from the strategic level to the tactical edge. Joint Warfighting Cloud Capability is a multiple award contract.

87 Benjamin Verdi, "The Coming Cyber-Industrial Complex: A Warning for the New US Administration," *Geopolitical Monitor*, Nov. 22, 2020, accesssed Dec. 5, 2022, https://geopoliticalmonitor.com/the-coming-cyber-industrial-complex-a-warning-for-the-new-us-administration
88 "Funding NATO," accessed Dec. 5, 2022, https://nato.int/cps/en/natohq/topics_67655.htm
89 Valerie Insinna, "Biden Administration Kills Trump-era Nuclear Cruise Missile Program," *Breaking Defense*, Mar. 28, 2022, accessed Dec. 6, 2022, https://breakingdefense.com/2022/03/biden-administration-kills-trump-era-nuclear-cruise-missile-program
90 Aaron Mehta, "Milley Breaks with Cancelation of New Nuclear Cruise Missile," *Breaking Defense*, Apr. 5, 2022, accessed Dec. 6, 2022, https://breakingdefense.com/2022/04/milley-breaks-with-cancelation-of-new-nuclear-cruise-missile
91 Neta C. Crawford, "The US Budgetary Costs of the Post-9/11 Wars," Watson Institute, Brown University, Sept. 1, 2021, accessed Nov. 9, 2022, https://watson.brown.edu/costsofwar/files/cow/imce/papers/2021/Costs of War_U.S. Budgetary Costs of Post-9 11 Wars_9.1.21.pdf
92 "Defense Primer: Department of Defense Unfunded Priorities," Congressional Research Services, Nov. 9, 2021, https://sgp.fas.org/crs/natsec/IF11964.pdf
93 Frederico Bartels, "Changing Current 'Use It or Lose It' Policy Would Result in More Effective Use of Defense Dollars," Jun. 23, 2021, accessed Dec. 6, 2022, https://heritage.org/defense/report/changing-current-use-it-or-lose-it-policy-would-result-more-effective-use-defense
94 Oriana Pawlyk, "Air Force No Longer Spending $10,000 on Toilet Seats, Officials Say," *Military.com*, Jul. 11, 2018, accessed Dec. 6, 2022, https://military.com/defensetech/2018/07/11/air-force-no-longer-spending-10000-toilet-seats-officials-say.html
95 "Policy Basics: Where Do Our Federal Tax Dollars Go?" Center on Budget and Policy Priorities, updated Jul. 28, 2022, accessed Dec. 26, 2022, https://cbpp.org/research/federal-budget/where-do-our-federal-tax-dollars-go
96 *2021 Demographics Report: Profile of the Military Community*, Department of Defense, ODASD MC&FP, accessed Dec. 26, 2022, https://download.militaryonesource.mil/12038/MOS/Reports/2021-demographics-report.pdf
97 Ibid.
98 "Decade of Service 1990s," DAV, Sep. 28, 2017, accessed Dec. 26, 2022, https://dav.org/learn-more/news/2017/decade-service-1990s
99 Alex Hollings, "The 5 Best Military Movies of the 1990s," *Sandboxx*, 2021, accessed Dec. 26, 2022, https://sandboxx.us/blog/the-best-military-movies-of-the-1990s
100 Neal Freyman, "For the Defense Industry, 9/11 Changed Everything," *Morning Brew*, Sept. 5, 2021, accessed Jan. 2, 2023, https://morningbrew.com/daily/stories/2021/09/05/defense-industry-911-changed-everything
101 Amy Reinkober Drummet, Marilyn Coleman, and Susan Cable Drummet, "Military Families Under Stress: Implications for Family Life Education," *Family Relations*, 52(3), (July 2003): 279–287, https://jstor.org/stable/3700279

102 David Gray, "The Military as Peacemakers and Enforcers: Military Operations Other Than War in the 1990s," Foreign Policy Research Institute. Apr. 23, 2018, accessed Dec. 26, 2022, https://fpri.org/article/2018/04/the-military-as-peacemakers-and-enforcers-military-operations-other-than-war-in-the-1990s

103 Corie Weathers, "For those of you who served before 9/11: What was your military experience like before 9/11 happened … ?" *Corie Weathers | Lifegiver* Facebook page post, Dec. 26, 2022, https://www.facebook.com/CorieLPC/posts/pfbid0fhZcQHBoJVbx6RsL5jMtntLec3XWRfhdbKFr6WrBfnnx4WnCj4BvYPRoFpPxURRVl

104 Ibid.

105 Ibid.

106 Mady Wechsler Segal, "The Military and the Family as Greedy Institutions," *Armed Forces & Society*, 13(1), (Fall 1986): 9–38, https://doi.org/10.1177/0095327X8601300101

107 Weathers, "For those of you who served before 9/11," (n103).

108 "2023 Small Business Trends," *Guidant*, accessed Aug. 23, 2023, https://guidantfinancial.com/small-business-trends

109 Joel Searls, "The Military Influences of GI Joe Cartoons," *We Are the Mighty*, Dec 13, 2022, https://wearethemighty.com/mighty-history/the-military-influences-of-g-i-joe-cartoons

110 Kristen Soltis Anderson, "How the Post-9/11 Generation Views American Power," Ronald Reagan Presidential Foundation & Institute, accessed Jan. 2, 2023, https://reaganfoundation.org/reagan-institute/publications/how-the-post-911-generation-views-american-power

111 "Operation Iraqi Freedom," Naval History and Heritage Command, Nov. 7, 2022, accessed Jan. 3, 2023, https://history.navy.mil/browse-by-topic/wars-conflicts-and-operations/middle-east/operation-iraqi-freedom.html

112 "How Military Spending has Changed Since 9/11," National Priorities Project, accessed Jan. 2, 2023, https://nationalpriorities.org/campaigns/how-military-spending-has-changed

113 *DoD Financial Management Regulation*, Vol. 7A, Chap. 9 (February 2002), https://comptroller.defense.gov/Portals/45/documents/fmr/archive/07aarch/07a_09_200202.pdf

114 *Military Housing: Management Issues Require Attention as the Privatization Program Matures*, US Government Accountability Office, Report GAO-06-438, Apr. 28, 2006, accessed Jan. 3, 2023, https://gao.gov/assets/gao-06-438.pdf

115 *US Army Family Readiness Support Assistant FRSA Resource Guide*, US Department of Army, 2007, accessed Jan. 23, 2023, http://militarywives.com/images/Specific/army/02_US__Army_Family_Readiness_Support_Assistant_dated_2007.pdf

116 "Strong Bonds," US Army, Building Strong & Ready Teams, https://bsrt.army.mil/program

117 TRICAREdotMil Staff, "Delivering Readiness: The Evolution of TRICARE," Health.mil, Jul. 1, 2021, accessed Jan. 5, 2023, https://health.mil/News/Articles/2021/07/01/Delivering-Readiness-The-Evolution-of-TRICARE

118 "Modify TRICARE Enrollment Fees and Cost Sharing for Working-Age Military Retirees," *Options for Reducing the Deficit*, Congressional Budget Office, Dec. 13, 2018, accessed Jan. 5, 2023, https://cbo.gov/budget-options/2018/54763

119 Anderson, "How the Post-9/11 Generation Views American Power," (n110).

120 "Alaska's 'Hot Sauce' Mom Sentenced to 3 Years of Probation, Fine, for Child Abuse," *ABC News*, Aug. 29, 2011, accessed Jan. 7, 2023, https://abcnews.go.com/US/alaskas-hot-sauce-mom-sentenced-years-probation-child/story?id=14408795

121 Anderson, "How the Post-9/11 Generation Views American Power," (n110).

122 Darlene E. Stafford and Henry S. Griffis, "A Review of Millennial Generation Characteristics and Military Workforce Implications," The CNA Corp., May 2008, accessed Jan. 7, 2023, https://cna.org/archive/CNA_Files/pdf/d0018211.a1.pdf

123 Robert Hale, "Budgetary Turmoil at the Department of Defense from 2010 to 2014," Center for 21st Century Security and Intelligence at Brookings, Aug. 2015, accessed Jan. 7, 2023, https://brookings.edu/wp-content/uploads/2016/06/dod_budgetary_turmoil_final.pdf

124 "The Budget Control Act: Frequently Asked Questions," Congressional Research Service, updated Oct. 1, 2019, https://sgp.fas.org/crs/misc/R44874.pdf

125 Defense Secretary Robert Gates and Joint Chiefs of Staff Chairman Admiral Michael Mullen, Testimony, "Defense Department Fiscal Year 2012 Budget Request," *C-SPAN*, Mar. 2, 2011, accessed Jan. 7, 2021, https://c-span.org/video/?298247-1/defense-department-fiscal-year-2012-budget-request

126 Hale, "Budgetary Turmoil," (n123).

127 General James F. Amos, testimony, "House Armed Services Committee Holds Hearing on the Impact of Sequestration on the Defense Department," US Marine Corps, Feb. 13, 2013, https://www.hqmc.marines.mil/Portals/142/Docs/130213%20--%20HASC%20Hearing%20(Transcript).pdf

128 General Martin Dempsey, chairman of the Joint Chiefs of Staff, testimony, Ibid.

129 Tara Copp, The Associated Press, "The Death Toll for Rising Aviation Accidents: 133 Troops Killed in Five Years," *Military Times*, Apr. 8, 2018, https://militarytimes.com/news/your-military/2018/04/08/the-death-toll-for-rising-aviation-accidents-133-troops-killed-in-five-years

130 Alan W. Dowd, "The Sacrifices of Sequestration," *American Legion*, Jan. 19, 2018, accessed Jan. 9, 2023, https://legion.org/magazine/240775/sacrifices-sequestration

131 Ibid.

132 Hale, "Budgetary Turmoil," (n123).

133 Jennifer Herbek, comment, "For those who remember sequestration (2013+), where/when did you first feel its impact on your daily life?" *Corie Weathers | Lifegiver* Facebook page post, Jan. 7, 2023, https://www.facebook.com/CorieLPC/posts/pfbid0quxKHxKMi6FCxzVTq7q7TVs66XFwpRQZEamBSX5sVyXXxLUh4m2ipHxyA2JvaFxml

134 Amy Bushatz, "Officer 'Separations' Break More Than Years of Service," *The New York Times*, Jul. 7, 2014, accessed Jan. 8, 2023, https://archive.nytimes.com/atwar.blogs.nytimes.com/author/amy-bushatz

135 Pew Research Center, "Millennials in Adulthood," Mar. 7, 2014, accessed Jan. 8, 2023, https://pewresearch.org/social-trends/2014/03/07/millennials-in-adulthood

136 In 2010, Don't Ask, Don't Tell was repealed, allowing same-sex couples to openly serve in the military. In 2011, I met our female commander's wife as she attended her first-ever spouse event. They had been together for fourteen years. While other commanders' spouses were running FRG meetings, she was just beginning. Millennials had a hand in making that possible.

137 General Ray Odierno, former Army chief of staff, "The Value of Bold Leadership," JPMorgan Chase & Co., https://jpmorganchase.com/news-stories/gen-odierno-bold-leader

138 *Military Times* staff and Robert Burns, *The Associated Press*, "Odierno, Former Army Chief of Staff, Remembered as 'Loyal,' 'Extraordinary Leader,'" Oct. 10, 2021, https://militarytimes.com/news/your-military/2021/10/10/odierno-former-army-chief-of-staff-remembered-as-loyal-extraordinary-leader

139 *The Atlanta Journal-Constitution*, "Pension Loss Jolts Some Ex-Delta Pilots," Oct. 9, 2006, accessed Jan. 25, 2023, Aviation Pros, https://aviationpros.com/home/news/10396326/pension-loss-jolts-some-exdelta-pilots

140 Dave Philipps, "Army Cuts Hit Officers Hard, Especially Ones Up From Ranks," *The New York Times*, Nov. 12, 2014, accessed Jan. 25, 2023, https://nytimes.com/2014/11/13/us/cuts-in-military-mean-job-losses-for-career-staff.html

141 Ibid.

142 J.D. Leipold, "Odierno Warns 2016 Sequestration Could Result in 'Hollow' Army," US Army, Jan. 29, 2015, accessed Jan. 12, 2023, https://www.army.mil/article/141812/odierno_warns_2016_sequestration_could_result_in_hollow_army

143 Honorable John M. McHugh, secretary of the Army, and General Raymond T. Odierno, Army chief of staff, "On the Posture of the United States Army," Mar. 11, 2015, accessed Jan. 12, 2023, https://appropriations.senate.gov/imo/media/doc/hearings/DEF%20Secretary%20McHugh-General%20Ordierno%20Army%20Posture%20Statement%20031115.pdf

144 Dowd, "The Sacrifices of Sequestration," (n130).

145 "Department of Defense (DoD) Releases Fiscal Year 2017 President's Budget Proposal," US Department of Defense, Feb. 9, 2016, accessed Jan. 28, 2023, https://www.defense.gov/News/Releases/Release/Article/652687/department-of-defense-dod-releases-fiscal-year-2017-presidents-budget-proposal

146 M.B. Pell, "Air Force Landlord Falsified Records to Boost Income, Documents Show," *Reuters Investigates*, Jun. 18, 2019, accessed Jan. 27, 2023, https://reuters.com/investigates/special-report/usa-military-maintenance

147 Laurie Simmons, "Special Investigation: Military Families Say Dangerous Mold Is Taking Over Their Homes," *WTKR*, Nov. 21, 2011, accessed Jan. 29, 2023, https://wtkr.com/2011/11/21/special-investigation-military-families-say-dangerous-mold-is-taking-over-their-homes

148 Pell, "Air Force Landlord Falsified Records," (n146).

149 Ibid.

150 Ibid.

151 Military Family Advisory Network, *2019 Military Family Support Programming Survey Results*, https://mfan.org/research-reports/2019-military-family-support-programming-survey-results

152 Ray Sanchez, "Mold at Florida Military Housing Caused Mushrooms to Grow Out of Carpets, Lawsuit Says," *CNN*, Nov. 4, 2019, accessed Jan. 29, 2023, https://cnn.com/2019/12/04/us/macdill-air-force-base-housing-florida-lawsuit/index.html

153 Pell, "Air Force Landlord Falsified Records," (n146).

154 General Raymond T. Odierno, Army chief of staff, "Feb. 13, 2013—CSA Testimony Before the Senate Armed Services Committee," Feb. 21, 2013, accessed Jan. 28, 2023, https://www.army.mil/article/96868/feb_13_2013_csa_testimony_before_the_senate_armed_services_committee

155 *2013 Sequestration: Agencies Reduced Some Services and Investments, While Taking Certain Actions to Mitigate Effects*, US Government Accountability Office, Report GAO-14-244, Mar. 6, 2014, accessed Jan. 27, 2023, https://gao.gov/assets/gao-14-244.pdf

156 David Vergun, "Odierno: Brigade Readiness Half What It Should Be," US Army, Mar.12, 2015, accessed Jan. 28, 2023, https://www.army.mil/article/144302/odierno_brigade_readiness_half_what_it_should_be

157 Karen Jowers, "Raw Sewage, Mold, Vermin: Military Families Ask Court to Withhold Rent Until All Houses on These Two Bases Are Certified as Safe," *Military Times*, Dec. 6, 2019, https://militarytimes.com/pay-benefits/2019/12/06/raw-sewage-mold-vermin-military-families-ask-court-to-withhold-rent-until-all-houses-on-these-2-bases-are-certified-as-safe

158 James LaPorta, "Military Families Live in Housing with Mice and Mold, and Congress Wants 'Slumlords' and Top Brass Held Accountable," *Newsweek*, Dec. 4, 2019, https://newsweek.com/military-families-housing-mice-mold-slumlords-brass-congress-accountable-1475354

159 *Privatized Military Housing: Update on DoD's Efforts to Address Oversight Challenges*, US Government Accountability Office, Report GAO-22-105866, Mar. 31, 2022, accessed Jan. 28, 2023, https://gao.gov/assets/gao-22-105866.pdf

160 Robert Timmons, "Leaders Hear Residents' Housing Concerns," US Army, Jan. 19, 2023, accessed Feb. 12, 2023, https://www.army.mil/article/263356

161 Stony Brook University, "What's a Wicked Problem?" (n15).

162 In 2021, the Army shuttered the University of Foreign Military and Cultural Studies, home of "Red Team Training." Red Teaming taught individuals to mitigate groupthink, practice contrarian thinking, and mitigate cognitive bias as a component of the Combined Arms Center at Fort Leavenworth. The previous proponent for this course (Training and Doctrine Command—TRADOC—G2) has partnered with the Army Special Operations Forces (ARSOF) community to offer an "interim red team education element located at Fort Leavenworth [to] deliver one six-week red team qualification course per quarter beginning Q4 FY24," https://usacac.army.mil/organizations/red-team-program

163 Staff Sgt. Alexandra Hemmerly-Brown, "ARFORGEN: Army's deployment cycle aims for predictability," Nov. 19, 2009, accessed Jan. 30, 2023, https://www.army.mil/article/30668/arforgen_armys_deployment_cycle_aims_for_predictability

164 Ibid.

165 Service member on Twitter: "ARFORGEN destroyed two generations of officers and NCOs. It took away training management, and understanding how to train your formation with given resources, and how to mitigate risk across capability/resource gaps in your formation. It dictated everything."

166 Adam C. Resnick, Mireille Jacobson, Srikanth Kadiyala, et al., "How Deployments Affect the Capacity and Utilization of Army Treatment Facilities," *RAND Health Quarterly*, 4(3), https://ncbi.nlm.nih.gov/pmc/articles/PMC5396211

167 Russ Bynum, The Associated Press, "Fort Stewart Deployment Takes a Third of Ga. Town," *Star News*, Jan. 3, 2003, https://starnewsonline.com/story/news/2003/01/03/fort-stewart-deployment-takes-a-third-of-ga-town/30504310007

When our family moved to Fort Stewart in 2011, those stationed there since before the Global War on Terror spoke about how the deployment of 10,000 soldiers from the 3rd Infantry Division in 2003 drastically impacted the surrounding town of Hinesville. This was the first major deployment since the 1991 Gulf War; many families left and about half the businesses were forced to shutter.

168 Brock Bastian, Jolanda Jetten, and Laura J. Ferris, "Pain as Social Glue: Shared Pain Increases Cooperation," *Psychological Science*, 25(11), Sept. 5, 2014, accessed Feb. 3, 2023, https://doi.org/10.1177/0956797614545886

169 Arin Yoon, "A Military Spouse Reflects on Life Over Two Decades of War—and What Comes Next," *National Geographic*, Sept. 10, 2021, accessed Feb. 27, 2023, https://nationalgeographic.com/history/article/military-spouse-reflects-on-life-over-two-decades-of-war

170 Senior Airman Jason J. Brown, "AEF Next Represents Next Evolution in Air Force Expeditionary Operations," Joint Base Langley-Eustis, May 12, 2014, accessed Feb. 3, 2023, https://www.jble.af.mil/News/Article-Display/Article/844091/aef-next-represents-next-evolution-in-air-force-expeditionary-operations

171 David B. Larter, "The US Navy's Vaunted Deployment Plan Is Showing Cracks Everywhere," *Defense News*, Feb. 7, 2020, accessed Feb. 3, 2023, https://www.defensenews.com/naval/2020/02/07/the-us-navys-vaunted-deployment-plan-is-showing-cracks-everywhere

172 Rick Brennan Jr., Charles P. Ries, Larry Hanauer, et al., "Smooth Transitions? Lessons Learned from Transferring U.S. Military Responsibilities to Civilian Authorities in Iraq," RAND Corp., 2013, accessed Feb. 11, 2023, https://rand.org/pubs/research_briefs/RB9749.html

173 G. K. SteelFisher, A.M. Zaslavsky, R. J. Blendon, "Health-Related Impact of Deployment Extensions on Spouses of Active Duty Army Personnel," *Military Medicine*, 173(3), (Mar. 2008): 221-229. https://doi.org/10.7205/MILMED.173.3.221

174 Erin Sahlstein, Katheryn C. Maguire, and Lindsay Timmerman, "Contradictions and Praxis Contextualized by Wartime Deployment: Wives' Perspectives Revealed Through Relational Dialectics," *Communication Monographs*, 76(4), (Dec. 2009):421-442, https://doi.org/10.1080/03637750903300239

175 Brian Cafferky and Lin Shi, "Military Wives Emotionally Coping During Deployment: Balancing Dependence and Independence," *The American Journal of Family Therapy*, 43(3), (May 15, 2015):282-295, https://doi.org/10.1080/01926187.2015.1034633

176 Tova B. Walsh, Carolyn J. Dayton, Michael S. Erwin, et al., "Fathering after Military Deployment: Parenting Challenges and Goals of Fathers of Young Children," *Health & Social Work*, 39(1), (Feb. 2014):35-44. https://doi.org/10.1093/hsw/hlu005

177 Leanne K. Knobloch, Aaron T. Ebata, Patricia C. McGlaughlin, et al., "Depressive Symptoms, Relational Turbulence, and the Reintegration Difficulty of Military Couples Following Wartime Deployment," *Health Communication*, 28(8), (Oct. 2013):754–766, https://doi.org/10.1080/10410236.2013.800440

178 Stacy Ann Hawkins, Annie Condon, Jacob N. Hawkins, et al., "What We Know About Military Family Readiness: Evidence from 2007-2017," Research Facilitation Laboratory, Mar. 30, 2018, accessed Feb. 9, 2023, https://apps.dtic.mil/sti/pdfs/AD1050341.pdf

179 Lyndon A. Riviere, PhD, Julie C. Merrill, MS, Jeffrey L. Thomas, MSC USA, et al., "2003–2009 Marital Functioning Trends Among US Enlisted Soldiers Following Combat Deployments," *Military Medicine*, 177(10), (Oct. 2012):1169–1177, https://doi.org/10.7205/milmed-d-12-00164

180 Andrea L. Joseph and Tamara D. Afifi, "Military Wives' Stressful Disclosures to Their Deployed Husbands: The Role of Protective Buffering," *Journal of Applied Communication Research*, 38(4), (Oct. 2010):412-434, https://doi.org/10.1080/00909882.2010.513997

181 Hawkins, et al., "What We Know About Military Family Readiness," (n178).

182 Karen Jowers, "Two-Thirds of Military Teens Want to Follow in Their Parents' Footsteps, but These Kids 'are Not Okay,' Survey Finds," *Military Times*, Oct, 7, 2021, accessed Feb. 11, 2023, https://militarytimes.com/pay-benefits/2021/10/07/two-thirds-of-military-teens-want-to-follow-in-their-parents-footsteps-but-these-kids-are-not-okay-survey-finds

In addition to the primary concern of military teens' well-being, further investigation is needed: If our most reliable recruitment cohort is children of service members, and if those families believe that receiving a behavioral health diagnosis could be a discriminating factor against potential future military service, how does that affect their decision to seek out services for their teens?

183 Jim Garamone, "Sequestration Endangers Defense Strategy, Dempsey Says," US Department of Defense, Mar. 3, 2015, accessed Sept. 18, 2023, https://www.defense.gov/News/News-Stories/Article/Article/604209/sequestration-endangers-defense-strategy-dempsey-says

184 Andrew Feickert, "The Army's Regionally Aligned Readiness and Modernization Model" Congressional Research Service, *In Focus* IF11670 Version 3, Sept. 22, 2022, accessed Feb. 11, 2023, https://sgp.fas.org/crs/weapons/IF11670.pdf

185 James McConville, *Military Operations Force Generation—Sustainable Readiness*, Headquarters Department of the Army, Oct. 1, 2019, accessed Feb. 11, 2023, https://armypubs.army.mil/epubs/DR_pubs/DR_a/pdf/web/ARN9412_AR525_29_FINAL.pdf

186 "The Army's Sustainable Readiness Model (SRM)," Congressional Research Service, Mar. 13, 2017, accessed Feb. 11, 2023, https://www.everycrsreport.com/reports/IN10679.html

187 Clearinghouse Technical Assistance Team, "The Military Spouse Experience: Current Issues and Gaps in Service, Rapid Literature Review," Clearinghouse for Military Family Readiness at Penn State, Jun. 3, 2021, accessed Feb. 11, 2023, https://militaryfamilies.psu.edu/wp-content/uploads/2021/07/Military-Spouse-Experience_Current-Issues-and-Gaps-in-Service_3June2021.pdf

188 Blue Star Families, *2021 Military Family Lifestyle Survey Comprehensive Report*, Mar. 2022, accessed Sept. 18, 2023, https://bluestarfam.org/wp-content/uploads/2022/03/BSF_MFLS_Results2021_ComprehensiveReport_03_14.pdf

189 Headquarters, Deputy Chief of Staff, "Regionally Aligned Readiness and Modernization Model", US Army, Oct. 16, 2020, accessed Feb. 11 2023, https://www.army.mil/standto/archive/2020/10/16

190 *The Associated Press*, "Military Must Focus on Current Wars, Gates Says," *NBC News*, May 13, 2008, accessed Feb. 10, 2023, https://nbcnews.com/id/wbna24600218

191 US Army War College. "Panel III and Featured Speaker: LTG Keith Walker, ARCIC Director," video, *YouTube*, Apr 17, 2012, accessed Feb. 10, 2023, https://youtube.com/watch?v=BQup5zpmH9g

192 Definition of "enslaved," *Oxford English Dictionary*, retrieved Feb. 10, 2023.

193 Christine Chmielewski, "The Importance of Values and Culture in Ethical Decision Making," NACADA Clearinghouse of Academic Advising, 2004, accessed Sept. 18, 2023, http://nacada.ksu.edu/Resources/Clearinghouse/View-Articles/Values-and-culture-in-ethical-decision-making.aspx

194 William Hojnacki, "Three Rules of Management," *Managerial Decision Making*, 2004 graduate course conducted in the School of Public and Environmental Affairs, Indiana University South Bend.

195 Jeffrey Bradfield and Cole Clark, "Seven Principles for Effective Change Management," Deloitte US, accessed Sept.18, 2023, https://www2.deloitte.com/us/en/pages/public-sector/articles/effective-change-management-higher-education.html

196 Yvonne Doll and Billy Miller, "Leading and Making a Transformational Change," Command and General Staff College, 2007, accessed Sept.18, 2023, https://usacac.army.mil/sites/default/files/documents/cace/DCL/DCL_LeadingMakingTransChange.pdf

197 Pat Williams and Karen Kingsbury, *Forever Young: Ten Gifts of Faith for the Graduate*, (Faith Communications, 2005), 156.

198 Elizabeth Dixon, LISW-CP, "Breaking the Chains of Generational Trauma," *Psychology Today*, Jul. 3, 2021, accessed Feb. 27, 2023, https://psychologytoday.com/us/blog/the-flourishing-family/202107/breaking-the-chains-generational-trauma

199 *2021 Demographics Profile of the Military Community*, US Department of Defense, Office of the Deputy Assistant Secretary of Defense for Military Community and Family Policy, accessed Jul. 15, 2023, https://download.militaryonesource.mil/12038/MOS/Reports/2021-demographics-report.pdf

200 "Don't Ask, Don't Tell Repeal Act of 2010," National Archives Foundation, accessed Nov. 26, 2022, https://archivesfoundation.org/documents/dont-ask-dont-tell-repeal-act-2010

201 "The Hidden Financial Costs of Military Spouse Unemployment," Hiring Our Heroes, First Command Financial Services, accessed Nov. 23, 2022, https://hiringourheroes.org/resources/hidden-financial-costs-military-spouse-unemployment

202 National Academies of Sciences, Engineering, and Medicine; Div. of Behavioral and Social Sciences and Education; Board on Children, Youth, and Families; Committee on the Well-Being of Military Families; Kenneth W. Kizer and Suzanne Le Menestrel, editors, "Family Well-Being, Readiness, and Resilience," *Strengthening the Military Family Readiness System for a Changing American Society*, (Washington, DC: National Academies Press, 2019), https://ncbi.nlm.nih.gov/books/NBK547609

203 Daniel Brown, "Russian-backed Separatists Are Using Terrifying Text Messages to Shock Adversaries—and It's Changing the Face of Warfare," *Business Insider*, Aug.14, 2018, https://businessinsider.com/russians-use-creepy-text-messages-scare-ukrainians-changing-warfare-2018-8

204 "Breaking Barriers in History: Martha Washington's Active Role During the American Revolution," George Washington's Mount Vernon, accessed Sept.18, 2023, https://www.mountvernon.org/education/for-students/national-history-day-2020/martha-washington-breaking-barriers

205 Nancy Shea, *The Army Wife*, Third Revised Edition, (New York: Harper & Brothers, 1954)

206 Jennifer Mittelstadt, "Welfare's Last Stand: Long in Retreat in the US, the Welfare State Found a Haven in an Unlikely Place—the Military, Where it Thrived for Decades," Sept. 21, 2015, accessed Nov. 21, 2022, *Aeon*. https://aeon.co/essays/how-the-us-military-became-a-welfare-state

207 Serena Covkin, "A Short History of US Army Wives, 1776-1983." *US History Scene*, Apr. 29, 2019, https://ushistoryscene.com/article/a-short-history-of-u-s-army-wives-1776-1983

208 John Worsencroft, "Family Matters: The United States Army, Family, and the Search for Stability 1980-1984," *Army Heritage Center Foundation*, accessed Sept. 18, 2023. https://armyheritage.org/wp-content/uploads/2020/06/FamilyMattersBarnes.pdf

209 Covkin, "A Short History of US Army Wives," (n207).

210 General John A. Wickham Jr., *The Army Family*, white paper, Office of the Army Chief of Staff, Aug. 15, 1983.

211 Cynthia Enloe, *Maneuvers: The International Politics of Militarizing Women's Lives*, (Berkeley: University of California Press, 2000); 3, 293.

212 Mittelstadt, "Welfare's Last Stand," (n206).

213 "Emotionally Unavailable Partners and The Highly Sensitive Person," Eggshell Therapy and Coaching, Feb. 28, 2022, https://eggshelltherapy.com/emotionally-unavailable

214 *2021 Demographics Profile of the Military Community*, (n199).

215 Henrietta C. McGowan, "The Military Experience: Perceptions from Senior Military Officers' Wives," dissertation, Capella University, Sept. 2008, https://proquest.com/openview/d72fff3202831c670409ef8e0f43ca4f/1?pq-origsite=gscholar&cbl=18750

216 Richard Tedeschi, Lawrence G. Calhoun, and Jessica M. Groleau, "Clinical Applications of Posttraumatic Growth," *Positive Psychology in Practice*, (John Wiley & Sons, Inc, 2015), Ch. 30, 503–518, https://doi.org/10.1002/9781118996874.ch30

217 Jaia Barrett, *LinkedIn* post, https://linkedin.com/in/jaia-barrett

218 Rebekah F. Cole, Rebecca G. Cowan, Hayley Dunn, et al., "Military Spouses' Perceptions of Suicide in the Military Spouse Community," *The Professional Counselor* (11)2, Apr. 1, 2021, accessed

Sept. 18, 2023, https://tpcjournal.nbcc.org/military-spouses-perceptions-of-suicide-in-the-military-spouse-community

219 Yoon, "A Military Spouse Reflects," (n169).

220 Military Family Advisory Network, *2021 Military Family Support Programming Survey Executive Summary*, accessed Sept. 18, 2023, https://mfan.org/wp-content/uploads/2022/07/Executive-Summary-MFAN-Programming-Survey-Results-2021.pdf

221 CPI Inflation Calculator, US Bureau of Labor Statistics, https://bls.gov/data/inflation_calculator.htm

222 Rosalyn Oshmyansky, "Monica Lewinsky: 'I Was The First Person Destroyed by the Internet,'" *Entertainment Tonight*, Oct. 20, 2014, accessed Feb. 15, 2023, https://etonline.com/news/152767_monica_lewinsky_was_the_first_person_destroyed_by_the_internet

223 Verity Jennings, "A Brief History of Email: Dedicated to Ray Tomlinson," *Phrasee*, Mar. 10, 2016, accessed Feb. 16, 2023, https://phrasee.co/blog/a-brief-history-of-email

224 Louis Anslow, "Here's What the Internet Looked Like in the Hours After 9/11," *Timeline*, Sept. 11, 2016, accessed Feb. 18, 2023, https://timeline.com/internet-looked-like-9-11-dd4afad61330

225 Pew Research Center, "One Year Later: September 11 and the Internet," Sept. 5, 2002, https://pewresearch.org/internet/2002/09/05/one-year-later-september-11-and-the-internet

226 Kate Moran, "Social Media Natives: Growing Up with Social Networking," Nielsen Norman Group, Aug 28, 2016, accessed Feb. 18, 2023, https://nngroup.com/articles/social-media-natives

227 Jon Loomer, "Detailed History of Facebook Changes 2004-12," May 6, 2012, updated Sept. 15, 2023, https://jonloomer.com/history-of-facebook-changes

228 Blue Star Families, *2013 Military Family Lifestyle Survey Comprehensive Report*, Mar. 2013, https://bluestarfam.org/wp-content/uploads/2020/02/2013-Blue-Star-Families-annual-Military-Family-Lifestyle-Survey-Comprehensive-Report.pdf

229 *Doctrine Man* Facebook page, accessed Sept. 18, 2023, https://fb.com/DoctrineMan

230 Blue Star Families, *2013 Military Family Lifestyle Survey*, (n228).

231 Michelle D. Sherman, PhD, Lynne M. Borden, PhD, Octavia Cheatum, BA, et al., *Social Media Communication with Military Spouses*, University of Minnesota, Feb. 2015, accessed Feb. 19, 2023, https://reachfamilies.umn.edu/sites/default/files/rdoc/Social%20Media%20Communication%20with%20Military%20Spouses.pdf

232 Christine Barton, Lara Koslow, and Christine Beauchamp, "How Millennials Are Changing the Face of Marketing Forever," Boston Consulting Group, Jan. 15, 2014, accessed Feb. 19, 2023 https://bcg.com/publications/2014/marketing-center-consumer-customer-insight-how-millennials-changing-marketing-forever

233 Amelia Tait, "How Instagram Changed Our World," *The Guardian*, May 3, 2020, accessed Feb. 19, 2023, https://theguardian.com/technology/2020/may/03/how-instagram-changed-our-world

234 Zara Abrams, "How Can We Minimize Instagram's Harmful Effects?" *Monitor on Psychology*, 53(2), (Dec. 2021): 30, https://apa.org/monitor/2022/03/feature-minimize-instagram-effects

235 Kate Nelson, "I Blame '90s Diet Culture for My Adult Body Image Issues," *Saveur*, Dec. 1, 2022, accessed Feb. 22, 2023, https://saveur.com/culture/90s-diet-culture-bipoc

236 Tait, "How Instagram Changed Our World," (n233).

237 Hannah Messinger, "Dis-like: How Social Media Feeds into Perfectionism," *Penn Medicine*, Nov. 19, 2019, accessed Feb. 20, 2023, https://pennmedicine.org/news/news-blog/2019/november/dis-like-how-social-media-feeds-into-perfectionism

238 William R. Miller and Stephen Rollnick, *Motivational Interviewing: Helping People Change*, Third Edition, (The Guilford Press, 2013).

239 General Robert B. "Abe" Abrams, "Social Media: Senior Leaders Need to Get on the Bus," *From the Green Notebook*, Oct. 8, 2019, accessed Feb. 22, 2023, https://fromthegreennotebook.com/2019/10/08/social-media-senior-leaders-need-to-get-on-the-bus

240 Christina Newberry, "2023 Facebook Algorithm: How to Get Your Content Seen," Hootsuite, Feb. 22, 2023, https://blog.hootsuite.com/facebook-algorithm

241 Brianna Wiest, "Millennials Hate Phone Calls, And They Have A Point." *Forbes*, Nov. 4, 2019, https://forbes.com/sites/briannawiest/2019/11/04/millennials-hate-phone-calls-they-have-a-point

242 Ashley Fantz, "As ISIS Threats Online Persist, Military Families Rethink Online Lives," *CNN*, Mar. 23, 2015, https://cnn.com/2015/03/23/us/online-threat-isis-us-troops/index.html

243 Michael Hoffman, "Military Spouses Threatened by ISIS Affiliate on Twitter, Facebook," *Military.com*, Feb. 10, 2015, https://military.com/daily-news/2015/02/10/military-spouses-threatened-by-isis-affiliate-on-twitter-facebo.html

244 Andrew Perrin, "Social Media Usage: 2005-2015," Pew Research Center, Oct. 8, 2015, accessed Feb. 18, 2023, https://pewresearch.org/internet/2015/10/08/social-networking-usage-2005-2015

245 Pat Matthews-Juarez, Paul D. Juarez, and Roosevelt T. Faulkner, "Social Media and Military Families: A Perspective," *Journal of Human Behavior in the Social Environment*, 23(6), (Jun. 2013):769–776. https://doi.org/10.1080/10911359.2013.795073

246 Noam Lapidot-Lefler and Azy Barak, "Effects of Anonymity, Invisibility, and Lack of Eye-Contact on Toxic Online Disinhibition," *Computers in Human Behavior*, 28(2), (Mar. 2012):434–443. https://doi.org/10.1016/j.chb.2011.10.014

247 Jennifer Kristine Rea, "The Role of Social Networking Sites in the Lives of Military Spouses," Unpublished master's thesis, 2014, North Carolina State University, Raleigh, NC, https://repository.lib.ncsu.edu/bitstream/handle/1840.16/9630/etd.pdf

248 Stacy Jo Dixon, "Daily Time Spent on Social Media by US Active-Duty Military Members 2020," *Statista*, Apr. 28, 3022, accessed Sept. 18, 2023, https://statista.com/statistics/1130021/hours-spent-social-media-us-active-duty-military-members

249 MandatoryFunDay, https://mandatoryfunday.carrd.co

250 Austin (Host), "Leave Packets, Hands in Pockets, and Gaining 100k+ Followers While Active Duty with Austin from MandatoryFunDay," audio podcast, *The Veteran (Semi) Professional*, Mar. 25, 2023, Ep. 185, accessed Aug. 8, 2023, https://buzzsprout.com/1084379/12512653

251 John Hughel, "Popular Military Social Media Influencer Takes Part in 'Hope in the Trenches' Podcast," Defense Visual Information Distribution Service, Apr. 28, 2023, accessed Sept. 18, 2023, https://dvidshub.net/news/443584/popular-military-social-media-influencer-takes-part-hope-trenches-podcast

252 Brady Shearer, "The Biggest Communication Shift in the Last 500 Years," *Pro Church Tools*, Season 4 Ep. 5, Jan. 12, 2018, accessed Feb. 16, 2023, https://prochurchtools.com/biggest-communication-shift-last-500-years-ep-005

253 "The Gutenberg Printing Press", *Vaia*, accessed Feb. 17, 2023, https://hellovaia.com/explanations/history/protestant-reformation/the-gutenberg-printing-press

254 "The Incredible Growth of the Internet Since 2000," *Pingdom*, Oct. 22, 2010, accessed Feb. 17, 2023, https://pingdom.com/blog/incredible-growth-of-the-internet-since-2000

255 Lisa Zyga, "Internet Growth Follows Moore's Law Too", *Phys.org*, Jan. 14, 2009, accessed Feb. 17, 2023, https://phys.org/news/2009-01-internet-growth-law.html

256 Amitabh Ray, "Human Knowledge Is Doubling Every 12 Hours," *LinkedIn*, Oct. 22, 2020, accessed Feb. 17, 2023 https://linkedin.com/pulse/human-knowledge-doubling-every-12-hours-amitabh-ray

257 Abby McCain, "How Fast Is Technology Advancing? [2023]: Growing, Evolving, and Accelerating at Exponential Rates," *Zippia*, Jan. 11, 2023, accessed Feb. 19, 2023, https://zippia.com/advice/how-fast-is-technology-advancing

258 "How Does ChatGPT Work?" *Atria Innovation*, Jan. 5, 2023, accessed Mar. 20, 2023, https://atriainnovation.com/en/how-does-chat-gpt-work

259 Leading Generations Training Event, 5th Special Forces Group, Dec. 8, 2022, Fort Bragg, NC.

260 Kim Parker and Ruth Igielnik, "On the Cusp of Adulthood and Facing an Uncertain Future; What We Know About Gen Z So Far," Pew Research Center, May 14, 2020, accessed Feb. 26, 2023, https://pewresearch.org/social-trends/2020/05/14/on-the-cusp-of-adulthood-and-facing-an-uncertain-future-what-we-know-about-gen-z-so-far-2

261 Kyle Fitzgerald, "Activists Use Google Maps to Send Messages to Russian People," *The National*, Mar. 1, 2022, accessed Feb. 26, 2023, https://thenationalnews.com/world/us-news/2022/02/28/activists-use-google-maps-to-send-messages-to-russian-people

262 Jeffrey M. Jones, "More Parents, Children Fearful for Safety at School," Gallup, Aug. 24, 2018, accessed Feb, 13, 2023, https://news.gallup.com/poll/241625/parents-children-fearful-safety-school.aspx

263 "Millennials Grew Up With the Internet—But Now They're Aging Out of It", *YPulse*, Aug. 9, 2022, accessed Feb. 16, 2023, https://ypulse.com/newsfeed/2022/08/09/millennials-grew-up-with-the-internet-but-now-theyre-aging-out-of-it

264 Riad Chikhani, "The History of Gaming: An Evolving Community," *TechCrunch*, Oct. 31, 2015, accessed Feb. 20, 2023, https://techcrunch.com/2015/10/31/the-history-of-gaming-an-evolving-community

265 Mike Wadhera, "The Information Age is Over; Welcome to the Experience Age," *TechCrunch*, Mar. 9, 2016, accessed Feb. 26, 2023, https://techcrunch.com/2016/05/09/the-information-age-is-over-welcome-to-the-experience-age

266 Chikhani, "The History of Gaming," (n264).

267 Dom Barnard, "History of VR: Timeline of Events and Tech Development," *Virtual Speech*, Jun. 14, 2023, accessed Feb. 26, 2023, https://virtualspeech.com/blog/history-of-vr

268 Sol Rogers, "2019: The Year Virtual Reality Gets Real," *Forbes*, Jun. 21, 2019, accessed Feb. 26, 2023, https://forbes.com/sites/solrogers/2019/06/21/2019-the-year-virtual-reality-gets-real

269 Susan Katz Keating, "Army Leaders Benefit from Engaging on Social Media," Association of the United States Army (AUSA), Oct. 14, 2019, accessed August 18th, 2023, https://ausa.org/news/army-leaders-benefit-engaging-social-media

270 Meghann Myers, "Retired Sergeant Major, US Army W.T.F. Moments Admin Reveals Himself at AUSA," Washington Military Resource Directory, accessed Feb. 21, 2023, https://wamilitary.com/retired-sergeant-major-us-army-wtf-moments-admin-reveals-himself-at-ausa

271 Corey Dickstein, "Former Fort Benning Commander Retires with No Reprimand After Army Investigates His Social Media Use," *Stars and Stripes*, Jan. 3, 2023, accessed Mar. 20, 2023, https://stripes.com/branches/army/2023-01-03/army-general-twitter-fox-carlson-8633939.html

272 Katz Keating, "Army Leaders Benefit from Engaging," (n269).

273 Amber Ferguson, "Nonstop Worship Service at Kentucky College Set to End After Attracting Thousands," *The Washington Post*, Feb. 19, 2023, accessed Feb. 26, 2023, https://washingtonpost.com/religion/2023/02/18/asbury-university-revival-kentucky

274 Kenneth Ramos, "Happy Friday Nation!" *US Army W.T.F! Moments* Facebook page, Jan. 20, 2023, accessed Feb. 26, 2023, https://fb.com/usawtfm/photos/10161243424713606

275 C. Todd Lopez, "Trust Bedrock of Army Profession," US Army, Dec. 16, 2016, accessed Jul. 28, 2023, https://www.army.mil/article/179601/trust_bedrock_of_army_profession

276 Stef W. Kight, "Generation V for Virus," *Axios*, Apr. 11, 2020, accessed Mar. 12, 2023, https://axios.com/2020/04/11/generation-z-coronavirus

277 Charlie Dunlap, JD, "Mackenzie Eaglen on '5 Lessons the US Military Learned from the Pandemic,'" *Lawfire*, Jun. 17, 2020, accessed Mar. 9, 2023, https://sites.duke.edu/lawfire/2020/06/17/mackenzie-eaglen-on-5-lessons-the-u-s-military-learned-from-the-pandemic

278 Carol R. Byerly, PhD, "The US Military and the Influenza Pandemic of 1918-1919," *Public Health Rep*, 125(3), (Apr. 2010):82-91, https://ncbi.nlm.nih.gov/pmc/articles/PMC2862337

279 "Travel Restrictions for DoD Components in Response to Coronavirus Disease 2019," Secretary of Defense, Mar. 11, 2020, accessed Mar. 9, 2023, https://documentcloud.org/documents/6808531-Travel-Restrictions-for-DoD-Components-in.html

280 Mark F. Cancian and Adam Saxton, "COVID-19 and the Military; Maintaining Operations While Supporting Civil Society," Center for Strategic & International Studies, Feb.12, 2021, accessed Mar. 9, 2023, https://csis.org/analysis/covid-19-and-military-maintaining-operations-while-supporting-civil-society

281 Sarah Streyder, "How COVID-19 is Affecting Military Families," *Secure Families Initiative*, Mar. 16, 2020, accessed Mar. 9, 2023, https://securefamiliesinitiative.org/how-covid-19-is-affecting-military-families

282 Ibid.

283 William M. Donnelly (ed.), Jamie L. H. Goodall (ed.), Kendall E. Cosley, et al., *The United States Army and the COVID-19 Pandemic January 2020–July 2021*, Center of Military History, US Army, accessed Mar. 9, 2023, https://history.army.mil/html/books/070/70-133/cmhPub_70-133-1.pdf

284 Brianna Keilar and Catherine Valentine, "More Than a Third of Military Families Said They Have No One to Ask for a Favor, Survey Finds," *CNN*, Feb. 26, 2020, accessed Mar. 9, 2023, https://cnn.com/2020/02/26/politics/homefront-military-family-isolation/index.html

285 National Military Family Association, "Military Families During the COVID-19 Pandemic: Lack of Child Care, Military Spouse Unemployment, and Income Loss," accessed Mar. 9, 2023, https://militaryfamily.org/wp-content/uploads/Military-Families-During-the-Covid-19-Pandemic.pdf

286 Blue Star Families, *2020 Military Family Lifestyle Survey Comprehensive Report*, Oct. 2021, https://bluestarfam.org/wp-content/uploads/2021/10/BSF_MFLS_CompReport_FULL.pdf

287 Ibid.

288 Ibid.

289 "About DoDEA," Department of Defense Education Activity, accessed Mar. 10, 2023, https://dodea.edu/about/about-dodea

290 Jennifer S. Thomas, PhD, Amanda Trimillos, EdD, and Stacy Allsbrook-Huisman, "Military Adolescent Pandemic Study 2021: MAPS21," *Journal of School Health*, 92(11), (Nov. 2022):1051-1061, https://doi.org/10.1111/josh.13277

291 Kathy Broniarczyk and Shelly Wadsworth, PhD, "Measuring Our Communities: The State of Military and Veteran Families in the United States," Purdue University, Sept. 2022, https://mfri.purdue.edu/wp-content/uploads/2022/09/2022_MeasuringCommunitiesReport-lores.pdf

292 Marcela Bombardieri, "COVID-19 Changed Education in American—Permanently," *Folitico*, Apr. 15, 2021, https://politico.com/news/2021/04/15/covid-changed-education-permanently-479317

293 Apoorva Mandavilli, "The Coronavirus May Spread From Corpses, Scientists Report," *The New York Times*, Dec. 15, 2022, https://nytimes.com/2022/12/15/health/covid-dead-bodies.html

294 Broniarczyk and Wadsworth, "Measuring Our Communities," (n291).

295 Ibid.

296 Ibid.

297 Ibid.

298 "Understanding the Effects of Social Isolation on Mental Health," Tulane University, Dec. 8, 2020, https://publichealth.tulane.edu/blog/effects-of-social-isolation-on-mental-health

299 Amy Morin, "Quarantine Can Leave You Traumatized. Here's What That Means, and How to Address It," *Business Insider*, May 16, 2020, accessed Mar. 8, 2023, https://businessinsider.com/can-quarantine-leave-you-traumatized-heres-what-it-means-2020-5

300 Corie Weathers, "Working on a chapter about the impact of COVID/2020 on our community. Anyone else still feel the trauma in their body? ... ," *Corie Weathers | Lifegiver* Facebook page post, Mar. 9, 2023, https://fb.com/CorieLPC

301 Thomas J Walsh, "Military Kids Are Resilient, But Far From Immune to Pandemic Effects," *Health.mil*, Apr. 28, 2021, accessed Mar. 9, 2023, https://health.mil/News/Articles/2021/04/28/Military-kids-are-resilient-but-far-from-immune-to-pandemic-effects

302 Ibid.

303 Ibid.

304 Alexandra Jones, "The Five Phases of Lockdown, from Tiger King to Sourdough Starters," *The Independent*, Sept. 23, 2020, accessed Mar. 12, 2023, https://independent.co.uk/life-style/lockdown-boris-johnson-nhs-joe-wicks-tiger-king-normal-people-banana-masturbation-houseparty-b549943.html

305 Elizabeth M. Collins, "Experts Explain Mental State of Military Children," *Soldiers*, May 6, 2015, https://www.army.mil/article/147786/experts_explain_mental_state_of_military_children

306 Ibid.

307 *Health of the Army Family*, Defense Centers for Public Health, Aberdeen, Dec. 2021, accessed Mar. 12, 2023, https://phc.amedd.army.mil/Periodical%20Library/haf-2021-report.pdf

308 Anne Kniggendorf, "A New Survey Found Isolation and Frequent Moves Are Harming Military Teens' Mental Health," The American Homefront Project, Oct. 7, 2021, https://americanhomefront.wunc.org/news/2021-10-07/a-new-survey-found-isolation-and-frequent-moves-are-harming-military-teens-mental-health

309 Patricia Kime, "Soldiers Need Friends and Family to Help Amid Mental Health Provider Shortage, Army Leaders Say," *Military.com*, Oct. 12, 2022, https://military.com/daily-news/2022/10/12/soldiers-need-friends-and-family-help-amid-mental-health-provider-shortage-army-leaders-say.html

310 Thomas W Travis, MD and David L Brown, MD, "Human Performance Optimization: A Framework for the Military Health System," *Military Medicine*, 188(1), (Mar/Apr. 2023):44-48. https://doi.org/10.1093/milmed/usac411

311 Paula Sumner, "10 of the Lowest-Paying Majors (Plus How to Increase Salary)," *Indeed*, Jul. 27, 2023, accessed Apr. 10, 2023, https://indeed.com/career-advice/finding-a-job/lowest-paying-majors

312 K. Marksberry, "The Holmes-Rahe Stress Inventory," The American Institute of Stress, accessed Mar. 8, 2023, https://stress.org/holmes-rahe-stress-inventory

313 Angela Duckworth, *Grit: The Power of Passion and Perseverance*, (New York: Scribner, 2016).

314 Corie Weathers, "Service families know that the past year has only added stress …," *Corie Weathers | Lifegiver* Facebook page, Mar. 7, 2023, https://fb.com/CorieLPC/videos/488759552115508

315 Blue Star Families, *2020 Military Family Lifestyle Survey*, (n286).

316 Josephine Joly, Luke Hurst, and David Walsh, "Four-Day Week: Which Countries Have Embraced It and How's It Going So Far?" *Euronews*, Jun. 21, 2023, https://euronews.com/next/2023/02/23/the-four-day-week-which-countries-have-embraced-it-and-how-s-it-going-so-far

317 Nathalie Grogan, "Schools, Communities Need to Make Sure Military Children Don't Get Lost During COVID-19," *Military.com*, Dec. 3, 2020, accessed Mar. 10, 2023, https://military.com/daily-news/opinions/2020/12/03/schools-communities-need-make-sure-military-children-dont-get-lost-during-covid-19.html

318 Audrey Watters, "A Brief History of Calculators in the Classroom," *Hack Education*, Mar. 12, 2015, accessed Mar. 15, 2023, https://hackeducation.com/2015/03/12/calculators

319 Frederick Peck and David Erickson, "The Rise—and Possible Fall—of the Graphing Calculator," *The Conversation*, Jun. 13, 2017, accessed Mar. 15, 2023, https://theconversation.com/the-rise-and-possible-fall-of-the-graphing-calculator-78017

320 Watters, "A Brief History of Calculators," (n318).

321 Peck and Erickson, "The Rise—and Possible Fall," (n319).

322 Jean M. Twenge, *iGen: Why Today's Super-Connected Kids Are Growing Up Less Rebellious, More Tolerant, Less Happy—and Completely Unprepared for Adulthood—and What That Means for the Rest of Us*, (New York: Atria Books, 2017).

323 Steven Bell, "Farewell Print Textbook Reserves: A COVID-19 Change to Embrace," *Educause*, Jan. 14, 2021, accessed Mar. 10, 2023, https://er.educause.edu/articles/2021/1/farewell-print-textbook-reserves-a-covid-19-change-to-embrace

324 Ingrid Noguera Fructuoso, "How Millennials Are Changing the Way We Learn," *Revista Iberoamericana de Educación a Distancia*, 18(1), (2015):45-65, ProQuest, accessed Mar. 16, 2023, https://proquest.com/docview/1649109655

325 Cory Armes, MEd, "2022 Education Trends That Might Excite You," *Carnegie Learning*, Jan. 3, 2023, accessed Mar. 15, 2023, https://carnegielearning.com/blog/2022-education-trends

326 Christina Counts, "Five of the Biggest Education Trends in 2023," *eSchool News*, Jan. 30, 2023, https://eschoolnews.com/innovative-teaching/2023/01/30/5-of-the-biggest-education-trends-in-2023

327 Sarah Galasso,"One Stone High School: The Teacher-Student Connection," *Carnegie Learning*, Mar. 26, 2019, accessed Mar. 14, 2023, https://carnegielearning.com/blog/carnegie-learning-and-one-stone-teacher-student-connection-2

328 Kimberlee R. Mendoza, "Engaging Generation Z: A Case Study on Motivating the Post-Millennial Traditional College Student in the Classroom," *US-China Foreign Language*, 17(4), (Apr. 2019):157-166, https://davidpublisher.com/Public/uploads/Contribute/5cd383fedd216.pdf

329 Cynthia Zimmer, "Getting to Know Gen Z: Exploring Middle and High Schoolers' Expectations for Higher Education," *Barnes & Noble College*, Oct. 2015, accessed Mar. 15, 2023, https://next.bncollege.com/wp-content/uploads/2015/10/Gen-Z-Research-Report-Final.pdf

330 Mien Company, "5 Benefits of Esports in School," accessed Mar. 15, 2023, https://miencompany.com/5-benefits-of-esports-in-schools

331 Jamie McCrary, "10 Trends in the Era of Generation Z College Students," *Lead Squared*, Dec. 28, 2022, https://leadsquared.com/industries/education/generation-z-college-students

332 Irina Sidorcuka and Anna Chesnovicka, "Methods of Attraction and Retention of Generation Z Staff," *CBU International Conference Proceedings 2017*, 5, (2017):807-814, https://doi.org/10.12955/cbup.v5.1030

333 Shelly Haslam-Ormerod, "'Snowflake Millennial' Label Is Inaccurate and Reverses Progress to Destigmatise Mental Health," *The Conversation*, Jan. 11, 2019, https://theconversation.com/snowflake-millennial-label-is-inaccurate-and-reverses-progress-to-destigmatise-mental-health-109667

334 Ibid.

335 Greg Lukianoff and Jonathan Haidt, *The Coddling of the American Mind: How Good Intentions and Bad Ideas Are Setting Up a Generation for Failure*, (New York: Penguin Press, 2018).

336 Jennifer McKenzie, "Gen Z: Redefining Authority in the Workplace," *Bespoke Careers*, May 31, 2019, https://www.bespokecareers.com/uk/articles/12/gen-z:-redefining-authority-in-the-workplace

337 Ibid.

338 MasterClass, "Daniel Pink Teaches Sales and Persuasion," 2022, https://masterclass.com/classes/daniel-pink-teaches-sales-and-persuasion

339 Madi Hammond, "Living as Gen Z: From Fear to Positive Change," *TEDx Talks*, May 3, 2019, accessed Mar. 17, 2023, https://youtube.com/watch?v=DQcraSYkakk&t=6s

340 Peter Docker, *Leading From The Jumpseat: How to Create Extraordinary Opportunities by Handing Over Control*, (Why Not Press, 2021)

341 Jon Michail, "Strong Nonverbal Skills Matter Now More Than Ever in This 'New Normal,'" *Forbes*, Aug. 24, 2020, accessed Mar.17, 2023, https://forbes.com/sites/forbescoachescouncil/2020/08/24/strong-nonverbal-skills-matter-now-more-than-ever-in-this-new-normal

342 N. J. Dortch, "Does Your Seat Choice Say Anything About You?" *The Sandbox News*, St Petersburg College, Aug. 2015, accessed 18 March 2023, https://sandbox.spcollege.edu/index.php/2015/08/does-your-seat-choice-say-anything-about-you

343 Madeline Will, "Teachers Are Ready for Systemic Change. Are Schools?" *Education Week*, Sept. 6, 2022, https://edweek.org/leadership/teachers-are-ready-for-systemic-change-are-schools/2022/09

344 Quote attributed to Jerry Lewis.

345 Sam Sumac, *Piss & Vinegar*, (Black Rose Writing, 2020).

346 Karl Monger, "Memorial Day Remarks by Grant McGarry, 1st Ranger Battalion Veteran," *GallantFew*, May 25, 2015, https://gallantfew.org/memorial-day-remarks-by-grant-mcgarry

347 Danielle DeSimone,"Why 9/11 Inspired These Service Members to Join the Military," USO, Sept. 7, 2021, https://uso.org/stories/2849-why-9-11-inspired-these-patriots-to-join-the-military

348 Bradley Johnson, "My Life as a Soldier in the 'War on Terror,'" *The Journal of Military Experience*, accessed Mar. 20, 2023, https://encompass.eku.edu/cgi/viewcontent.cgi?article=1030&context=jme

349 Carmen Sanchez and David Dunning, "Research: Learning a Little About Something Makes Us Overconfident," *Harvard Business Review*, Mar. 29, 2018, accessed Mar. 20, 2023, https://hbr.org/2018/03/research-learning-a-little-about-something-makes-us-overconfident

350 Inspire Solutions, "War Is Not a Video Game – Or Is It?" Jun. 19, 2013, accessed Mar. 25, 2023, https://inspire.dawsoncollege.qc.ca/2013/06/19/war-is-not-a-video-game-or-is-it/index.html

351 Johnson, "My Life as a Soldier," (n348).

352 Rachel Martin, "A Former Marine Details the Chaotic Exit from Afghanistan—and How We Should Mark It," *NPR*, Aug. 6, 2022, accessed Jun. 13, 2023, https://npr.org/2022/08/06/1115666703/afghanistan-marine-war-exit-departure-united-states

353 Mark Benjamin, "Post-Traumatic Futility Disorder," *Salon*, Dec. 21, 2006, accessed Mar. 22, 2023, https://salon.com/2006/12/21/ptsd_6

354 Thom Shanker, "Army Is Worried by Rising Stress of Return Tours to Iraq," *The New York Times*, Apr. 6, 2008, accessed Mar. 23, 2023, https://nytimes.com/2008/04/06/washington/06military.html

355 Service member anonymous; Terrorism statistic: Bruce Hoffman and Jacob Ware, "The Terrorist Threats and Trends to Watch Out for in 2023 and Beyond," Combatting Terrorism Center, *CTC Sentinel*, 15(11), (Nov.-Dec. 2022), https://ctc.westpoint.edu/the-terrorist-threats-and-trends-to-watch-out-for-in-2023-and-beyond

356 William Deverell, "Warren Harding Tried to Return America to 'Normalcy' After WWI and the 1918 Pandemic. It Failed," *Smithsonian*, May 19, 2020, https://smithsonianmag.com/history/warren-harding-back-to-normalcy-after-1918-pandemic-180974911

357 Robert Longley, "The Lost Generation and the Writers Who Described Their World," *ThoughtCo*, updated Mar. 2, 2022, accessed Mar. 22, 2023, https://thoughtco.com/the-lost-generation-4159302

358 US Department of Defense, "Casualty Status," https://www.defense.gov/casualty.pdf

359 Ibid.

360 Daniel Brown and Azmi Haroun, "The Wars in Iraq and Afghanistan Have Killed at Least 500,000 People, According to a Report that Breaks Down the Toll," *Business Insider*, Aug. 26, 2022, https://businessinsider.com/how-many-people-have-been-killed-in-iraq-and-afghanistan

361 "PTSD: National Center for PTSD," US Department of Veterans Affairs, accessed Mar. 23, 2023, https://ptsd.va.gov/understand/common/common_veterans.asp

362 Marie-Louise Sharp, Nicola T. Fear, Roberto J. Rona, et al., "Stigma as a Barrier to Seeking Health Care Among Military Personnel With Mental Health Problems," *Epidemiologic Reviews*, 37(1), (2015):144-162, https://doi.org/10.1093/epirev/mxu012

363 Ibid.

364 J. Connor Stull, "Is 'Up or Out' Holding Us Back?" US Army, Jun. 22, 2021, accessed Mar. 24, 2023, https://www.army.mil/article/247749/is_up_or_out_holding_us_back

365 Tony L. Dedmond, "Sixteen Things I Wish I Could Tell My Senior Rater," *From the Green Notebook*, Jan. 23, 2023, accessed Mar. 24 2023, https://fromthegreennotebook.com/2023/01/23/sixteen-things-i-wish-i-could-tell-my-senior-rater

366 Scott R. Johnson, "US Army Evaluations, a Study on Inaccurate and Inflated Reporting," US Marine Corps Command and Staff College, Apr. 2012, https://apps.dtic.mil/sti/tr/pdf/ADA602724.pdf

367 Claudia Grisales, "Military Experts, Leaders Say 'Up or Out' Promotion System Outdated," *Stars and Stripes*, Jan. 24, 2018, accessed Mar. 25, 2023, https://stripes.com/news/military-experts-leaders-say-up-or-out-promotion-system-outdated-1.508382

368 Howard Risher, "The Real Problem With Inflated Performance Evaluations," *Government Executive*, Jun. 13, 2016, accessed Mar. 25, 2023, https://govexec.com/management/2016/06/real-problem-inflated-performance-evaluations/129037

369 The Athena Project is improving the situation, especially with the introduction of peer reviews at each career educational juncture. The BCAP & CCAP program identifies lieutenant colonels for battalion command, based on assessments (psychological, written, verbal, blind interviews).

370 Leonard Wong and Stephen J. Gerras, *Lying to Ourselves: Dishonesty in the Army Profession*, (US Army War College Press, 2015), https://press.armywarcollege.edu/monographs/466

371 Ibid.

372 Leonard Wong and Stephen J. Gerras, "Still Lying to Ourselves: A Retrospective Look at Dishonesty in the Army Profession," *The Journal of Character & Leadership Development*, (Spring 2022):56-62, accessed Mar. 26, 2023, https://jcldusafa.org/index.php/jcld/article/view/10/10

373 "US Navy Suspends 'Up-or-Out' Policy in a Bid to Boost Retention," *The Maritime Executive*, Dec. 27, 2022, accessed Mar. 28 2023, https://maritime-executive.com/article/u-s-navy-suspends-up-or-out-policy-to-boost-retention

374 "Opt In to Promotion," *US Army Talent Management*, https://talent.army.mil/opt

375 Devon L. Suits, "Changes to Promotion Process Provide Officers More Career Flexibility," *Military News*, Feb. 28, 2020, accessed Mar. 28, 2023, https://militarynews.com/peninsula-warrior/news/army_news/changes-to-promotion-process-provide-officers-more-career-flexibility/article_32925a9c-58dc-11ea-90e5-eb8b8e98722b.html

376 Michelle Tan, "Putting People First: McConville Looks to Revolutionize How Soldiers Serve," Association of the United States Army (AUSA), Oct. 3, 2019, accessed Sept. 23, 2022, https://ausa.org/articles/putting-people-first-mcconville-looks-revolutionize-how-soldiers-serve

377 Haley Britzky, "The Army Chief of Staff Wants You to Have Work-Life Balance. Seriously," *Task & Purpose*, Oct, 15, 2020, https://taskandpurpose.com/news/army-chief-mcconville-people-priority

378 Michael A. Grinston, James C. McConville, and Ryan D. McCarthy, *The Army People Strategy*, Oct. 2019, https://www.army.mil/e2/downloads/rv7/the_army_people_strategy_2019_10_11_signed_final.pdf

379 Jennifer McDonald, "Army Human Resources Command's Online Talent Management System Is a Bust for Military Families," *Military Times*, Dec 28, 2019, accessed Mar, 27, 2023, https://militarytimes.com/opinion/commentary/2019/12/28/human-resources-commands-online-talent-management-system-is-a-bust-for-military-families

380 Subcommittee on Military Personnel, "Fort Hood 2020: The Findings and Recommendations of the Fort Hood Independent Review Committee," Hearing, Dec. 9, 2020, accessed Mar. 27, 2023, https://congress.gov/116/chrg/CHRG-116hhrg42928/CHRG-116hhrg42928.pdf

381 Ibid.

382 Monica Hassall, RN, and Barbara Hunter, MEd., "Fight, Flight, Freeze … or Fib?" *ADDitude*, updated Sept. 13, 2023, https://additudemag.com/why-lie-adhd-fight-flight-freeze

383 Ross Andel, Michael Crowe, Ingemar Kåreholt, Jonas Wastesson, and Marti G. Parker, "Indicators of Job Strain at Midlife and Cognitive Functioning in Advanced Old Age," *The Journals of Gerontology Series B*, 66B(3), (May 2011):287-291, https://doi.org/10.1093/geronb/gbq105

384 Daniel Kahneman, *Thinking, Fast & Slow* (Farrar, Straus and Giroux 2013).

The inability to self-regulate behavior is tied to an exhausted System 2 part of the brain in charge of slow, methodical, deep-think. As we move from an army at war back to a "zero-defect garrison army," the exhausted System 2 gives way to the impulses of System 1, the fast, reactive thinking part of the brain.

385 Mazarine Treyz, "Are You Trauma-Bonded to Your Job?," *MazarineTreyz.com*, Aug, 15, 2022, accessed Mar. 28, 2023, https://mazarinetreyz.com/are-you-trauma-bonded-to-your-job

386 Nicole McDermott and Stacey Diane Arañez Litam, PhD, "Trauma Bonding: What You Need To Know—And How To Get Help," *Forbes*, Mar. 31, 2023, accessed Jun. 13, 2023, https://forbes.com/health/mind/what-is-trauma-bonding

387 K. J. Ramsey, *The Lord Is My Courage: Stepping Through the Shadows of Fear Toward the Voice of Love*, (Zondervan, 2022).

388 Watson Institute for International & Public Affairs, "US Veterans & Military Families," Brown University, updated Aug. 2021, https://watson.brown.edu/costsofwar/costs/human/veterans

389 Ibid.

390 Christoph Mikulaschek, Saurabh Pant, and Beza Tesfaye, "Winning Hearts and Minds in Civil Wars: Governance, Leadership Change, and Support for Violent Groups in Iraq," *American Journal of Political Science*, 64(4) (Oct. 2020): 773-790, https://doi.org/10.1111/ajps.12527

391 Kristi Keck, "US Must Win Afghan Hearts and Minds, Commander Says," *CNN*, Sept. 8, 2009, accessed Mar. 27, 2023, https://cnn.com/2009/POLITICS/09/28/afghanistan.obama/index.html

392 Sam Wilkins, "The Rise and Fall of Village Stability Operations in Afghanistan: Lessons for Future Irregular Warfare Campaigns," *Modern War Institute*, Aug. 8, 2022, accessed Mar. 30, 2023, https://mwi.usma.edu/the-rise-and-fall-of-village-stability-operations-in-afghanistan-lessons-for-future-irregular-warfare-campaigns

393 "Green Beret, Operation Pineapple Express, Evacuated Afghan Allies from Kabul, Last Man Out, Scott Mann," *Combat Story* podcast, Jan. 7, 2023, https://podcasts.apple.com/us/podcast/combat-story/id1537933985?i=1000592970191

394 Daniel R. Green, "It Takes a Village to Raze an Insurgency," *Defense One*, Aug. 31, 2017, accessed Mar. 30, 2023, https://defenseone.com/ideas/2017/08/it-takes-village-raze-insurgency/140663

395 Nick Schifrin and Dan Sagalyn, "Was the War in Afghanistan Worth Fighting? 3 Veterans Weigh In," *PBS News Hour*, Jul 6, 2021, accessed Mar. 27, 2023, https://pbs.org/newshour/show/was-the-war-in-afghanistan-worth-fighting-3-veterans-weigh-in

396 Tom Bowman and Renee Montagne, "Rules of Engagement Are a Dilemma for US Troops," *NPR Morning Edition*, Dec. 11, 2009, accessed Mar. 30, 2023, https://www.npr.org/templates/story/story.php?storyId=121330893

397 Adam, Linehan, "What Mattis' New Rules Of Engagement Mean For The War In Afghanistan," *Task & Purpose*, Oct. 4, 2017, accessed Mar. 30, 2023, https://taskandpurpose.com/news/what-mattis-new-rules-of-engagement-mean-for-the-war-in-afghanistan

398 Katherin Schaeffer, "A Year Later, A Look Back at Public Opinion About the US Military Exit from Afghanistan," Pew Research Center, Aug. 17, 2022, https://pewresearch.org/fact-tank/2022/08/17/a-year-later-a-look-back-at-public-opinion-about-the-u-s-military-exit-from-afghanistan

399 Cami Mondeaux, "Timeline of Terror: How Biden's Afghanistan Withdrawal Disaster Unfolded," *Washington Examiner*, Aug. 17, 2022, accessed Mar. 28, 2023, https://washingtonexaminer.com/policy/how-afghanistan-withdrawal-disaster-unfolded

400 Ruby Mellen, "Two Weeks of Chaos: A Timeline of the US Pullout of Afghanistan," *The Washington Post*, Aug. 15, 2022, accessed Sept. 18, 2023 https://washingtonpost.com/world/2022/08/10/afghanistan-withdrawal-timeline

401 Martin, "A Former Marine Details the Chaotic Exit," (n352).

402 Jim Garamone, "US Honors Service Members Involved in Afghan Evacuation," *DoD News*, Aug. 31, 2022, https://defense.gov/News/News-Stories/Article/Article/3146236/us-honors-service-members-involved-in-afghan-evacuation

403 Air Mobility Command Public Affairs, "AMC Leaders to Present 96 Distinguished Flying Crosses, 12 Bronze Star Medals, Gallant Unit Citation for Operation Allies Refuge," US Air Force, Oct. 24, 2022, accessed Mar. 30, 2023, https://www.af.mil/News/Article-Display/Article/3196767/amc-leaders-to-present-96-distinguished-flying-crosses-12-bronze-star-medals-ga

404 Charles Pope, "One Year Later, Historic Afghan Airlift Inspires Pride and Reflection Across the Air Force," US Air Force, Aug. 30, 2022, https://www.af.mil/News/Article-Display/Article/3144426/one-year-later-historic-afghan-airlift-inspires-pride-and-reflection-across-the

405 Bessel van der Kolk, MD, *The Body Keeps the Score: Brain, Mind, and Body in the Healing of Trauma*, (New York: Penguin Books, 2015).

406 Khaleda Rahman, "Veterans Angry Over 'Wasted Years' in Afghanistan as Taliban Take Power," *Newsweek*, Aug. 16, 2021, accessed Mar. 30, 2023, https://newsweek.com/veterans-angry-over-wasted-years-afghanistan-taliban-take-power-1619646

407 Corie Weathers, "It's Time for Veterans and Military Families to Let Go of Afghanistan—But Not Forget," *Military.com*, Aug 17, 2021, https://military.com/daily-news/opinions/2021/08/17/its-time-veterans-and-military-families-let-go-of-afghanistan-not-forget.html

408 The Veterans and Citizens Initiative, "After Kabul: Veterans, America, and the End of the War in Afghanistan," Nov. 2021, accessed Mar. 30, 2023, https://moreincommon.com/media/vvjpdo2k/after-kabul-more-in-common.pdf

409 James R. Webb and Andrea Scott, "Marine Relieved for Viral Video Now Says He's Resigning His Commission," *Marine Corps Times*, Aug. 30, 2021, https://marinecorpstimes.com/news/your-marine-corps/2021/08/30/marine-relieved-for-viral-video-now-says-hes-resigning-his-commission

410 Philip Athey, "Viral Marine Who Criticized Afghanistan Withdrawal Charged with 6 Violations," *Marine Corps Times*, Oct. 6, 2021, https://marinecorpstimes.com/news/your-marine-corps/2021/10/06/lt-col-stuart-scheller-charged-with-6-ucmj-violations

411 Randi Stenson, "Leaders Take Note: The Army's Counseling Form Gets a Much-Needed Update," Mission Command Center of Excellence Public Affairs, Mar. 31, 2023, https://www.army.mil/article/264745/leaders_take_note_the_armys_counseling_form_gets_a_much_needed_update

412 Oriana Pawlyk, "Air Force to Pause Operations After Spike in Suicides," *Military.com*, Aug. 1, 2019, https://military.com/daily-news/2019/08/01/air-force-pause-operations-after-spike-suicides.html

413 Thomas Howard Suitt, III, PhD, "High Suicide Rates Among United States Service Members and Veterans of the Post9/11 Wars," Watson Institute, Brown University, Jun. 21, 2021, https://watson.brown.edu/costsofwar/files/cow/imce/papers/2021/Suitt_Suicides_Costs%20of%20War_June%2021%202021.pdf

414 Caitlyn M. Kenney, "Active-Duty Suicide Rate Hit Record High in 2020," *Defense One*, Oct. 6, 2021, https://defenseone.com/threats/2021/10/active-duty-suicide-rate-hits-record-high/185882

415 Loryana L. Vie, PhD and Adam D. Lathrop, MLIS, "Department of the Army Career Engagement Survey: Second Annual Report," Headquarters, Department of the Army Deputy Chief of Staff and Office of the Assistant Secretary of the Army (Manpower & Reserve Affairs), Sept. 2022, https://talent.army.mil/wp-content/uploads/2023/01/DACES-Second-Annual-Report_FINAL.pdf

416 Dedmond, "Sixteen Things," (n365).

417 Caitlin Doornbos, "Woke Army General Rebuked for Twitter-trolling Tucker Carlson, Right-Wingers," *New York Post*, Oct. 5, 2022, accessed Mar. 5, 2023, https://nypost.com/2022/10/05/gen-patrick-donahoe-rebuked-for-twitter-trolling-tucker-carlson

418 Jeff Schogol, "Army General Investigated for Defending Female Troops Online Retires Honorably," *Task & Purpose*, Jan. 3, 2023, accessed Apr. 2 2023, https://taskandpurpose.com/news/army-general-patrick-donahoe-retires

419 Kelly Sanders, "Operation Allies Refuge: Ramstein AB Transforms into Major Evac Hub," US Air Force 86th Airlift Wing Public Affairs, Aug. 26, 2021, https://www.af.mil/News/Article-Display/Article/2747949/operation-allies-refuge-ramstein-ab-transforms-into-major-evac-hub

420 Anthony D. Kauders, "From Particularism to Mass Murder: Nazi Morality, Antisemitism, and Cognitive Dissonance," *Holocaust and Genocide Studies*, 36(1), (Spring 2022):46-59, https://doi.org/10.1093/hgs/dcac011

421 Ibid.

422 Caroll Tavris and Elliot Aronson, *Mistakes Were Made (But Not By Me)*, (Mariner Books, 2020), 29; Joel M. Cooper, *Cognitive Dissonance*, (SAGE Publications Ltd, 2007), 75; and James F. Laird, *Feelings: The Perception of Self* (Oxford University Press, 2007), 160.

423 Kauders, "From Particularism to Mass Murder," (n420).

424 Viktor Frankl Institute of Logotherapy in Israel, "From Dissonance to Meaning," Apr. 4, 2023, https://themeaningseeker.org/from-dissonance-to-meaning; and Viktor E. Frankl, *The Will to Meaning: Foundations and Applications of Logotherapy*, (Plume, 2014), 62.

425 "Toxic Positivity," *Psychology Today*, accessed Apr. 4, 2023, https://psychologytoday.com/us/basics/toxic-positivity

426 Suzanne Degges-White, PhD, "Ditch Toxic Positivity for Tragic Optimism," *Psychology Today*, May 2, 2022, accessed Apr. 4, 2023, https://psychologytoday.com/us/blog/lifetime-connections/202205/ditch-toxic-positivity-tragic-optimism

427 Current military families are often surprised to hear that some of the issues they are advocating for today are the same issues presented in Wickham's white paper in 1983 (n210), including safer housing, childcare buildings, schools, employment challenges. Telling the stories of the past helps us evaluate progress. Not sharing these stories creates a perceived lack of transparency, creating more mistrust of the institution. When generations leave and retire, stories, progress, advocacy, and experiences can leave with them.

428 This is a powerful example of the successful use of campaigns in the world wars. Consistency, patience, and action make a difference, even though people may not understand or see it at first.

429 Wickham, *The Army Family*, (n210).

430 "Military Leaders Apologize for Substandard Living Conditions at Family Housing," *CBS News*, Mar. 7, 2019, accessed Apr. 6, 2023, https://cbsnews.com/news/military-housing-hearing-leaders-apologize-substandard-living-conditions

431 Laura Geller, "Are We Not Worth Safe Houses?" *WUSA9*, updated Nov. 21, 2019, https://wusa9.com/article/news/investigations/are-we-not-worth-safe-houses-military-spouses-create-their-own-advocacy-group-amid-concerns-about-mold-remediation-fort-belvoir/65-aef68db7-b416-4ed7-8485-6dc62b48f009

432 Major Stoney L. Portis, "Unit Reconstitutions: Combat Stress as an Indicator of Unit Effectiveness," US Army Command and General Staff College, Jun. 13, 2014, https://apps.dtic.mil/sti/pdfs/ADA611627.pdf

433 Stephen M.R. Covey, *The Speed of Trust*, (New York: Simon & Schuster, 2006).

434 Edgar H. Schein and Peter A. Schein, *Humble Inquiry: The Gentle Art of Asking Instead of Telling*, Second Edition, (Oakland, CA: Berrett-Koehler Publishers 2021).

435 Staff Sgt. Michael Reinsch, "In a War for Talent—Recruiting, Retention and Opportunity: Army Leaders Work to Grow the Army of 2030," *Army News Service*, Oct. 14, 2022, accessed Apr. 5, 2023, https://www.army.mil/article/261158/in_a_war_for_talent_recruiting_retention_and_opportunity_army_leaders_work_to_grow_the_army_of_2030

436 Jim Absher, "The Blended Retirement System Explained," *Military.com*, Dec. 19, 2022, Accessed: April 5, 2023, https://military.com/benefits/military-pay/blended-retirement-system.html

437 Jim Garamone, "DoD Official Cites Widening Military-Civilian Gap," US Department of Defense, May 16, 2019, accessed Oct. 8, 2022, https://defense.gov/News/News-Stories/Article/Article/1850344/dod-official-cites-widening-military-civilian-gap

438 Wickham, *The Army Family*, (n210).

439 Caitlyn M. Kenney, "Fewer Military Families Would Recommend Uniformed Service, Survey Finds," *Defense One*, Jul. 14, 2022, accessed April 5, 2023, https://defenseone.com/policy/2022/07/fewer-military-families-would-recommend-uniformed-service-survey-finds/374481

440 Ibid.

441 "There's a Hole in My Bucket," accessed Apr. 5, 2023, https://nurseryrhymescollections.com/lyrics/theres-a-hole-in-my-bucket.html

442 Allie Townsend, "All-Time 100 Greatest Toys," *Time*, Feb. 16, 2011, accessed Mar. 13, 2023, https://content.time.com/time/specials/packages/0,28757,2049243,00.html

443 "We Are Your Defense," accessed Mar. 13, 2023, https://www.defense.gov/About

444 Charles L. Anspach, "CAVU: A Salute to Those Graduating from College," *Peabody Journal of Education*, 26(2), (Sept. 1948):102–106, accessed Sept.18, 2023, http://jstor.org/stable/1489990

ACKNOWLEDGMENTS

IT IS ONE THING TO SACRIFICE YOUR OWN TIME and effort for something you feel passionate about. It is quite humbling when others are willing to join you.

Matthew, you added kindling to this spark years ago. Thank you for sacrificing our time and your time to graciously send me on a "deployment" in search of the truth. Thank you for your servant heart, proofreading and barista skills, endless patience, and for championing me to the finish line with a shared heart and concern for our community. Thank you to my boys who similarly sacrificed their mom for a season and a cause difficult to understand. Thank you for teaching me to respect and value your generation with an open mind and new perspective. You're my people and more important than any mission.

Thank you to Rebecca Brown for your eagerness to support, for your talents in editing and research, and for your gentle guidance when I had writer's block. You willingly jumped in and I'll never forget it.

Thank you to those who gave a considerable amount of time and thought. Jon and Evie and Brad and Sue Brown for reading, digesting, challenging, and always encouraging. Thank you to Brad for letting me share your words and leadership example as it continues to inspire. Thank you to Mac and Sally Thornberry for similarly lending your time and energy. You had earned a quiet retirement and instead, you continue to serve and inspire through wisdom and truth.

Thank you to those who were willing to read, review, and share the message of this book. Healing this community is not on the shoulders of one, it is something we must all do together to make a difference. May my gift of words give you the inspiration to begin change where you are.

Thank you to the individuals, families, and groups I have served and continue to serve. Your stories, some with incredible pain, are not in vain. My biggest hope is that you feel your story was used with the most care possible. Thank you for trusting me to walk with you through some of the most vulnerable moments in your life.

Thank you to the Elva Resa Publishing team, especially Karen, for believing in me and this message, and for sacrificing more than I will ever know to publish this beast. To the team of editors and designers, I'm sorry, and thank you. You gave an incredible amount of focus and energy to make this a reality. I hope you will share in the joy of bringing healing and hope to this community. We did this together and I wouldn't have had it any other way.